The LabVIEW Student Edition User's Guide

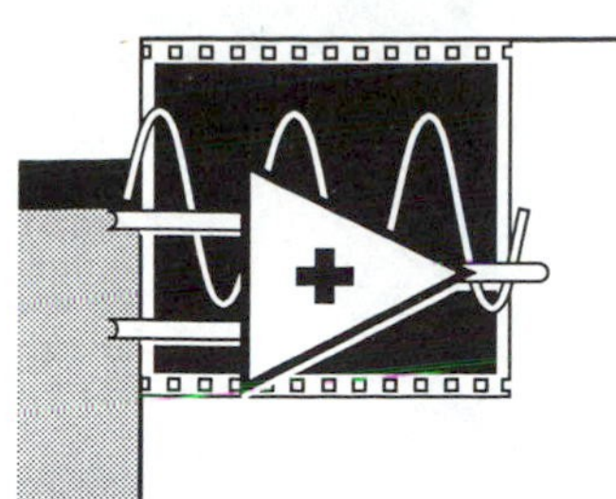

The LabVIEW Student Edition User's Guide

Lisa K. Wells
National Instruments Corporation

PRENTICE HALL, Englewood Cliffs, New Jersey 07632

Library of Congress Cataloging-in-Publication Data

Wells, Lisa K.
The LabVIEW student edition : user's guide / Lisa K. Wells.
p. cm.
Includes index.
ISBN 0-13-210683-3
1. LabVIEW. 2. Scientific apparatus and instruments--Computer simulation. 3. Computer graphics. I. Title.
Q185.W45 1994
006--dc20 94-44174
CIP

Acquisitions editor: Linda Ratts
Production editor: Bayani Mendoza de Leon
Copy editor: Brenda Melissaratos
Cover designer: Trevoris Morgan
Buyers: Dave Dickey / Bill Scazzero
Editorial assistant: Naomi Goldman

A Simon & Schuster Company
Englewood Cliffs, New Jersey 07632

The author and publisher of this book have used their best efforts in preparing this book. These efforts include the development, research, and testing of the theories and programs to determine their effectiveness. The author and publisher make no warranty of any kind, express or implied, with regard to these programs or the documentation contained in this book. The author and publisher shall not be liable in any event for incidental or consequential damages in connection with, or arising out of, the furnishing, performance, or use of these programs.

Printed in the United States of America
10 9 8 7 6 5 4 3 2 1

ISBN 0-13-210683-3

Prentice-Hall International (UK) Limited, *London*
Prentice-Hall of Australia Pty. Limited, *Sydney*
Prentice-Hall Canada Inc., *Toronto*
Prentice-Hall Hispanoamericana, S.A., *Mexico*
Prentice-Hall of India Private Limited, *New Delhi*
Prentice-Hall of Japan, Inc., *Tokyo*
Simon & Schuster Asia Pte. Ltd., *Singapore*
Editora Prentice-Hall do Brasil, Ltda., *Rio de Janeiro*

I would like to dedicate the LabVIEW Student Edition to my family for all of their support.

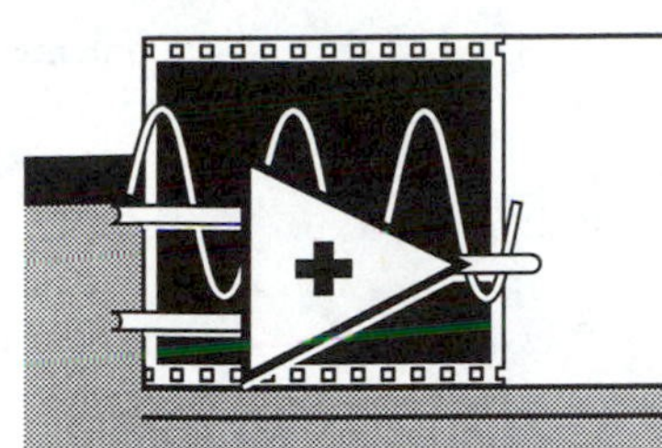

Contents

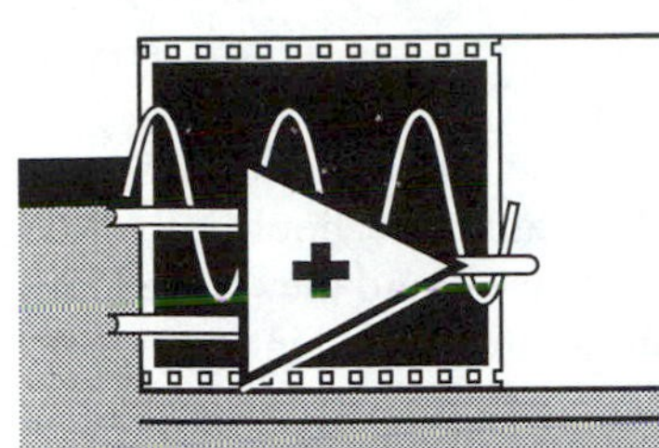

Foreword

The room is filled with students consulting their notes and one another, downloading data and studying the information carefully, designing experiments and analyzing data, composing text and animations to present their results, integrating audio and video with their data to give depth and strength to their presentations. They are not synthesizing bits of predigested knowledge—they are engineering their own works. They are not merely demonstrating their knowledge and understanding of a particular topic; they are applying and synthesizing it, then analyzing and evaluating the results.

This is not only the educator's dream. It is the dream of American business people, engineers, and scientists, all of whom are anxious to employ young people who are eager, self-confident problem-solvers, experienced in cooperating with others and in using the latest technological tools. The publication of the *LabVIEW Student Edition* is a giant step toward achieving this dream.

Incorporating LabVIEW into the classroom has been a goal of mine for many years. We designed interfaces for oxygen, temperature, and pH probes; for amplifiers acquiring EKGs and action potentials from nerves and muscles; for interfaces with pressure transducers and strain gauges. With the use of LabVIEW, maintaining and modifying these labs became easier. The data analysis, for instance, could be changed on the spot. Training the instructors to man the labs became easier. Administrators were happy, for we started throwing out equipment that was expensive to maintain.

Strange things began to happen. Because we could prototype courseware rapidly in LabVIEW, some faculty members became interested. They began to work on projects and to think more about what they were teaching and how they were teaching it. LabVIEW offered an excellent tool for developing courseware that would provide student-centered activities and challenging problem-solving environments. Simulations of population genetics, action potentials, and others were introduced into the labs. Students and faculty members clustered around computers, discussing the models. Education had suddenly become a personal enterprise.

Nor is the appeal of LabVIEW limited to students at the university level. Several years ago I began to demonstrate some of our courseware to secondary schools. LabVIEW captured the imagination of several students from local high schools, and they came to my lab after school to learn how to use it, picking it up easily and soon working independently with very little supervision.

One Saturday afternoon I observed a student using LabVIEW to simulate the lift and drag on a wing. (We even allow engineering projects in our biology labs.) I suggested that a graph of the wing cross section could be used as data input rather than a cumbersome array of coordinates. The lights turned on for him, and he reworked his project, going on to win several awards and scholarships. I received a note from his parents (one of whom is an engineer), thanking me for introducing him to the software that was "a little addictive." It is, of course, addictive. A tool that, with relative ease, allows its user to become an independent creator could not keep from becoming an addiction. All education should be that addictive!

Is it too difficult a tool for most students to learn to use? No. LabVIEW can hold the interest of innercity school students last period on a Friday afternoon. Why? Because students know when they are being shown a valid tool, one that could make a difference to their own performance ability.

LabVIEW is used in many prominent research labs, from Berkeley to NASA. (It was recently used for data collection on the space shuttle.) It is used in manufacturing processes. (Ben and Jerry's new plant in Vermont will be controlled by LabVIEW.) Shouldn't it be widely used in our schools as well? Any software that can make engineers and research scientists more effective and businesses more competitive is probably an important educational tool.

Providing our students with the tools to be competitive is the goal of all education, but it is a constant battle for schools to keep up with technology. Money and training time are the problems. The *LabVIEW Student Edition* addresses these difficulties, placing the price of the newest technology at a level where every school can afford it.

Clark Gedney, Ph.D.
Instructional Computing
Department of Biological Sciences
Purdue University

Acknowledgments

Many wonderful people at National Instruments and at Prentice Hall provided invaluable assistance in the creation of the *LabVIEW Student Edition* software and user's guide. Without their hard work, the student edition would still be just an idea.

I would like to recognize Greg Richardson and Gregg Fowler, LabVIEW engineers extraordinaire, for all of their hard work creating the student edition. A big thank-you goes to Jeff Kodosky, Monnie Anderson, Crystal Doubrava, and the rest of the LabVIEW development team for their never-ending commitment to LabVIEW and support for the student edition.

Special appreciation is also in order for the invaluable folks who handled the sales, marketing, and publications aspects of the student edition. Without the knowledge and text material provided by Ann Arnold and the rest of the National Instruments Tech Pubs group, we would not have a student manual to read. An extra measure of thanks goes to Roxanne Green and Sandy Bartnett for all of their valuable contributions to the style, organization, and flow of the book. Ray Almgren and Dan Phillips provided invaluable assistance keeping the project on track and providing the support of their marketing and sales departments. Finally, I'd like to acknowledge Don Fowley and Linda Ratts-Engelman, executive editors at Prentice Hall, for their devotion to a timely and successful *LabVIEW Student Edition.*

Lisa K. Wells

About the Author

Lisa Wells has extensive experience both using and teaching LabVIEW. During the two years she has taught LabVIEW classes, Lisa has become attuned to the unique learning needs of the LabVIEW beginner. Lisa first learned LabVIEW as a student; she also wrote her undergraduate thesis on using LabVIEW for monitoring electrical power quality. She joined the National Instruments (Austin, TX) team in August 1992, after earning a B.S. in electrical engineering and a B.A. in liberal arts from the University of Texas at Austin. Her first 18 months at the company were focused on technical support for LabVIEW, and teaching both basic and advanced LabVIEW courses. In early 1994, Lisa diverted her attentions to developing and promoting LabVIEW and related products for use by students and instructors in the classroom and laboratory. When she's not playing with LabVIEW, Lisa enjoys traveling, snow skiing, water sports, and camping.

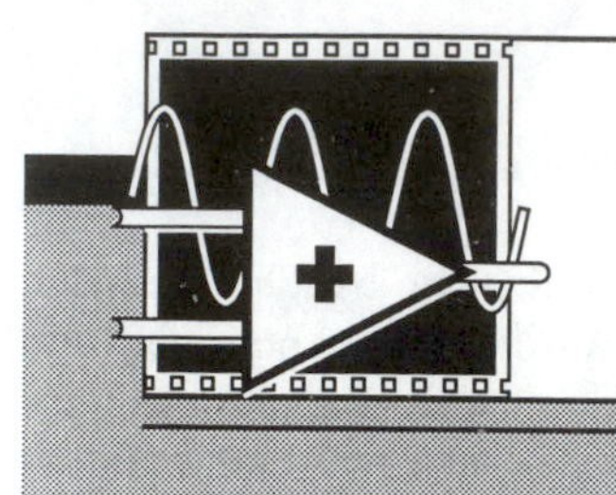

Preface

To the Instructor

LabVIEW, or Laboratory Virtual Instrument Engineering Workbench, is a graphical programming language that has been widely adopted throughout industry, academia, and government labs as the standard for data acquisition and instrument control software. Its intuitive user interface makes learning exciting and fun! Now, LabVIEW is available to students at an affordable price so that they can learn valuable skills on their own, at home as well as in the classroom laboratory. LabVIEW is ideal for a countless number of science and engineering applications. With his or her own copy of LabVIEW, the student has an opportunity for creativity, independent project development, and self-pacing that cannot always be found in the lab.

Computer usage in labs has skyrocketed. Computers are much more flexible than standard instruments, and with LabVIEW, creating your own *virtual instrument (VI)* is simple. We created the *LabVIEW Student Edition* to give students early exposure to the many uses of graphical programming. LabVIEW not only helps reinforce basic scientific, mathematical, and engineering principles, but it encourages students to explore advanced topics as well. Students can run LabVIEW programs designed to teach a specific topic, or they can use their skills to develop their own applications. Either by itself or coupled with LabVIEW curriculum series texts, LabVIEW provides a real-world, hands-on experience that will greatly enhance the learning process.

We've worked closely with the academic community to make sure LabVIEW meets the needs of today's students. The student edition is easy to use, inexpensive, and runs on lower-end computers, yet retains all of the functionality most students will ever need. We also offer academic discounts on data acquisition and instrument control hardware as well as on the professional version of LabVIEW. Classroom laboratories can be easily and cost effectively configured for computer automated data acquisition, analysis, and presentation.

To the Student

LabVIEW, or Laboratory Virtual Instrument Engineering Workbench, is a powerful and flexible instrumentation and analysis software system for PCs running Microsoft Windows and Apple Macintosh computers. LabVIEW departs from the sequential nature of traditional programming languages and features an easy-to-use graphical programming environment, including all of the tools necessary for data acquisition, analysis, and presentation. With this graphical programming language, called "G," you can program using a block diagram method that compiles into machine code. Using LabVIEW, you can solve many types of problems in only a fraction of the time and hassle it would take to write "conventional" code. LabVIEW integrates data acquisition, analysis, and presentation in one system. With LabVIEW, your software *becomes* your instrument, since you define the functionality you need.

This manual expects you to have basic knowledge of your operating system, either Mac OS or Windows. If you don't have much computer experience, you may want to spend a little time with your Mac or Windows manual, familiarizing yourself with your computer. For example, you should know how to access menus, open and save files, make backup disks, and use a mouse. You will also find previous programming experience helpful, although it is not necessary.

Because LabVIEW is a general-purpose programming tool, it can be used in a wide variety of industries and applications. A few examples include:

- Simulating heart functions
- Controlling an ice cream-making process
- Detecting Hydrogen gas leaks on the space shuttle
- Monitoring feeding patterns of baby ostriches
- Modeling power systems to analyze power quality
- Measuring physical effects of exercise in lab rats
- Motion control of servo and stepper motors
- Testing electronic circuit boards in computer and electronic devices

Objectives of the *LabVIEW Student Edition User's Guide*

After reading this manual and working through the exercises, you should be able to

- Write LabVIEW programs, called *virtual instruments*, or VIs, which are comprised of front panels, block diagrams, and icons and connectors.
- Use LabVIEW to create your laboratory applications.
- Employ various debugging techniques.
- Manipulate both built-in LabVIEW functions and library VIs.
- Create and save your own VIs so you can use them as subVIs.
- Design custom graphical user interfaces (GUIs)
- Build applications that use GPIB or serial instruments.

- Create applications that use plug-in data acquisition (DAQ) boards.
- Use built-in analysis functions to process your data.

Organization of the Student Edition User's Guide

The *LabVIEW Student Edition User's Guide* teaches you to make optimum use of LabVIEW to develop your applications. The manual is divided into ten body chapters, each covering a topic or a set of topics, plus detailed Examples and Function Reference sections. Each chapter consists of:

- **Introduction,** which describes the main ideas covered in that chapter.
- **Overview, goals, and key terms** to guide your learning process.
- **Discussion** of the featured topics.
- **Demonstration exercises** to reinforce the information presented in the discussion.
- **Summary**, which outlines important concepts and skills taught in the lesson.
- **Additional exercises** to reinforce the new material.

Chapter 1 introduces you to LabVIEW.

In Chapter 2, you will get an overview on how data acquisition, GPIB, serial port communication, and data analysis are performed with LabVIEW.

In Chapter 3, you will get acquainted with the LabVIEW environment, including the essential parts of a VI, the LabVIEW menus, run and edit modes, the Help window, and the use of subVIs (LabVIEW-ese for subroutine).

In Chapters 4 and 5, you will become familiar with the basic LabVIEW principles—using controls and indicators (such as numerics, Booleans, and strings), creating and saving VIs, editing and debugging VIs, creating subVIs, and documenting VIs. You will also begin to understand dataflow programming.

Chapter 6 describes the basic programming structures of LabVIEW: While Loops, For Loops, shift registers, Case structures, Sequence structures, and Formula Nodes.

In Chapter 7, you will learn how to use arrays and clusters in your programs.

Chapter 8 details the variety of charts and graphs available in LabVIEW and teaches you how to use them for colorful and informative data presentation.

Chapter 9 discusses string data types, string functions, and how to save data to a file.

Chapter 10 covers miscellaneous topics such as how to configure the appearance of your VI, how to control the timing of your VI, popping up dialog boxes, and printing front panels from within the VI.

The Examples section will explain all of the software examples that ship with the student edition. Many of these examples are functional LabVIEW programs that you can use in your applications.

The Function Reference section describes in detail every LabVIEW function and its parameters.

At the end of this text you will find a complete glossary, index, and appendices.

Conventions Used in this Manual

The following table describes the conventions used in this manual:

Convention	Definition
bold	Bold text denotes VI names, function names, menus, and menu items. In addition, bold text denotes VI input and output parameters. For example, **File**.
bold italic	Bold italic text denotes a note, caution, or warning. For example, **Note:** ***Be sure to save your work frequently during thunderstorms.***
italics	Italic text denotes emphasis, a cross reference, or an introduction to a key term or concept. For example, VIs have three main parts: the *front panel*, the *block diagram*, and the *icon/connector.*
Sans Serif	Sans Serif type denotes text or characters that you enter using the keyboard, and folder or directory names. Sections of code, programming examples, syntax examples, and messages and responses that the computer automatically prints to the screen also appear in this font. For example, an instruction such as "type Digital Indicator inside the bordered box."
< >	Angle brackets enclose names of keys on the keyboard, for example <shift>.

The Professional Version of LabVIEW and its Add-On Toolkits

The *LabVIEW Student Edition* is a compact version of the professional version. We have designed the student edition to be easier to use and to have lower system requirements than the professional version

With the *LabVIEW Student Edition*, you can learn LabVIEW programming techniques without encountering some of the more advanced features. You can run the student edition without a math coprocessor (though we recommend one), and it has smaller memory requirements than the professional version.

While the *LabVIEW Student Edition* is extremely powerful, the professional version has even more built-in functions and features to make programming easy for you. In addition, you can use the following special add-on toolkits with the professional version to further increase your flexibility and capability.

Application Builder

The LabVIEW Application Builder is an add-on package for creating stand-alone applications. When accompanied by the Application Builder, a LabVIEW system can create VIs that operate as stand-alone applications. You can run the executable, but cannot edit it. Applications minimize RAM and disk requirements to run by saving only those resources needed for execution. The application contains the same execution resources as those in the professional version, so VIs will execute at the same high performance rates.

Test Executive Toolkit

The LabVIEW Test Executive Toolkit is a multipurpose LabVIEW add-on for automated test execution. The Test Executive Toolkit gives you the capability to control the execution of sequences of tests for production and manufacturing test applications. The toolkit delivers the Test Executive in block diagram source code with high-level modules, making it easy to modify to meet your application needs.

JTFA Toolkit

The Joint Time-Frequency Analysis (JTFA) Toolkit is a LabVIEW add-on for precise signal analysis of data whose frequency content changes with time. The toolkit features the Gabor Spectrogram, which is ideal in applications of speech processing, sound analysis, sonar, radar, vibration analysis, and dynamic signal monitoring.

Picture Control Toolkit

The LabVIEW Picture Control Toolkit is a versatile graphics add-on package for creating arbitrary front panel displays. The toolkit adds the Picture control and a library of VIs to your LabVIEW system. You can create diagrams using a set of VIs to describe the drawing operations to build these images dynamically. With the toolkit, you can create new front panel displays such as specialized bar graphs, pie charts, and Smith charts. You can also display and animate arbitrary objects such as robot arms, test equipment, a unit under test, or a two-dimensional display of a real-world process.

SPC Toolkit

The LabVIEW SPC Toolkit is a VI library for statistical process control (SPC) applications. Using the SPC Toolkit, you can use SPC methods to analyze and track process performance. In addition to subVIs that perform the SPC computations, the toolkit contains an extensive collection of example and graph VIs that demonstrate how to incorporate typical

SPC methods and displays into LabVIEW applications. The toolkit addresses three areas of SPC—control charts, process statistics, and Pareto analysis.

PID Control Toolkit

The PID Control Toolkit adds sophisticated control algorithms to LabVIEW. With this package, you can quickly build data acquisition and control systems for your own control application. By combining the PID Control Toolkit with the math and logic functions in LabVIEW, you can quickly develop programs for control.

SQL Toolkit–DatabaseVIEW

National Instruments is distributing DatabaseVIEW from Ellipsis, Inc. (Boston, MA), as the SQL Toolkit–DatabaseVIEW. It is a collection of LabVIEW VIs for direct interaction with a local or remote database. High-level Access VIs simplify database access by intelligently encapsulating common database operations into easy-to-use VIs. Low-level Interface VIs directly access the database, operating on columns and records in database tables.

Upgrading to the Professional Version

The LabVIEW professional version runs on the following platforms: Macintosh, Power Macintosh, computers running Windows, Sun SPARCstations, and HP 9000/700-series workstations running HP-UX. If you would like information on upgrading, please write to

National Instruments
att.: Academic Sales
6504 Bridge Point Parkway
Austin, TX 78730-5039

Technical Support

Student users of the *LabVIEW Student Edition* receive technical support from their instructors. If you run into trouble, first read the appropriate sections of this book to try to solve your problem. If you still have questions, consult with your instructor.

Replacement of Defective Disks

Contact Prentice Hall at (201) 592-3096 if you have a defective disk. Prentice Hall will replace damaged disks free of charge to registered users only.

Registration

Make sure to fill out and mail in the registration card that comes with your *LabVIEW Student Edition* package. Registered users receive replacement of defective disks at no charge.

Limited Warranty

The software and the documentation are provided "as is" without warranty of any kind, and no other warranties, either expressed or implied, are made with respect to the software. National Instruments does not warrant, guarantee, or make any representations regarding the use or the results of the use of the software or the documentation in terms of correctness, accuracy, reliablility, or otherwise and does not warrant that the operation of the software will be uninterrupted or error free. This software is not designed with components and testing for a level of reliability suitable for use in the diagnosis and treatment of humans or as critical components in any life-support systems whose failure to perform can reasonably be expected to cause significant injury to a human. National Instruments expressly disclaims any warranties not stated herein. Neither National Instruments nor Prentice Hall shall be liable for any direct or indirect damages.

The entire liability of National Instruments and its dealers, distributors, agents, or employees are set forth above. To the maximum extent permitted by applicable law, in no event shall National Instruments or its suppliers be liable for any damages including any special, direct, indirect, incidental, exemplary, or consequential damages, expenses, lost profits, lost savings, business interruption, lost business information, or any other damages arising out of the use or inability to use the software or the documentation even if National Instruments has been advised of the possibility of such damages.

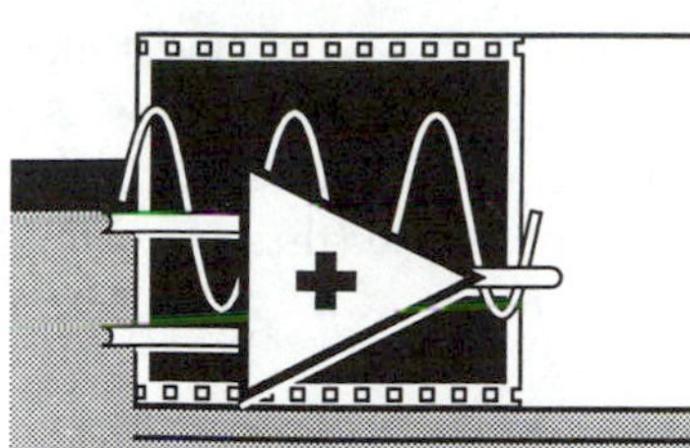

Installation

Before You Take the Plunge

Checking Out Your Package

Your *LabVIEW Student Edition* package should contain several items:

- This handy manual
- 3 Mac disks or 4 PC disks containing the *LabVIEW Student Edition* software and examples
- Registration card

Checking System Configuration

PC running Windows

- 6 MB memory, 8 MB recommended (you can use virtual memory to meet this requirement)
- Minimum 386sx (any speed, but faster is always better)
- Floating-point unit recommended but not necessary
- Microsoft Windows, version 3.1 or later
- 10 MB free hard disk space

Mac

- 6 MB memory, 8 MB recommended (you can use virtual memory to meet this requirement)
- 68020 processor or later
- Floating-point unit recommended but not necessary

- The student edition runs well on low-end Macs such as the Macintosh SE/30, LC, II, and IIsi. It does not run on the Classic, Plus, SE, or Powerbook 100 (these machines contain antiquated 68000 processors).
- 10 MB free hard disk space

Backing Up Your Disks

Before you go any further, you should make backup copies of the disks that came with this manual. If you don't know how to back up disks, please look at your Windows or Macintosh manuals for instructions. Store the originals in a safe place and work with the copies. Then when (notice we don't say "if") something happens to your backup disks, you can make new ones. Remember, the frequency of catastrophe is inversely proportional to the number of precautions you take!

Installing the LabVIEW Student Edition

How to Do It

Note

Some virus detection programs may interfere with the LabVIEW installer. Check the distribution disks for viruses before you begin installation. Then turn off the automatic virus checker and run the installer. After installation, check your hard disk for viruses again and turn the virus checker back on.

Windows

You can install LabVIEW from the Windows File Manager or with the **Run...** command from the **File** menu of the Program Manager. If you are unfamiliar with basic Windows operation, please see your Windows manual for instructions.

Insert Disk 1 and run the SETUP.EXE program on Disk 1 from Windows using one of the following methods:

1. Select **Run...** from the **File** menu of the Program Manager. A dialog box appears.
2. Type A:\SETUP (where A is the proper drive designation).

OR

1. Launch the File Manager.
2. Click on the drive icon that contains Disk 1.
3. Find SETUP.EXE in the list of files on that disk and double-click on it.

Follow the instructions that appear on the screen. The student edition will install into a directory called LVSE on the top level of the drive where Windows is located. To change this location, use the "Set Location" button. At the end of the installation, the installer asks you if you want to create a Windows program group with icons for LabVIEW. Click the YES button. Once you have installed LabVIEW, it is ready to do your bidding.

If your computer doesn't have a coprocessor, you will need to relaunch Windows to make sure the new configuration parameters take effect.

Watch Out!
The LabVIEW Student Edition installer does not install the hardware drivers you will need to use DAQ or GPIB. These drivers are available from your instructor on a separate disk. For more information, please see Appendix A, DAQ and GPIB Configuration.

Mac

1. Insert Disk 1 and double-click on the student edition Installer icon.
2. Drag the LabVIEW Student Edition icon to the destination hard disk. It creates a `LabVIEW Student Edition` folder on the top level of the selected drive and places LabVIEW there. If the folder already exists, it will overwrite the old LabVIEW and supporting files. If you wish to keep them, you should rename the folder to something other than `LabVIEW Student Edition`.
3. Follow the instructions that appear on the screen. Once you have completely installed LabVIEW, it is ready to do your bidding.

If you wish to install only certain files, you can click on the Show Other Installations button at the top of the installation screen. Choose the items you wish to install and drag them to the desired hard drive, where they will be installed in a `LabVIEW Student Edition` folder.

If you are unfamiliar with any of these Mac operations, please see your Mac manual for instructions.

Understanding What Installation Does—What Did I Get?

Windows

The installation process should install the following files in your `LVSE` directory (or whichever directory you specify):

`VI.LIB <dir>`	A directory that contains LabVIEW's built-in functions. Do not modify!
`EXAMPLES <dir>`	A directory that contains useful example programs you can run or modify.
`LABVIEW.EXE`	LabVIEW executable
`LVDEVICE.DLL`	Part of LabVIEW's device driver interface
`LABVIEW.RSC`	LabVIEW resource file
`SERPDRV`	Serial Port driver
`DAQDRV`	DAQ driver
`GPIBDRV`	GPIB driver

WEMU387.386	Floating-point emulation file that is installed only if your machine doesn't have an FPU. Goes in WINDOWS\SYSTEM directory.

Mac

On the Macintosh, the installation process places LabVIEW, the VI.LIB folder, and the Examples folder into the LabVIEW Student Edition folder. VI.LIB contains LabVIEW's built-in functions—you should not change the contents of this folder in any way or you risk causing yourself much suffering. The Examples folder contains great example programs that you can use as is or modify to suit your needs. It also contains programs that you will look at when you do exercises, and even solutions to the exercises!

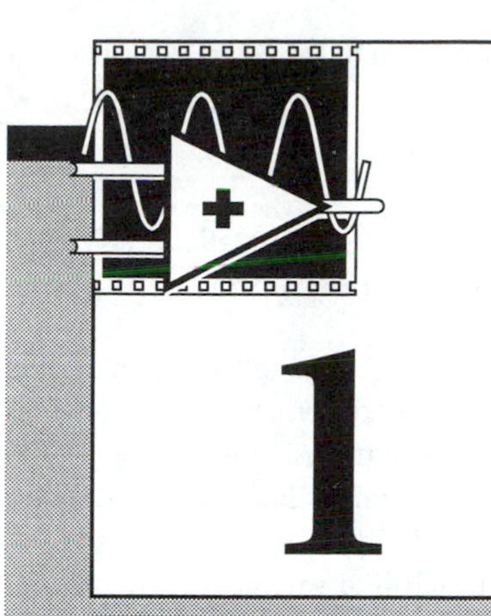

1

An Introduction to LabVIEW

Overview

Welcome to the world of LabVIEW! This chapter gives you a basic overview of LabVIEW and its capabilities. It also explains how this book is organized and the things you will be doing in your quest to become a LabVIEW expert.

Your Goals

- Develop an idea of *what* LabVIEW really *is*
- Peruse the introductory examples

Key Terms

LabVIEW
virtual instrument (VI)
front panel
block diagram
icon
connector
palette
hierarchy
modular programming

1.1 What Is LabVIEW? The Cure for the Common Code...

Now that you have checked your package, backed up your disks, and installed LabVIEW, you'd probably like to know what exactly LabVIEW is—what you can do with it and what it can do for you.

LabVIEW®is short for **Lab**oratory **V**irtual **I**nstrument **E**ngineering **W**orkbench. It is a powerful and flexible instrumentation and analysis software system that runs on PCs using Microsoft Windows, Apple Macintoshes, Sun SPARCstations, and HP 9000/700 series workstations running HP-UX. Currently, the *LabVIEW Student Edition* runs only on Macs and computers running Windows.

LabVIEW is a program development application, much like various commercial C or BASIC development systems. However, LabVIEW is different from those applications in one important respect. While other programming systems use text-based languages to create lines of code, LabVIEW uses a graphical programming language, G, to create programs in a flowchart-like form called a block diagram, eliminating a lot of the syntactical details. The following figure shows a simple LabVIEW user interface and the code behind it.

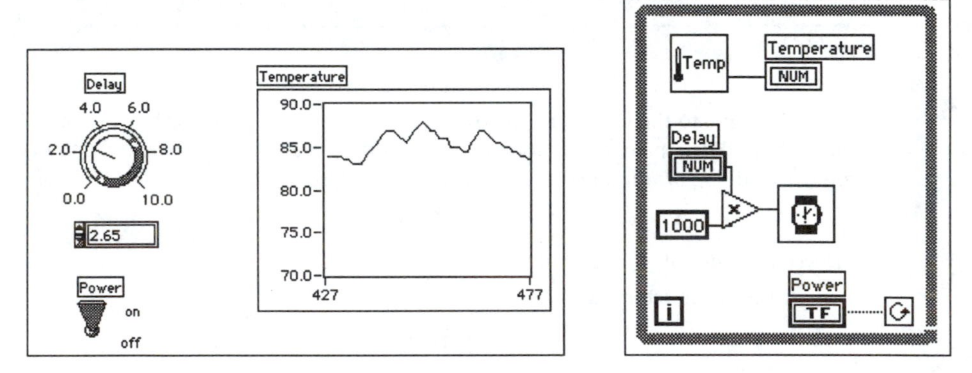

User Interface **Graphical Code**

LabVIEW uses terminology, icons, and ideas familiar to scientists and engineers. It relies on graphic symbols rather than textual language to describe programming actions. You can learn LabVIEW even if you have little or no programming experience, but you will find knowledge of programming fundamentals helpful. If you are brand new to programming or have forgotten some things, you may want to go over some basic concepts in a good beginning programming guide.

What Can LabVIEW Do for YOU?

LabVIEW tries to make your life as hassle free as it can. It has extensive libraries of functions and subroutines to help you with most programming tasks. LabVIEW also contains application-specific libraries for data acquisition, GPIB and serial instrument control, data analysis, data presentation, and data storage. The student edition Analysis Library contains a multitude of functions for signal generation, signal processing, filters, windows, statistics, regression, linear algebra, and array arithmetic.

LabVIEW includes conventional program development tools with which you can set breakpoints, single-step through the program, and animate the execution so you can observe the flow of data.

Because of LabVIEW's graphical nature, it is inherently a data presentation package. Output appears in any form you desire. Charts, graphs, and user-defined graphics comprise just a fraction of your output options. This text will show you how to present your data in all of these forms.

With its acquisition, analysis, and presentation tools, LabVIEW is functionally complete. Any computation possible in a conventional programming language is possible (and generally simpler) using the LabVIEW virtual instrument approach.

1.2 How Does LabVIEW Work?

LabVIEW programs are called *virtual instruments (VIs)* because their appearance and operation imitate actual instruments. However, behind the scenes they are analogous to main programs, functions, and subroutines from popular programming languages like C or BASIC. VIs have both an interactive user interface and a source code equivalent, and you can pass data between them. A VI has three main parts:

- The *front panel* is the interactive user interface of a VI, so named because it simulates the panel of a physical instrument. The front panel can contain knobs, push buttons, graphs, and many other controls (user inputs) and indicators (program outputs). You input data using a mouse and keyboard, and then view the results produced by your program on the screen. An example of a front panel is shown in the following figure.

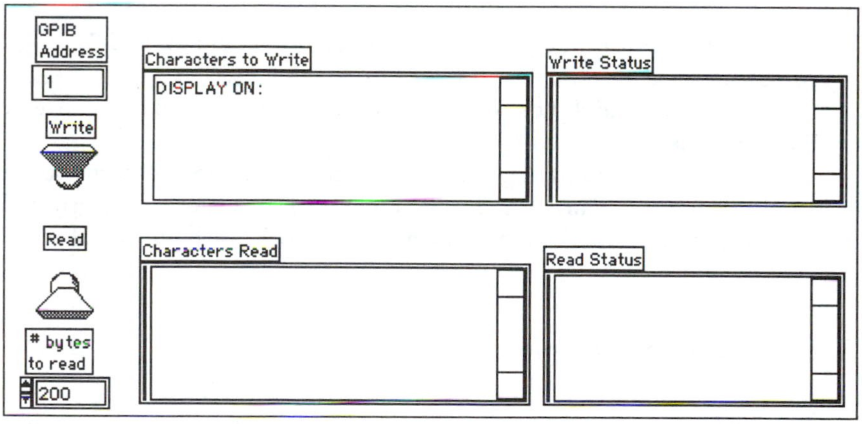

Front Panel

- The *block diagram*, shown below, is the VI's source code, constructed in LabVIEW's graphical programming language, G. *The block diagram, pictorial though it appears, is the actual executable program.* The components of a block diagram, *icons*, represent lower-level VIs, built-in functions, and program control structures. You draw wires to connect the icons together, indicating the flow of data in the block diagram.

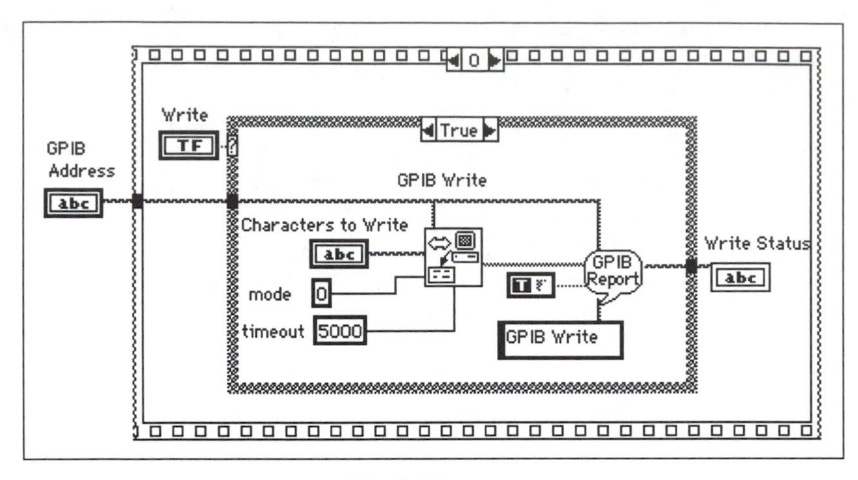

Block Diagram

- The *icon* and *connector* of a VI allow other VIs to pass data to the VI. The icon represents a VI in the block diagram of another VI. The connector defines the inputs and outputs of the VI. VIs are *hierarchical* and *modular*. You can use them as top-level programs, as subprograms within other programs, or even within other subprograms. A VI used within another VI, analogous to a subroutine, is called a *subVI*.

icon **connector**

With these features, LabVIEW promotes the concept of *modular programming*. First, you divide an application into a series of simple subtasks. Next, you build a VI to accomplish each subtask and then combine those VIs on a top-level block diagram to complete the larger task.

Modular programming is a plus because you can execute each subVI by itself, making debugging easy. Furthermore, many low-level subVIs often perform tasks common to several applications and can be used independently by each individual application.

1.3 Demonstration Examples

You will open existing LabVIEW programs and run them to get an idea of how LabVIEW works.

Note

Throughout this book, use the left mouse button (if you have one) unless we specifically tell you to use the right one.

EXERCISE 1.1
Temperature System Demo

Open and run the virtual instrument, or VI, called **Temperature System Demo.vi** by following these steps:

1. Launch LabVIEW by double-clicking on the LabVIEW icon. After a few moments, a blank, untitled Panel window appears. You can safely ignore it for this exercise.
2. Select **Open** from the **File** menu.
3. Next, open the Examples folder or directory by double-clicking on EXAMPLES. Then double-click on APPS and select TEMPSYS.LLB by double-clicking on it. Finally, double-click on **Temperature System Demo.vi** to open it. After a few moments, the Temperature System Demo front panel window appears, as shown in the next illustration. The front panel contains numeric controls, Boolean switches, slide controls, knob controls, charts, graphs, and a thermometer indicator.

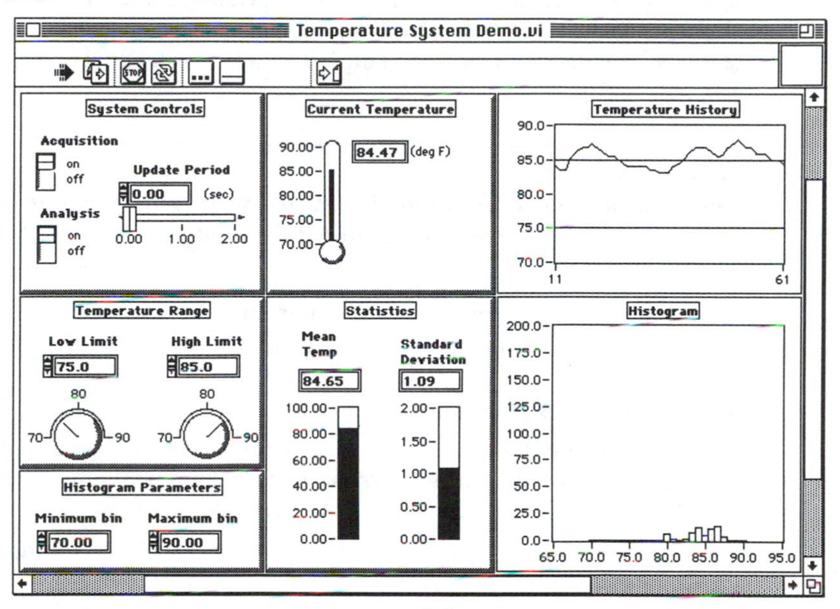

4. Run the VI by clicking on the run button. The button changes appearance to indicate that the VI is running. The palette, the row of icons on the top bar of the screen, also changes as the VI switches from edit mode to run mode (more on these modes later).

   Notice also that the stop button appears in the panel palette.

   Temperature System Demo.vi simulates a temperature monitoring application. The VI takes temperature readings and displays them in the thermometer indicator and on the chart. The readings are simulated in this case, but it is easy to modify the program to take real data. The Update Period slide controls how fast the VI acquires the new temperature readings. LabVIEW also plots high and low temperature limits on the chart; you can change these limits using the Temperature Range knobs. If the current temperature reading is out of the set range, LEDs light up next to the thermometer.

   This VI continues to run until you click the Acquisition switch to *off*. You can also turn the data analysis on and off. The analysis section shows you a running calculation of the mean and standard deviation of the temperature values and a histogram of the temperature values.

Tweaking Values

5. Use the cursor, which looks like the Operating tool while the VI is running, to change the values of the high and low limits. If you are in run mode, you will automatically have the Operating tool. If you are in edit mode, select the Operating tool by clicking on it in the palette at the top of the window. Once you have the Operating tool, highlight the old high or low value, either by clicking twice on the value you want to change, or by clicking and dragging across the value with the Operating tool. Then type in the new value and click on the enter button, located next to the run button on the Tools palette.
6. Change the Update Period slide control by placing the Operating tool on the slider, then clicking and dragging it to a new location.

   You can also operate slide controls using the Operating tool by clicking on a point on the slide to snap the slider to that location, by clicking on a scroll button to move the slider slowly toward the arrow, or by clicking in the slide's digital display and entering a number.

Note

LabVIEW does not accept values in digital displays until you press the enter button or click the mouse in an open area of the window.

---

7. Try adjusting the other controls in a similar manner.
8. Stop the VI by clicking on the Acquisition switch.

Examine the Block Diagram

The block diagram shown below represents a complete LabVIEW application. You don't need to understand all of these block diagram elements right now—we'll deal with that later. Just get a feel for the nature of a block diagram.

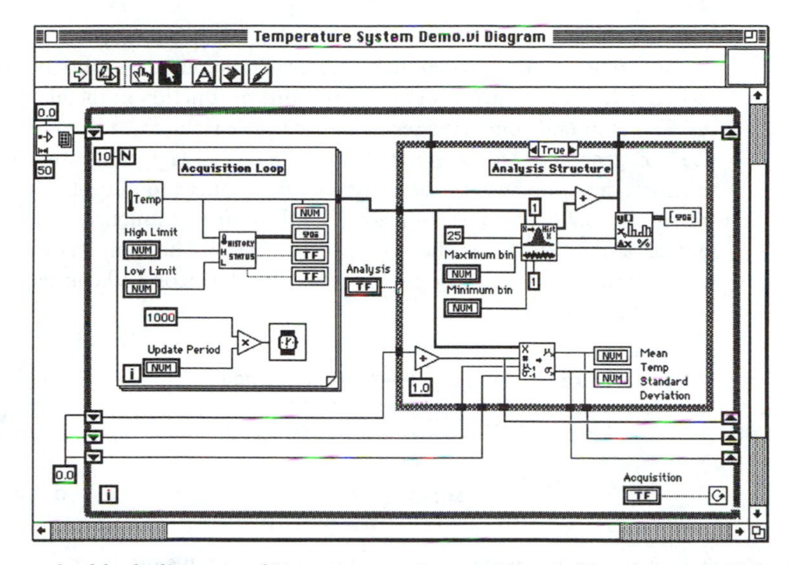

9. Open the block diagram of **Temperature System Demo.vi** by choosing **Show Diagram** from the **Windows** menu.
10. Examine the different objects in the diagram window. Each front panel has an accompanying block diagram, which is the VI equivalent of a program. You build the block diagram using the graphical programming language G. You can think of the block diagram as LabVIEW's source code. The components of the block diagram represent program nodes such as For Loops, Case structures, and arithmetic functions. The components are wired together to show the flow of data within the block diagram. Don't panic at the detail shown here! These structures are explained step by step later in this book.

Hierarchy

LabVIEW's power lies in the hierarchical nature of its VIs. After you create a VI, you can use it as a subVI in the block diagram of a higher-level VI, and you can have an essentially unlimited number of layers in the hierarchy. To demonstrate this versatile ability, look at a VI that **Temperature System Demo.vi** uses as a subVI in its block diagram.

11. Open the **Temperature Status** subVI by double-clicking on the subVI icon. The front panel shown in the following illustration springs to life.

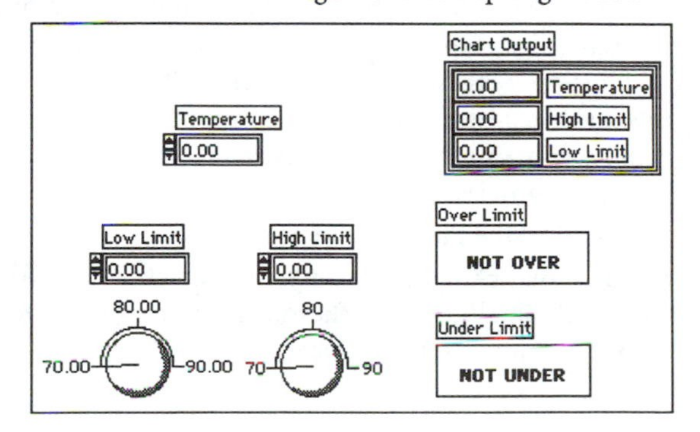

Icon/Connector

The icon/connector provides the graphical representation and parameter definitions needed if you want to use a VI as a subroutine or function in the block diagrams of other VIs. The icon and connector reside in the upper right corner of the VI's panel window. The icon graphically represents the VI in the block diagram of other VIs. The connector terminals determine where you must wire the inputs and outputs on the icon. These terminals are analogous to parameters of a subroutine or function. They correspond to the controls and indicators on the front panel of the VI. The icon sits on top of the connector pattern until you choose to view the connector. The icon and connector of the **Temperature Status** subVI are shown.

icon **connector**

By creating subVIs, you can make your block diagrams modular. This modularity makes VIs easy to debug, understand, and maintain.

12. Select **Close** from the **File** menu of the **Temperature Status** subVI. Do not save any changes.
13. Select **Close** from the **File** menu of **Temperature System Demo.vi,** and do not save any changes.

Note
Selecting Close from the File menu of a VI diagram closes the diagram window only. Selecting Close on a panel window closes both the panel and the diagram.

EXERCISE 1.2
Frequency Response Example

This example measures the frequency response of an unknown "black box." A function generator supplies a sinusoidal input to the black box (hint: it contains a bandpass filter). A digital multimeter measures the output voltage of the black box. Notice that while this VI uses subVIs to simulate a function generator and a digital multimeter, real instruments could easily be hooked up to a real black box to provide real-world data. The subVIs would then use the GPIB or serial port to bring in or send out data instead of simulating it.

Open, run, and observe the VI.

1. Select **Open** from the **File** menu to open the VI.
2. Double-click on EXAMPLES. Next, double-click on APPS; then double-click on FREQRESP.LLB. Finally, double-click on **Frequency Response.vi**. The front panel shown in the next illustration should appear.

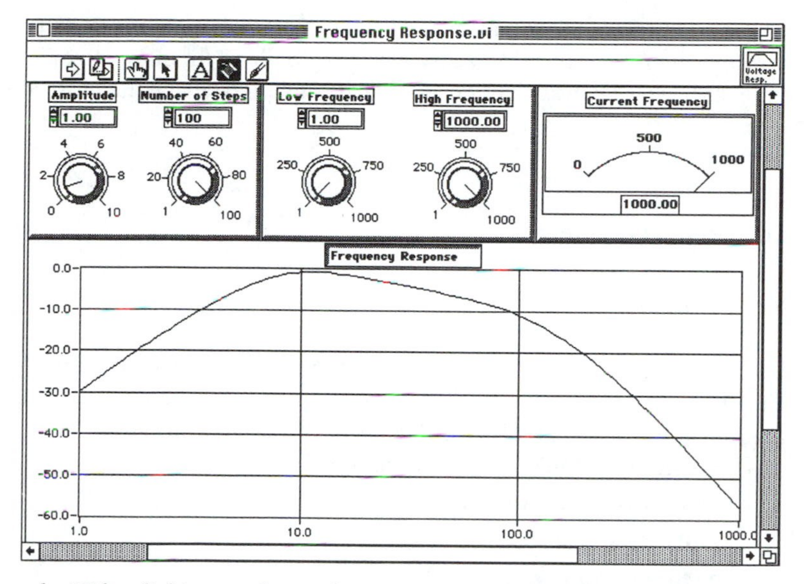

3. Run the VI by clicking on the run button. You can determine the amplitude of the input sine wave and the number of steps the VI uses to find the frequency response by changing the Amplitude control and the Number of Steps control and clicking the run button. You can also determine the frequency sweep by specifying the upper and lower limits with the Low Frequency and High Frequency knobs. Play with these controls and observe the effect they have on the output of the "black box;" then close the VI.
4. Open and examine the block diagram by choosing **Show Diagram** from the **Windows** menu.

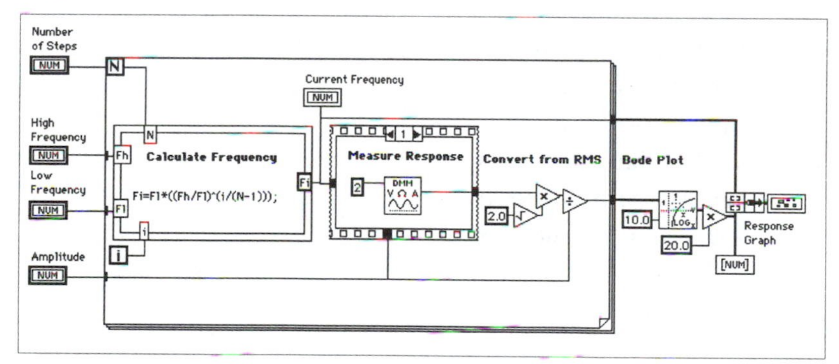

5. Close the VI by selecting **Close** from the **File** menu.

1.4 Wrap It Up!

LabVIEW is a powerful and flexible instrumentation and analysis software system. It uses the graphical programming language G to create programs called *virtual instruments*, or

VIs, in a flowchart-like form called a *block diagram.* The user interacts with the program through the *front panel.* LabVIEW has many built-in functions to facilitate the programming process. The next chapters will teach you how to make the most of LabVIEW's many features.

1.5 Additional Exercises

More Examples

Look through and run other VIs in the EXAMPLES directory or folder to get an idea of the LabVIEW environment and what you can do with it. The top-level examples (those that are not subroutines) are listed below. You will also find the solutions to every exercise in this book in the EXAMPLES directory or folder.

LabVIEW <--> GPIB.vi
Fluke 8840A.vi
AI Single Point.vi
AI Multi Point.vi
AO Single Point.vi
AO Multi Point.vi
Digital Read.vi
Digital Write.vi
Serial Communication.vi
Pulse Demo.vi
Frequency Response.vi
Temperature System Demo.vi
Building Arrays.vi
Temperature Analysis.vi
Separate Array Values.vi
Array to Bar Graph Demo.vi
Charts.vi
Waveform Graph.vi
X vs. Y graph.vi
Build String.vi
Extract Numbers.vi
Parse String.vi
Array Exercise.vi
Debug Exercise.vi
Editing Exercise.vi
Shift Register Example.vi
Pulse Demo

Amplitude Spectrum Example
Curve Fit Example
Extract the Sine Wave.vi
Integral & Derivative Example
IIR Filter Design
Global Switch.vi
Using Globals.vi
ASCII to Character.vi
Character to ASCII.vi

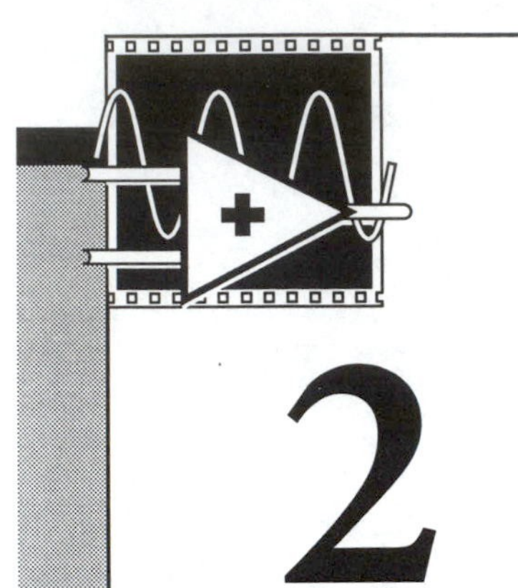

2

LabVIEW and Data Acquisition, Transfer, and Analysis

Overview

This chapter explains how to communicate with the outside world using LabVIEW. LabVIEW can command plug-in data acquisition (DAQ) boards to acquire or generate analog and digital signals. For example, you might use data acquisition boards and LabVIEW to monitor a temperature, send TTL signals to an external system, or determine the frequency of an unknown signal. LabVIEW also facilitates data transfer over the General Purpose Interface Bus (GPIB) and through the serial port. The GPIB bus is frequently used to communicate with oscilloscopes, scanners, and film recorders, and to drive instruments from remote locations. Once you have acquired or received your data, you can use LabVIEW's many analysis VIs to process and manipulate it. This chapter gives a very brief overview; you will learn more about DAQ, GPIB, serial communication, and analysis in the Examples section, Appendix A, and in the appropriate chapters of the Function Reference section.

Your Goals

- Understand the nature of data acquisition and GPIB
- Be able to describe the components of a typical data acquisition or GPIB system
- Learn about your computer's built-in serial port
- Understand the usefulness of analysis functions

Key Terms

DAQ
GPIB
IEEE 488
serial port

2.1 What Is Data Acquisition?

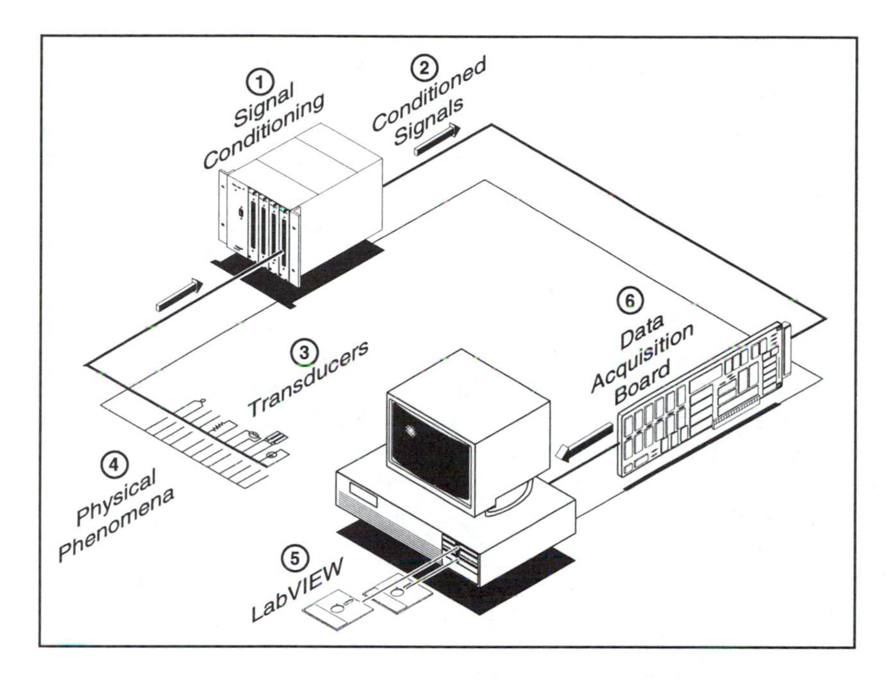

DAQ System

Data acquisition is simply the process of bringing a real-world signal, such as a voltage, into the computer for processing, analysis, storage, or other data manipulation. The above illustration shows the components of a *DAQ* system. *Physical phenomena* represent the real-world signals you are trying to measure, such as speed, temperature, humidity, pressure, flow, pH, start-stop, and so on. *Transducers* sense physical phenomena and produce electrical signals. For example, thermocouples, a type of transducer, convert temperature into a voltage that an A/D (analog to digital) converter can measure. Other examples of transducers include strain gauges, flowmeters, and pressure transducers, which measure force, rate of flow, and pressure, respectively. In each case, the electrical signal produced by the transducer is directly related to the phenomenon it monitors.

LabVIEW can command DAQ boards to read analog input signals (A/D conversion), generate analog output signals (D/A conversion), read and write digital signals, and manipulate the on-board counters for frequency measurement, pulse generation, etc. The voltage data goes into the plug-in DAQ board in the computer, which sends the data into computer memory for storage, processing, or other manipulation.

Signal conditioning modules "condition" the electrical signals generated by transducers so they are in a form that the DAQ board can accept. For example, you would want to isolate a high-voltage input lest you fry both your board and your computer. Signal conditioning modules can apply many different types of conditioning—amplification, linearization, filtering, isolation, and so on. Not all applications will require signal conditioning, but you should pay attention to your specifications to make sure.

2.2 What Is GPIB?

Hewlett Packard developed the General Purpose Interface Bus, or GPIB in the late 1960s to facilitate communication between computers and instruments. A bus is simply the means by which computers and instruments transfer data, and the GPIB provided a much-needed specification and protocol to govern this communication. The Institute of Electrical and Electronic Engineers (IEEE) standardized the GPIB in 1975, and the GPIB became known as the *IEEE 488* standard. The GPIB's original purpose was to provide computer control of test and measurement instruments. However, its use has expanded beyond these applications into other areas, such as computer-to-computer communication and control of scanners and film recorders.

The GPIB is a digital, 24-conductor parallel bus. It consists of eight data lines, five bus management lines (ATN, EOI, IFC, REN, and SRQ), three handshake lines, and eight ground lines. The GPIB uses an eight-bit parallel, byte-serial, asynchronous data transfer scheme. In other words, whole bytes are sequentially handshaked across the bus at a speed determined by the slowest participant in the transfer. Because the GPIB sends data in bytes (1 byte = 8 bits), the messages transferred are frequently encoded as ASCII character strings. Your computer can do GPIB only if it has a GPIB board and the proper drivers installed.

You can have many instruments and computers connected to the same GPIB bus. Every device, including the computer interface board, must have a unique GPIB address between 0 and 30, so that the data source and destinations may be specified. Address 0 is normally assigned to the GPIB interface board. The instruments on the GPIB can use addresses 1 through 30. The GPIB has one *Controller*, usually your computer, that controls the bus. To transfer instrument commands and data on the bus, the Controller *addresses* one *Talker* and one or more *Listeners*. The data strings are then handshaked across the bus from the Talker to the Listener(s). The LabVIEW GPIB VIs automatically handle the addressing and most other bus management functions, saving you the hassle of low-level programming. The following illustration shows a typical GPIB system.

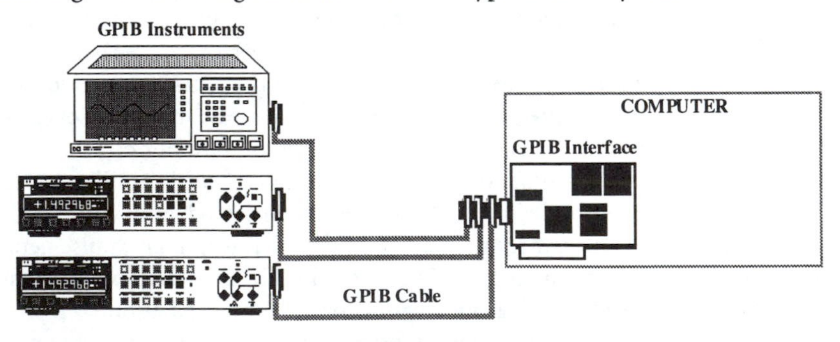

GPIB System

While the GPIB is one way to bring data into a computer, it is fundamentally different from data acquisition using DAQ boards that plug into the computer. Using a special protocol, the GPIB handshakes data into the computer that has been acquired by another computer or instrument, while data acquisition involves connecting a signal directly up to a DAQ board in the computer.

2.3 Communication Using the Serial Port

Serial communication is another popular means of transmitting data between a computer and another computer, or a peripheral device such as a programmable instrument. It uses the built-in *serial port* in your computer. Serial communication uses a transmitter to send data one bit at a time over a single communication line to a receiver. You can use this method when data transfer rates are low, or when you must transfer data over long distances. It is slower and less reliable than the GPIB, but you do not need a board in your computer to do it and your instrument does not need to conform to the GPIB standard. The following illustration shows a typical serial communication system.

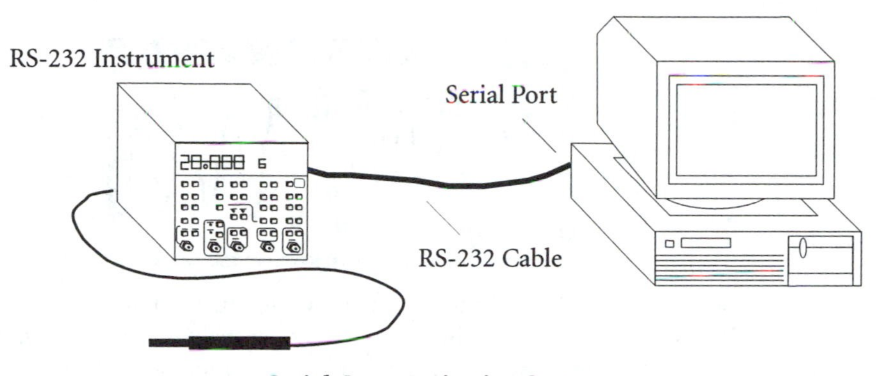

Serial Communication System

Serial communication is handy because most computers have one or two serial ports built in—you can send and receive data without buying any special hardware. Many GPIB instruments also have built-in serial ports. However, unlike GPIB, a serial port can communicate with only one device, which can be limiting for some applications. Serial port communication is also painstakingly slow.

The LabVIEW **Serial** library contains functions used for serial port operations.

2.4 Real-World Applications: Why We Analyze

Once you get data into your computer, you may want to process your data somehow. Modern, high-speed floating-point numerical and digital signal processors have become increasingly important to real-time and analysis systems. A few of the many possible applications include biomedical data processing, speech synthesis and recognition, and digital audio and image processing.

The importance of integrating analysis libraries into engineering stations is that the raw data collected from your DAQ board or GPIB instrument does not always immediately convey useful information, as shown below. Often you must transform the signal, remove noise perturbations, correct for data corrupted by faulty equipment, or compensate for environmental effects such as temperature and humidity.

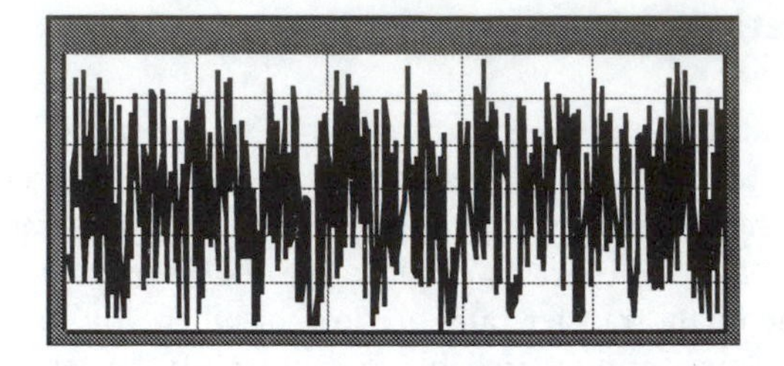

By analyzing and processing the digital data, you can extract the useful information from the noise and present it in a form more comprehensible than the raw data. The processed data looks more like this:

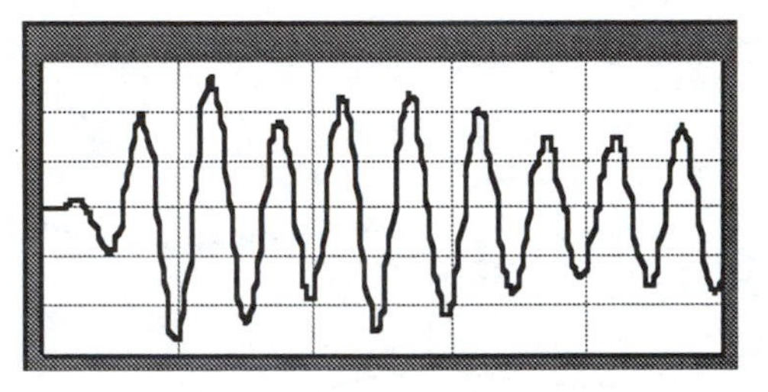

The LabVIEW block diagram programming method and the extensive set of LabVIEW analysis VIs simplify the development of analysis applications. The following example block diagram illustrates the LabVIEW programming concept.

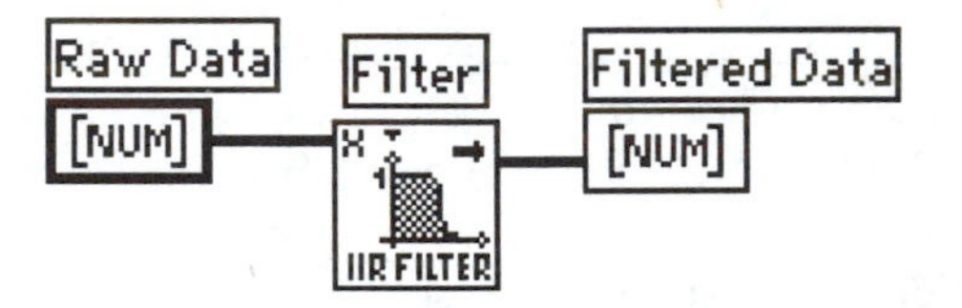

Because LabVIEW analysis VIs give you popular data analysis techniques in discrete VIs, you can wire them together, as shown above, to analyze data. Instead of worrying about implementation details for analysis routines as you do in most programming languages, you can concentrate on solving your data analysis problems.

The LabVIEW analysis VIs efficiently process blocks of information represented in digital form. They cover the following major processing areas:

- Pattern generation
- Digital signal processing
- Digital filtering
- Smoothing windows
- Statistical analysis
- Curve fitting
- Linear algebra
- Numerical analysis
- Measurement-based analysis

2.5 Wrap It Up!

LabVIEW's built-in functions facilitate communication with external devices. You can use National Instruments *DAQ* boards, managed by LabVIEW, to read and generate analog input, analog output, and digital signals and also to perform counter/timer operations. LabVIEW can also control communication over the *GPIB* bus if you have a GPIB board. If you don't have any special hardware, LabVIEW can communicate with other devices through your computer's *serial port.*

GPIB and DAQ boards need special driver and configuration files. While these files come with the boards, you should install them using the *LabVIEW Student Edition* Driver Setup disk to make sure you have a version that is compatible with LabVIEW. You can obtain this disk from your instructor.

LabVIEW analysis VIs make it easy for you to process and manipulate data once you have brought it into your computer. Rather than working through tricky algorithms by hand or trying to write your own code, you can simply access the built-in LabVIEW function that suits your needs.

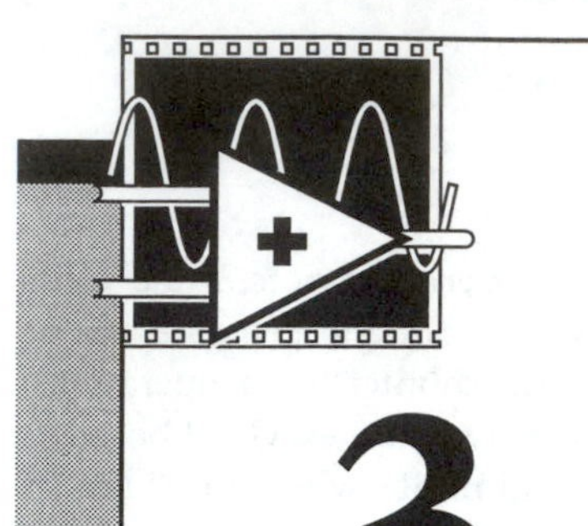

3

The LabVIEW Environment

Overview

In this chapter, you will investigate the LabVIEW environment and learn how its three parts—the *front panel, block diagram,* and *icon/connector*—work together. You will also learn about LabVIEW's menus, its two modes of operation (run mode and edit mode), the Help Window, and subVIs.

What is a Virtual Instrument, or VI? When all three main components—the front panel, the block diagram, and the icon/connector—are properly developed, you have a VI that can stand alone or be used as a subVI in another program.

Your Goals

- Understand and demonstrate the use of the front panel, block diagram, and icon/connector
- Learn the difference between the function and appearance of controls and indicators
- Understand the principle of dataflow programming
- Become familiar with LabVIEW menus, both pop-up and pull-down
- Know when to use run mode and when to use edit mode (hint: you can't make changes while you're in run mode)
- Learn what the Help window can do for you
- Understand what a subVI is and why it's useful
- Work through the exercises to get a feel for how LabVIEW works

Key Terms

control	terminal	icon	run mode
indicator	node	connector	edit mode
wire	dataflow	pop-up menus	Help window
subVI			

3.1 Front Panels

Simply put, the front panel is the window through which the user interacts with the program. When you run a VI, you must have the front panel open so you can give inputs to the executing program. You will also find the front panel indispensable if you want to see your program's output. The following illustration shows an example of a LabVIEW front panel.

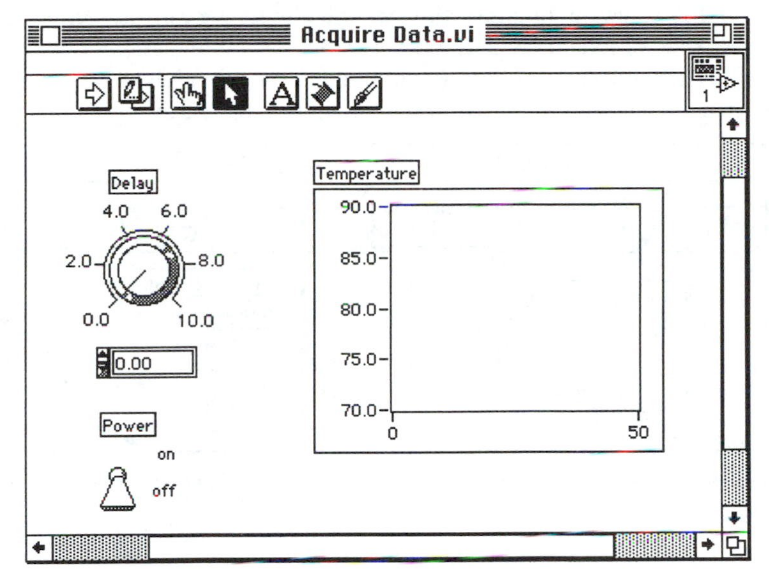

LabVIEW Front Panel

Controls and Indicators

The front panel is primarily a combination of *controls* and *indicators.* **Controls** simulate typical input devices you might find on a conventional instrument, such as knobs and switches. Controls supply data to the block diagram of the VI. **Indicators** simulate output displays that show data the program acquires or generates. Consider this simple way to think about controls and indicators:

Controls = Inputs

Indicators = Outputs

You "drop" controls and indicators onto the front panel by selecting them from the **Controls** menu at the top of the front panel window. Once an object is on the panel, you can easily adjust its size, shape, and position.

3.2 Block Diagrams

The block diagram window holds the graphical source code of a LabVIEW VI. LabVIEW's block diagram corresponds to the lines of text found in a popular language like C or BASIC—it is the actual executable code. You construct the block diagram by wiring together objects that perform specific functions. This pictorial representation, which looks kind of like the flowcharts used in conventional programming for preplanning, *is* the LabVIEW program. In this section, we will discuss the various components of a block diagram: *terminals*, *nodes*, and *wires*.

The following simple VI computes the sum of and difference between two numbers. Its diagram shows examples of terminals, nodes, and wires.

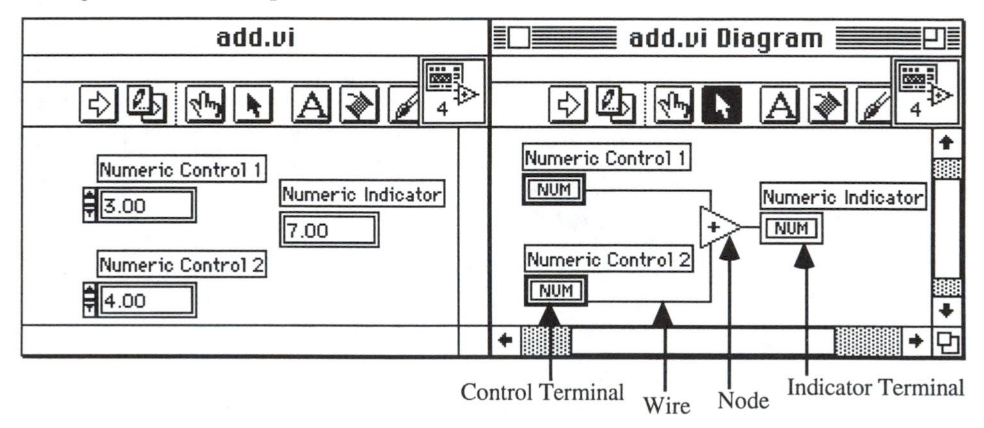

Terminals

When you place a control or indicator on the front panel, LabVIEW spontaneously creates a corresponding *terminal* on the block diagram. You cannot delete a terminal on the block diagram that belongs to a control or indicator, though you may try to your heart's content. The terminal disappears only when you delete its control or indicator on the front panel.

Note

Control terminals have THICK borders, while indicator terminal borders are THIN. It is VERY important to distinguish between the two since they are NOT functionally equivalent (Control=Input, Indicator=Output, so they are NOT interchangeable).

You can think of terminals as entry and exit ports, or as sources and destinations. Data that you enter into the Numeric Controls exits the front panel and enters the block diagram through the Numeric Control terminals on the diagram. As you may have noticed, the **Add** function icon also has a terminal. The data follows the wire and enters the **Add** function. When the **Add** function completes its internal calculations, it produces new data values at its exit terminal. The data flows to the Numeric Indicator terminal and reenters the front panel, where it is displayed for your enjoyment or torment, depending on how far along you are in the debugging process.

Nodes

A *node* is a program execution element. Nodes are analogous to statements, operators, functions, and subroutines in standard programming languages. The **Add** and **Subtract** functions represent one type of node. A structure is another type of node. Structures can execute code repeatedly or conditionally, similar to loops and case statements in traditional programming languages. LabVIEW also has special nodes, called Formula Nodes, which are useful for evaluating text-based formulas.

Wires

Wires are the data paths between source and destination terminals. You cannot wire a source terminal to another source, or a destination terminal to another destination, but you can wire one source to several destination terminals.

Note

This principle explains why controls and indicators are not interchangeable. Controls are source terminals, while indicators are sinks.

Each wire has a different style or color, depending on the data type that flows through the wire. The block diagram shown above depicts the wire style for a numeric scalar value—a thin, solid line. The following chart shows a few wires and corresponding types.

	Scalar	1D Array	2D Array	
Number				Orange
Boolean				Green
String				Purple

Basic Wires Styles Used in Block Diagrams

To avoid confusing your data types, simply match up the colors and styles!

Dataflow Programming—Going with the Flow

Since LabVIEW is not a text-based language, its code cannot execute "line by line." The principle that governs LabVIEW program execution is called *dataflow.* Stated simply, a node executes only when data arrives at all its input terminals; the node supplies data to all of its output terminals when it finishes executing; and the data passes immediately from source to sink, or destination, terminals. Dataflow contrasts strikingly with the control flow method of executing a text-based program, in which instructions are executed in the sequence in which they are written. This difference may take some getting used to. While traditional execution flow is instruction driven, dataflow execution is data driven or *data dependent.*

3.3 The Icon and the Connector

When your VI operates as a subVI inside another VI, its controls and indicators receive data from and return data to the VI that calls it. A VI's *icon* represents it in the block diagram of another VI. An icon can be a pictorial representation or a textual description of the VI, or a combination of both.

The VI's *connector* is a set of terminals that correspond to its controls and indicators. The connector is much like the parameter list of a function call; the connector terminals act like little graphical parameters to pass data to and from the subVI. Each terminal corresponds to its very own control or indicator on the front panel. A subVI's connector receives data at its input terminals and passes the data to the subVI code via the subVI controls. It also receives the results from the subVI indicators at its output terminals.

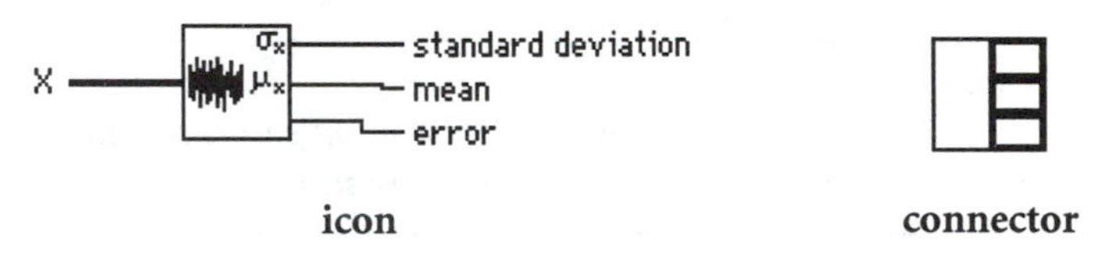

An icon and its underlying connector

Every VI has a default icon, which is displayed in the icon pane in the upper right corner of the Panel and Diagram windows. The default icon is depicted in the following illustration.

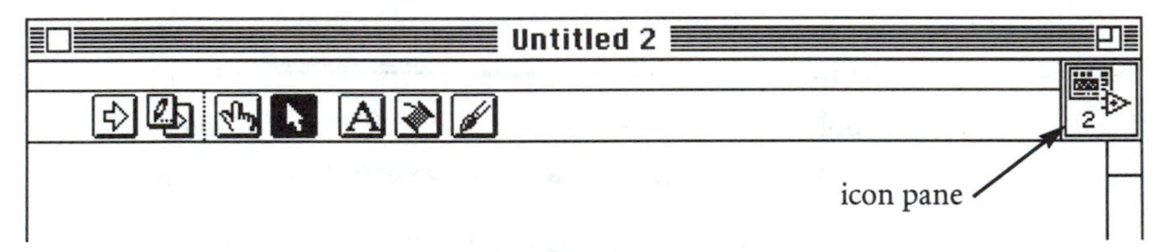

Every VI also has a connector, which you access by choosing **Show Connector** from the front panel icon pane pop-up menu. When you show the connector for the first time, LabVIEW helpfully suggests a connector pattern that has one terminal for each control or indicator on the front panel. You can select a different pattern if you desire, and you can assign up to 20 terminals before you run out of space on the connector.

EXERCISE 3.1
Getting Started

Okay, you've read enough. Now it's time to get some hands-on experience. Go ahead and launch LabVIEW. You will step through the creation of a simple LabVIEW VI that generates a random number and plots its value on a waveform chart.

1. After you've launched LabVIEW, you should have an "Untitled 1" front panel on your screen. Go to the **Controls** menu and select "waveform chart" from the **Array & Graph** palette. You will notice that, as you run the cursor over the icons in the **Array & Graph** palette, that the selected icon's name appears at the bottom of the palette.

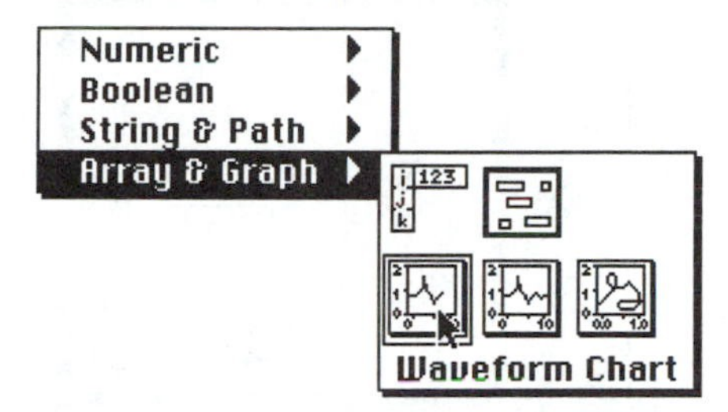

The chart will appear on your front panel. Now select the Positioning tool by clicking on its icon in the row of buttons at the top of the window (called the tool palette). Move the chart to the center of the panel window by clicking on it with the Positioning tool and dragging it where you want it.

2. Go back to the **Controls** menu and select "vertical toggle switch" from the **Boolean** palette.

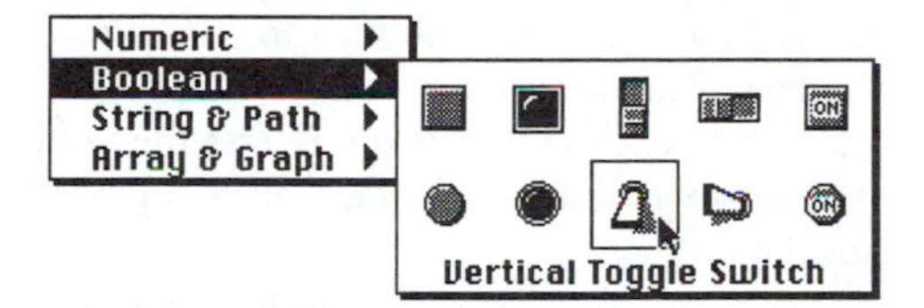

Move it next to the chart as shown in the following illustration.

3. Change the scale on the chart by selecting the Operating tool and dragging it across the number "10." The "10" should be highlighted. Now type in 1.0 and click on the enter button that appears in the panel palette.

4. Switch to the block diagram by selecting **Show Diagram** from the **Windows** menu. You will build the diagram shown in the following illustration. You should see two terminals already there.

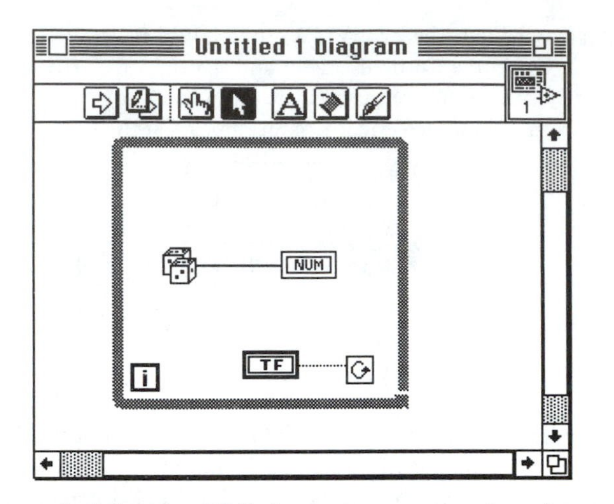

5. Now put the terminals inside a While loop to repeat execution of a segment of your program. Go to the **Structs & Constants** palette in the **Functions** menu and select the While loop from the top row of icons. Your cursor will change to a little loop icon. Now enclose the NUM and TF terminals by using the cursor. Click and hold down the mouse button while you drag the cursor from the upper left to the lower right corners of the objects you wish to enclose. When you release the mouse button, the dashed line that is drawn as you drag will change into the While Loop border. Make sure to leave some extra room inside the loop.
6. Go to the **Functions** menu and select **Random Number (0-1)** from the bottom row of the **Arithmetic** palette.

7. Select the Positioning tool from the tool palette and arrange your diagram objects so that they look like the previous illustration.
8. Click on the wiring tool to select it. Now click once on the **Random Number** icon, drag the mouse over to the NUM terminal, and click again. You should now have a solid orange wire connecting the two icons. If you mess up, you can select the wire or wire fragment with the Positioning tool and then hit the <delete> key to get rid of it. Now wire the TF terminal to the conditional terminal of the While loop. The loop will execute while the switch on the front panel is TRUE (in the "up" position).
9. You should be about ready to run your VI. First, switch back to the front panel by selecting **Show Panel** from the **Windows** menu. Using the Operating tool, flip the switch to the "up" position. Now click on the run button to run your VI. You will see a series of

random numbers plotted continuously across the chart. When you want to stop, click the switch to the down position.

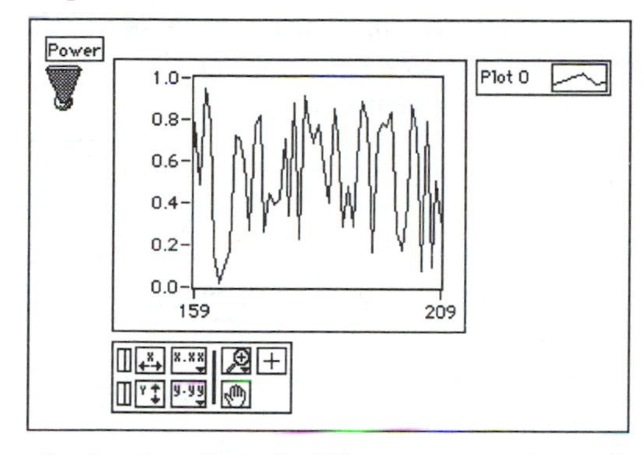

10. Save your VI by selecting **Save** from the **File** menu. Name it **Random Number.vi.**

~ ~ ~

3.4 Pull-Down Menus

LabVIEW has two types of menus, pull-down and pop-up, both of which you will use extensively in your program development. This section will cover pull-down menus. You used some of them in the last exercise, and now you will learn more about what they can do. You will probably find it helpful to look through the menus as we explain them.

The menu bar at the top of a VI window contains several pull-down menus. When you click on a menu bar item, a menu appears below the bar. The pull-down menus contain items common to many applications, such as **Open**, **Save**, **Copy**, and **Paste**, and many others particular to LabVIEW. Many menus also list shortcut keyboard combinations for you to use if you choose. The shortcuts shown in the example menus below work on the Mac. On the PC, the shortcuts are similar, with the <Command> key replaced by the <Control> key.

File
New ⌘N
Open... ⌘O
Close ⌘W
Save ⌘S
Save As...
Save A Copy As...
Save with Options...
Revert...
Page Setup...
Print Documentation...
Print Window... ⌘P
Get Info... ⌘I
Edit VI Library...
Mass Compile...
Quit ⌘Q

Edit
Undo ⌘Z
Cut ⌘X
Copy ⌘C
Paste ⌘V
Clear
Remove Bad Wires ⌘B
Panel Order...
Move Forward ⌘K
Move Backward ⌘J
Move To Front
Move To Back
Preferences...

Operate
Run ⌘R
Stop ⌘.
Change to Run Mode ⌘M
Make Current Values Default
Reinitialize All To Default

Controls
Numeric ▶
Boolean ▶
String & Path ▶
Array & Graph ▶

Note
Be sure to pay attention and learn these keyboard shortcuts, shown in the menus to the right of their corresponding commands. They are generally much faster and easier to use than the menus themselves.

File Menu

Pull down the **File** menu, which contains commands common to many applications such as **Save** and **Print**. You create new VIs or open existing ones from the **File** menu. You can also **Get Info** on a VI from this menu.

Edit Menu

Take a look at the **Edit** menu. It has some universal commands, like **Cut**, **Copy** and **Paste**, that let you edit your window. You can also change the layering of objects (that is, if two are in the same place, which one is on top?) and remove bad wires (the significance of bad wires will become apparent soon enough).

Operate Menu

You can run or stop your program from the **Operate** menu. You can also change a VI's default values and switch between LabVIEW's two modes, run mode and edit mode.

Controls Menu

Make sure your front panel window is active and click on the **Controls** menu, which actually contains both controls and indicators. Run your mouse through the palettes that appear when you highlight a **Controls** menu selection. From this menu, you select the control and indicator graphics that you want to appear on your front panel.

Note
The Controls menu is present only in the Front Panel window, NOT in the Block Diagram.

Another Note
The Controls menu is also accessible if you pop up in an empty area of the front panel. "Popping up" is defined as right-mouse-clicking on the PC and <Command>-clicking on the Mac . The advantage to accessing the Controls menu by popping up is that the selected control or indicator appears wherever you've put your cursor rather than at a location LabVIEW selects. It's also faster than pulling down menus. Try this now so you get in the habit quickly!

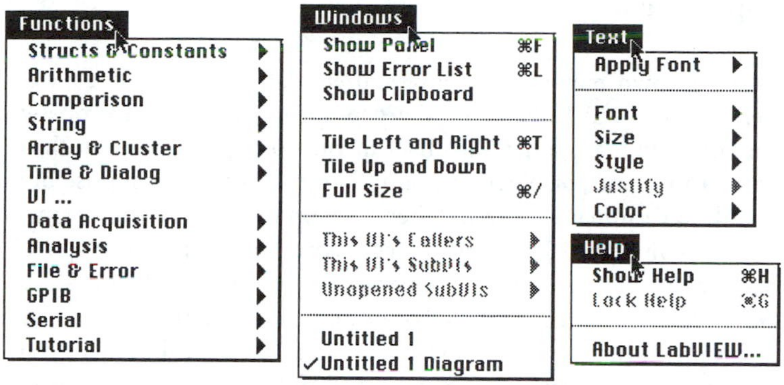

Functions
Structs & Constants
Arithmetic
Comparison
String
Array & Cluster
Time & Dialog
VI ...
Data Acquisition
Analysis
File & Error
GPIB
Serial
Tutorial

Windows
Show Panel ⌘F
Show Error List ⌘L
Show Clipboard
Tile Left and Right ⌘T
Tile Up and Down
Full Size ⌘/
This VI's Callers
This VI's SubVIs
Unopened SubVIs
Untitled 1
✓Untitled 1 Diagram

Text
Apply Font
Font
Size
Style
Justify
Color

Help
Show Help ⌘H
Lock Help ⌘G
About LabVIEW...

Functions Menu

Switch to the block diagram and observe the **Functions** menu, which allows you to choose the built-in functions you want to use in your VI.

Watch Out!
The Functions menu can be found only in the Block Diagram window!

Another Note
The Functions menu is also accessible if you pop up in an empty area of the block diagram.

Windows Menu

Pull down the **Windows** menu. Here you can toggle between the panel and diagram windows, show the error list and the clipboard, "tile" both windows so you can see them at the same time, and switch between open VIs. The **Windows** menu can also show you which subVIs are called by the current VI.

Text Menu

You can change the font, size, style, justification, and color of LabVIEW text using options in the **Text** menu.

Help Menu

You can show, hide, or lock the contents of the Help window using the **Help** menu. You can also get information about LabVIEW such as version number and available memory.

3.5 Pop-Up Menus

As if pull-down menus didn't give you enough to worry about, we will now discuss the other type of LabVIEW menu, the *pop-up menu.* You will probably use pop-up menus more often than any other LabVIEW menu. To pop up, position the cursor on the desired object, front panel, or block diagram; then click the right mouse button on a PC or hold down the <Command> key and then click on the Mac. A pop-up menu will appear.

Virtually every LabVIEW object, as well as empty front panel and block diagram space, has a pop-up menu of options and commands. *Options available in this pop-up menu depend on the object itself.* For example, a numeric control will have a very different pop-up menu than a graph indicator. You will find that instructions throughout the manual suggest that you select a command or option from an object pop-up menu, so try it now!

Note: How to *Pop Up*
PC: right mouse click on the object
Mac: <Command>-click on the object
Pop-up menus are ever-present in LabVIEW, so remember, when in doubt, POP UP!

Pop-Up Menu Things to Keep in Mind

Menu items that expand into submenus are called hierarchical menus and are denoted by a right arrowhead, as shown in the following illustration.

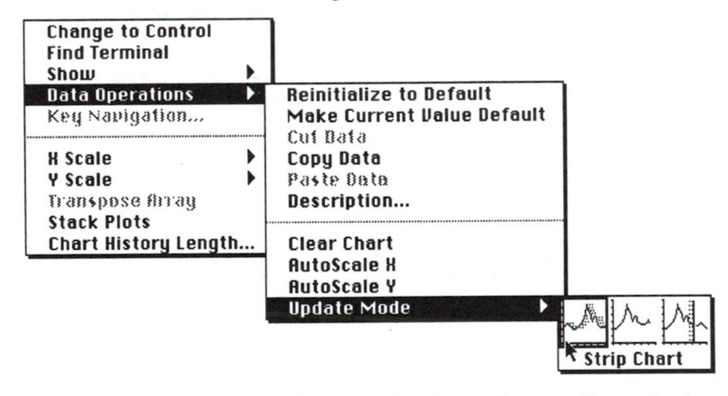

Hierarchical menus sometimes have a selection of mutually exclusive options. The currently selected option is denoted by a check mark for text-displayed options, or by a box for graphical options.

Some menu items pop up dialog boxes containing options for you to configure. Menu items leading to dialog boxes are denoted by ellipses (**...**).

Menu items without right arrowheads or ellipses are usually *commands* that execute immediately upon selection. A command usually appears in verb form, such as **Change to Indicator.** When selected, many commands are replaced in the menu by their inverse commands. For example, after you choose **Change to Indicator**, the menu selection becomes **Change to Control**.

Note

Sometimes different parts of an object have different pop-up menus. For example, if you pop up on an object's label, the menu contains only a "Size to Text" option. Popping up elsewhere on the object gives you a full menu of options. So if you pop up and don't see the menu you want, pop up elsewhere on the object.

Nifty Pop-Up Features

Pop-up menus allow you to specify many traits of an object. The following options appear in numerous pop-up menus, and we thought they were important enough to describe them. We'll let you to figure out the other options, since we would put you to sleep detailing them all. The following illustration shows an example of a pop-up menu.

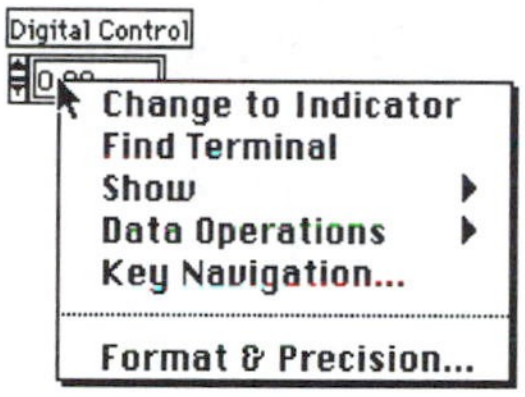

Digital Control
Change to Indicator
Find Terminal
Show
Data Operations
Key Navigation...
Format & Precision...

Change to Control and Change to Indicator

By selecting **Change to Control/Indicator**, you can turn an existing control (or input object), into an indicator (or output object) or vice versa. When an object is a control, its pop-up menu has the option, **Change to Indicator**. When it is an indicator, the pop-up menu reads **Change to Control**.

Watch Out!

Since Change to Control/Indicator is the first option in the pop-up menu, it is very easy to accidentally select it without realizing what you've done. The result may befuddle you because controls and indicators are not functionally interchangeable in a block diagram. Remember, a control terminal in the block diagram has a thicker border than an indicator terminal. Always pay attention to whether your objects are Controls or Indicators to avoid confusion!

Show Terminals/Show Icon

The **Show Terminals** or **Show Icon** pop-up option is available only on block diagram objects. If you want to see the connector terminals on a function or subVI, select **Show Terminals** from the pop-up menu. When you want to get the icon back, select **Show Icon**.

Show

Many items have **Show** menus with which you can show or hide certain cosmetic things like labels or scrollbars. If you select **Show**, you will get another menu off to the side, listing options of what can be shown. If an option has a check next to it, that option is currently visible; if it has no check, it is hidden. Release the mouse on an option to toggle its visibility.

Data Operations

The **Data Operations** pop-up menu has several handy options, so you can manipulate the data in a control or indicator.

Reinitialize to Default returns an object to its default value, while **Make Current Value Default** sets the default value to whatever is currently there.

Use **Cut Data**, **Copy Data**, and **Paste Data** to take data out or put data into a control or indicator.

Last but not least, **Description** allows you to enter or read a description of that control or indicator's function, one of LabVIEW's convenient documentation mechanisms.

3.6 Edit Mode and Run Mode

A VI can be in either *edit mode*, which lets you create or change a VI, or in *run mode*, where you can execute a VI. You cannot edit a VI in run mode. To draw a parallel with text-based languages, if a VI is in run mode, it has been successfully compiled and awaits your command to execute.

Watch Out!

The same object will have a different pop-up menu in run mode than it will in edit mode. If you cannot find a certain pop-up option, it is either not present for that object, you need to switch modes, or you should pop up elsewhere on the object.

If you want to switch to edit mode from run mode, click on the mode button or simply select **Change to Edit Mode** from the **Operate** menu. Better yet, use the keyboard shortcut: **<Ctrl><M>** under Windows, **<Cmd><M>** on the Mac. If you run the VI from run mode, the VI is still in run mode after executing.

Note

You can also run the VI from edit mode—it will convert itself to run mode, execute, and then return to edit mode when it is finished.

Tools

A *tool* is a special operating mode of the mouse cursor. You use tools to perform specific editing functions, similar to how you would use them in a standard paint program.

Edit Mode Palette

In edit mode, the editing tools become available on the edit mode palette, which you will find below the window menu bar.

Tool Palette—Edit Mode

The **Operating** tool lets you change values of front panel controls (and indicators if you are in edit mode). You can operate knobs, switches, and other objects with the operating tool, hence the name. It is the only tool available in run mode.

The **Positioning** tool selects, moves, and resizes objects.

A The **Labeling** tool creates and edits text labels.

The **Wiring** tool wires objects together on the block diagram and assigns controls and indicators on the front panel to terminals on the VI connector.

The **Coloring** tool brightens objects and backgrounds by allowing you to choose from a multitude of hues.

You change tools by clicking on the tool icon you want in the edit mode palette, or by pressing the tab key, which selects the next tool in sequence. You can also press the space bar to toggle between the Operating tool and the Positioning tool when the panel window is active, and between the Wiring tool and the Positioning tool when the diagram window is active.

The **warning** button appears if you have configured your VI to show warnings and you have any warnings outstanding. You can list the warnings by clicking on the button. A warning is not an error; it just alerts you that you are doing something strange (for example, you have a front panel control with nothing wired to it).

Run Mode Palette

When you are ready to test the VI, click on the mode button or select **Change to Run Mode** from the **Operate** menu. This command compiles your VI and puts the VI in run mode. The run mode palette is shown in the following illustration. In run mode, various debugging options become available, to help keep your frustration level low. Refer to Section 5.3, *Debugging Techniques*, for more information on using them.

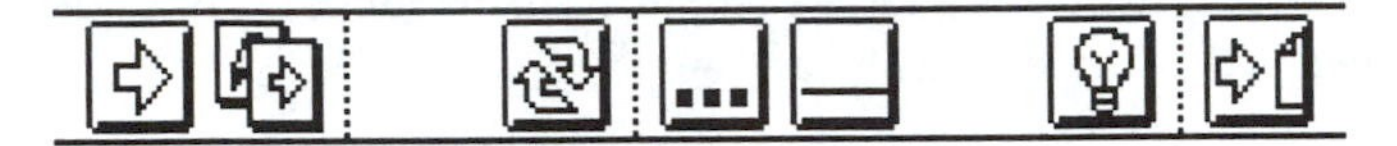

Tool Palette—Run Mode

The **run** button, which looks like an arrow, starts VI execution when you click on it. It changes appearance when a VI is actually running. When a VI won't compile, the **run** button is broken.

The **stop** button, easily recognizable because it looks like a tiny stop sign, appears after a VI begins to execute. You can click on this button to halt the VI.

Watch Out!

Using the stop button is like hitting the <break> key. Your program will stop immediately rather than coming to a graceful end, and data can be corrupted this way. You should code a more appropriate stopping mechanism into your program, as we will demonstrate later.

The **mode** button allows you to switch between run mode and edit mode.

The **continuous run** button causes the VI to execute over and over until you hit the **stop** button. It's kind of like a GOTO statement, so use it sparingly.

The **breakpoint** button, when set in a subVI, causes execution to suspend when the subVI is called. Then you can see what is going on and even change input values, another hassle-saving debugging mechanism.

The **step mode** button forces your VI to execute one step at a time. When you press it, the **single step** button appears. You click on the **single step** button each time you want the VI to execute another step.

The **execution highlighting** button causes the VI to highlight the flow of data as it passes through the diagram. When execution highlighting is on, you can see intermediate data values in your block diagram that would not otherwise appear.

The **print mode** button sends the front panel of a VI to the printer after the VI has completely finished executing.

3.7 Help

The LabVIEW *Help window* offers indispensable help information for functions, constants, subVIs, and controls and indicators. To display the window, choose **Show Help** from the **Help** menu or use the keyboard shortcut: **<Ctrl-H>** under Windows, **<Cmd-H>** on the Mac. You can move the Help window anywhere on your screen to keep it out of the way. The Help window is shown in the following illustration.

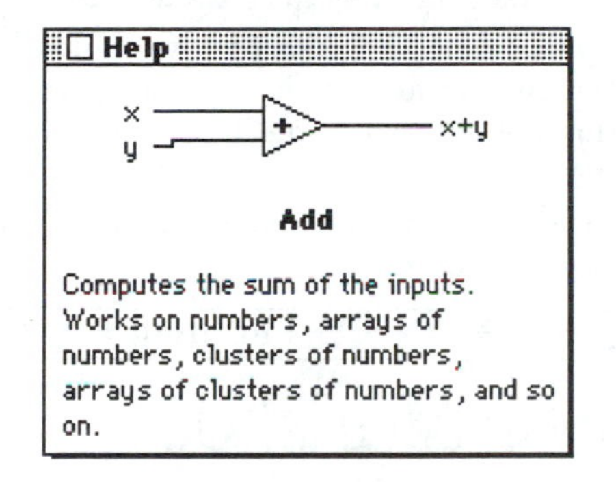

Help Window

When you slide a tool over a function, a subVI node, or a VI icon (including the icon of the VI you have open, at the top right of the VI window), the Help window shows the icon for the function or subVI with wires of the appropriate data type attached to each terminal. Input wires point to the left, and output wires point to the right. Terminal names appear beside each wire. If the VI has a description associated with it, this description is also displayed. For some subVIs or functions with many inputs, the Help window will show the names of the inputs you *must* wire in bold (often the default value can be used and you do not need to wire an input). You can lock the Help window so that its contents do not change when you move the mouse by selecting **Lock Help** from the **Help** menu. If you position the wiring tool over a specific node on a function or subVI, the Help window will flash the corresponding node so that you can make sure you are wiring to the right place.

3.8 SubVIs

You must understand and use the hierarchical nature of the VI if you want to take full advantage of LabVIEW's abilities. A *subVI* is simply a stand-alone program that is used by another program. After you create a VI, you can use it as a subVI in the block diagram of a higher-level VI. A LabVIEW subVI is analogous to a subroutine in C. Just as there is no limit to the number of subroutines you can use in a C program, there is no limit to the number of subVIs you can use in a LabVIEW program. You can also call a subVI inside of another subVI. If a block diagram has a large number of icons, you can group them into a lower-level VI to maintain the simplicity of the block diagram. This modular approach makes applications easy to debug, understand, and modify.

EXERCISE 3.2
Front Panel and Block Diagram Basics

Try to do the following basic things on your own. If you have any trouble, glance back through the chapter for clues.

1. Toggle between the front panel and block diagram. (Hint: Use keyboard shortcuts!)
2. Resize the windows so that both front panel and block diagram are visible simultaneously. You may need to move them around. (Hint: Do this using the standard resizing technique for your platform. If you don't know how, please look at your Windows or Mac instruction manuals.)
3. Drop a digital control, a string control, and a Boolean indicator on the front panel by selecting them from the **Controls** menu.

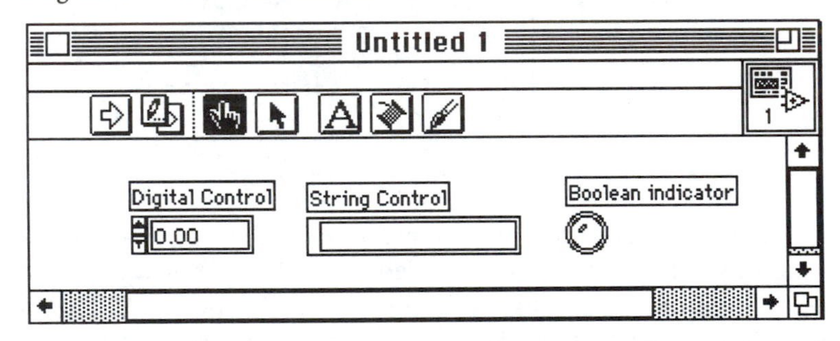

   To get the numeric control, pop up in a blank area of the front panel to get the **Controls** menu. Then select the **Numeric** hierarchical menu and move the mouse over the icons until you find the one called "Digital Control." Release the mouse button, and this icon will appear on your panel.

   Notice how LabVIEW creates corresponding terminals on the block diagram. Also notice that numeric terminals are orange, strings are pink, and Booleans are green. This color coding makes it easier for you to distinguish between data types.

4. Now pop up on the digital control (by right mouse-clicking on Windows or <Cmd>-clicking on Mac) and select **Change to Indicator** from the pop-up menu. Notice how the appearance of the numeric's front panel changes (the little arrows go away). Also notice how the terminal on the block diagram changes. Switch the object back and forth

between control and indicator until you can easily recognize the differences on both front panel and block diagram.

5. Change the VI from edit mode to run mode by clicking on the mode button; then change it back to edit mode.
6. Using the Positioning tool, select each object on the front panel and hit the <delete> key to remove it.
7. Drop another digital control from the **Numeric** palette of the **Controls** menu onto the front panel. If you don't click on anything, you should see a little box at the top of the control. Type `Number 1`, and you will see this text appear in the box. Click the enter button on the tool palette to enter the text. You have just created a label. Now create another digital control labeled `Number 2`, and a digital indicator labeled `N1+N2`.

   Use the Operating tool, click on the increment arrow of `Number 1` so that it contains the value "4." Give `Number 2` a value of "3."
8. Switch back to the diagram. Drop an **Add** function from the **Arithmetic** palette of the **Functions** menu in the block diagram. Pop up in a blank area of the diagram to get the **Functions menu**, select the **Arithmetic** hierarchical menu, and choose the **Add** function. Now repeat the process and drop a **Subtract** function.
9. Pop up on the **Add** function and select the **Show Terminals** option. Notice how the input and output terminals are arranged; then show the icon again by selecting **Show Icon.**
10. Bring up the Help window by using either the keyboard shortcut or the **Show Help Window** command from the **Windows** menu. Position the cursor over the **Add** function. The Help window provides valuable information about the function's use and wiring pattern.
11. You may have to use the Positioning tool to reposition some of the terminals. Then use the wiring tool to wire the terminals as shown in the following illustration. Click once on the NUM terminal and once on the appropriate terminal on the **Add** function to draw a wire. Wiring can be kind of tricky until you get the hang of it, so don't worry if it feels a little awkward right now!

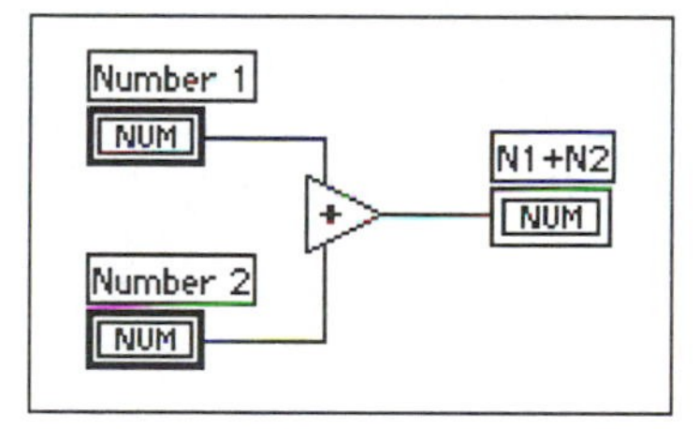

12. Switch back to the front panel and pop up on the icon pane (the little window in the upper right corner). Select **Show Connector** from the menu. Observe the connector that appears.

   Now pop up again on the connector and look at its menu to see the configuration options you have. Then show the icon by selecting **Show Icon**.
13. Run the VI by clicking on the run button. `N1+N2` should display a value of "7."

14. Save the VI by selecting **Save** from the **File** menu. Call it **Add.vi**.

Congratulations! You have now mastered several important basic LabVIEW skills!

~ ~ ~

3.9 Wrap It Up!

The LabVIEW environment has three main parts: the *front panel, block diagram*, and *icon/connector*. The front panel is the user interface of the program—you can input data through *controls* and observe output data through *indicators*. When you place an object on the front panel using the **Controls** menu, a corresponding terminal appears in the block diagram, making the front panel data available for use by the program. Wires carry data between nodes, which are LabVIEW program execution elements. A node will execute only when all input data is available to it, a principle called *dataflow*.

A VI should also have an *icon* and a *connector*. LabVIEW has two operating modes. When you use a VI as a subVI, its *icon* represents it in the block diagram of the VI you use it in. Its *connector*, usually hidden under the icon, defines the input and output parameters of the subVI.

LabVIEW has two types of menus: *pull-down* and *pop-up*. Pull-down menus are located at the top of a VI window under Windows and at the top of the screen on the Mac, while pop-up menus can be accessed by "popping up" on an object." **To pop up, right-mouse-click on machines running Windows and <Command>-click on the Mac.** Pull-down menus tend to have more universal commands, while pop-up menu commands affect only the object you pop up on. Remember, when in doubt about how to do something, pop up to see its menu options!

LabVIEW has two operating modes. When you are in *edit mode*, LabVIEW lets you make changes to your VI. When you want to run your VI, you can switch to *run mode* by clicking on the mode button on the palette.

The *Help window* provides priceless information about functions and how to wire them up; it can be accessed from the **Help** menu.

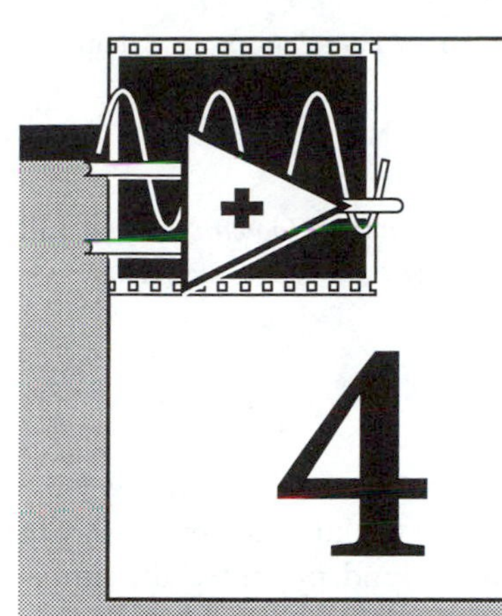

4

LabVIEW Foundations

Overview

Get ready to learn about LabVIEW's basic principles in this chapter. You will learn how to use LabVIEW's different data types and to fabricate, make changes to, wire, and run your own VIs. Make sure you understand these fundamentals before you proceed, because they are integral to all development you will do in LabVIEW.

Your Goals

- Learn the different types of controls and indicators, and the special options available for each
- Master the basics of creating a VI
- Be able to recognize and avoid the pitfalls that try to entrap you when you wire your VI
- Become comfortable with LabVIEW's editing techniques

Key Terms

numeric
string
Boolean
path
text ring

4.1 Basic Controls and Indicators and the Fun Stuff They Do

LabVIEW has four simple types of controls and indicators: *numeric, Boolean, string,* and the little-used *path.* You will also encounter a few more complex data types such as arrays, clusters, charts, and graphs, that we will expand on later.

Numeric Controls and Indicators

Numeric controls allow you to enter numeric values into your VIs; *numeric indicators* display numeric values you wish to see. LabVIEW has many types of numeric objects shown in the following illustration: knobs, slides, tanks, thermometers, and, of course, the simple digital display. All can be either controls or indicators, although each type has its own default. For example, a thermometer defaults to an indicator because you will most likely use it as one. By contrast, a knob appears on the front panel as a control because knobs are usually input devices.

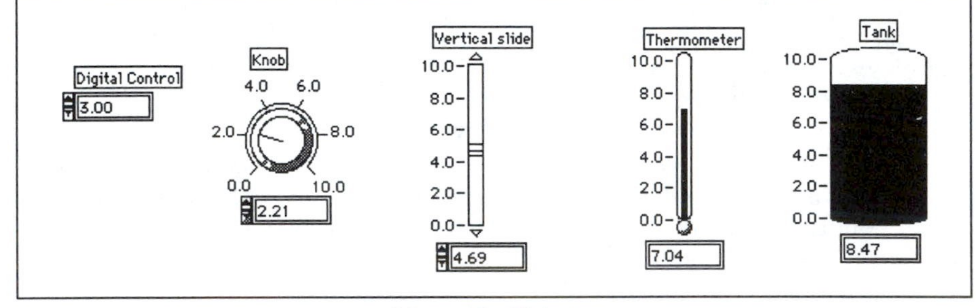

Numeric Objects

Numeric terminals on the block diagram appear as little orange boxes containing the letters, "NUM." Wires carrying numeric data are also orange.

Pop-Up Options. Pop up on a numeric control or indicator by right-mouse-clicking (on machines running Windows) or <Command>-clicking (on Macs) to view its pop-up menu.

You can specify the format and precision you want your numeric to display by selecting **Format & Precision** from its pop-up menu. A dialog box appears, and you can choose either scientific or floating-point notation. You can also enter the digits of precision to display (precision specifies the number of places to the right of the decimal point). More complex numerics, like knobs and thermometers, have the pop-up option to **Show Digital Display**, which brings up a digital window to display the precise numeric value.

Text Rings. A *text ring* is a special type of numeric that associates a number with a text message. You enter text into the ring, and the text becomes associated with a certain number (zero for the first text message, one for the next, and so on). You can see this number by selecting **Show ▶ Digital Display** from the text ring's pop-up menu.

If you want to add another number and corresponding message, select **Add Item After** or **Add Item Before** from the pop-up menu and a blank entry window will appear. If you click on a text ring with the Operating tool, you will see a list of all possible text messages, with the current one checked. Text rings are useful if you want a user to select an option (such as AC, DC, or Ohms), which will then correspond to a numeric value in the block diagram. Try dropping a text ring on the front panel; then show the digital display and add a few items.

Booleans

Boolean is just a fancy word for "on or off." Boolean data can have one of two states: true and false. LabVIEW provides a myriad of switches, LEDs, and buttons for your Boolean controls and indicators. You can change the state of a Boolean by clicking on it with the Operating tool. Like numerics, each type of Boolean has a default type based on its probable use (i.e. switches appear as controls, LEDs as indicators).

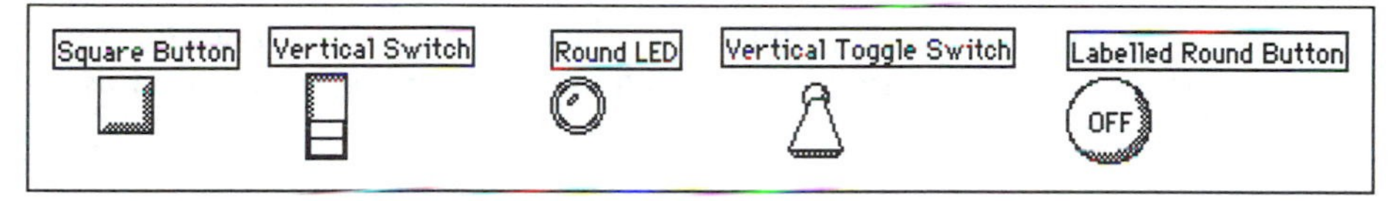

Boolean Objects

Note

Control terminals have THICK borders, while indicator terminal borders are THIN. It is VERY important to distinguish between the two since they are NOT functionally equivalent (Control=Input, Indicator=Output, so they are NOT interchangeable).

Boolean terminals appear green on the block diagram and contain the letters, "TF." [TF]

Labeled Buttons. LabVIEW has two buttons with text messages built into them—the labeled square button and the labeled round button. This text is merely informative for the user. Each labeled button can contain two text messages: one for the TRUE state and one for the FALSE state. When you first drop the button, the TRUE state says "ON" and the FALSE state says "OFF." You can then use the Labeling tool to change each message.

Mechanical Action. A Boolean control has a handy pop-up option called **Mechanical Action**, which lets you determine how the Boolean behaves when you click on it. Mechanical Action is covered in more detail in Chapter 8, *Charts and Graphs.*

Strings

String controls and *indicators* contain data in ASCII format, the standard way to store alphanumeric characters. They allow you to display text messages. String terminals and wires carrying string data appear pink on the diagram. The terminals contain the letters "abc."

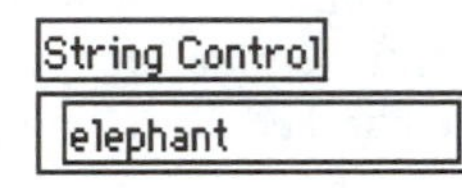

Watch Out!
While string controls and indicators can contain numeric characters, they do NOT contain numeric data. You cannot do any numerical processing on string data; that is, you can no more add an ASCII "9" character than you can an "A." If you need to use numeric information that is in string format, you must first convert it to numeric format using the appropriate functions (see Chapter 9, **Strings and File I/O***).*

String controls and indicators are fairly simple. Their pop-up menus contain few special options. If you choose the **Show Scrollbar** option from the string pop-up **Show** submenu, a vertical scrollbar appears on the string control or indicator. You can use this option to minimize the space taken up on the front panel by string controls that contain a large amount of text.

Note
If the Show Scrollbar menu item is dimmed, you must make the string icon taller to accommodate the scrollbar before you choose this option.

Entering and Viewing Nondisplayable Characters. If you choose **'\' Code Display** (instead of **Normal Display**) from the string pop-up menu, LabVIEW will interpret characters immediately following a backslash (\ ) as a code for nondisplayable characters. The following table shows how LabVIEW interprets these codes.

LabVIEW "\" Codes

Code	LabVIEW Interpretation
\00 - \FF	Hexadecimal value of an 8-bit character; alphabetical characters must be uppercase
\b	Backspace (ASCII BS, equivalent to \08)
\f	Formfeed (ASCII FF, equivalent to \OC)
\n	New Line (ASCII LF, equivalent to \OA)
\r	Return (ASCII CR, equivalent to \0D)
\t	Tab (ASCII HT, equivalent to \09)

\s	Space (equivalent to \20)
\\	Backslash (ASCII \, equivalent to \5C)

You must use uppercase letters for hexadecimal characters and lowercase letters for the special characters, such as formfeed and backspace. LabVIEW interprets the sequence \BFare as hex BF followed by the word "are," whereas LabVIEW interprets \bFare and \bfare as a backspace followed by the words "Fare" and "fare." In the sequence \Bfare, \B is not the backspace code, and \Bf is not a valid hex code. In a case like this, when a backslash is followed by only part of a valid hex character, LabVIEW assumes a zero follows the backslash, so LabVIEW interprets \B as hex 0B. Any time a backslash is not followed by a valid character code, LabVIEW ignores the backslash character.

Single Line Strings. If you choose **Limit to Single Line** from a string's pop-up menu, your string cannot exceed one line of text. If you hit <enter> or <return>, your text will be autmatically entered (rather than the cursor jumping to a new line to allow you to type more).

Paths

You use *path controls* and *indicators* to display paths to files, folders, or directories. If a function that is supposed to return a path fails, it will return <Not A Path> in the path indicator. Paths are a separate data type, and their terminals and wires appear bluish-green on the block diagram. A path is specified by *drivename* followed by *folder* or *directory names* and then finally the *file* itself. On a Mac, folder and file names are separated by a colon (:); on a computer running Windows, directory and file names are separated by a backslash (\).

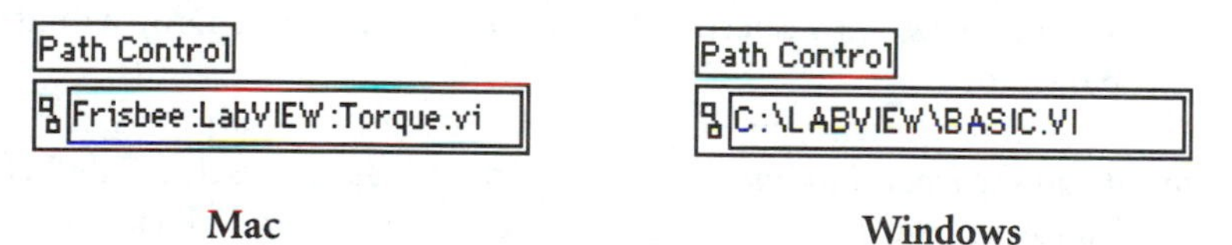

Mac **Windows**

Summary of Basic Controls and Indicators

Just to make sure you get your data types straight, we'll recap the four types of simple controls and indicators:

- *Numerics* contain standard numeric values.
- *Booleans* can have one of two states: on and off, true and false, one and zero.
- *Strings* contain ASCII data. Although they can contain numeric characters (such as 0–9), you must convert string data to numeric data before you can perform any arithmetic on it.
- *Paths* give you a platform-independent data type especially for file paths.

4.2 Creating VIs—It's Your Turn Now

We've gone over the basics of the LabVIEW environment, and now we're going to show you exactly how to build your own VIs. You should step through these instructions on your computer as you read them so you can learn the techniques more quickly.

Placing Items on the Front Panel

You will probably want to start by "dropping" controls and indicators on a front panel to define your user inputs and program outputs. First access the **Controls** menu, either by using the pull-down menu or by popping up in a blank area on the panel; then select the object you want with the mouse. The following illustration shows an item being selected.

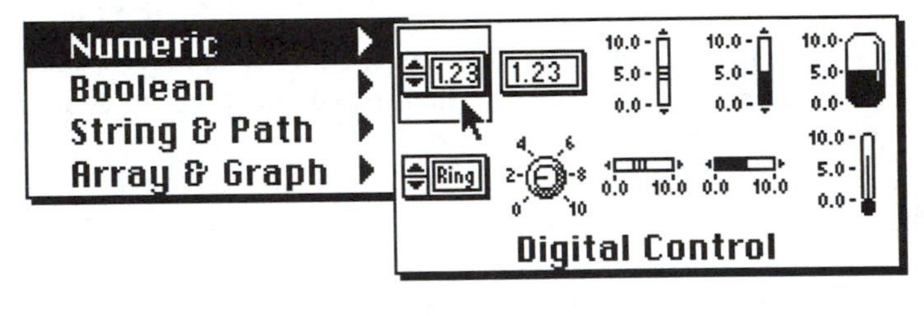

Note
If you use the pull-down Controls menu, LabVIEW will choose where to place the object, but if you pop up to get the menu (definitely the recommended method!), the item appears wherever your mouse is.

Create a new VI by selecting **New** from the **File** menu. Drop a digital control on your front panel now.

Remember, when you drop an item on the front panel, its corresponding terminal appears on the block diagram. You might find it helpful to select **Tile Left and Right** from the **Windows** menu so that you can see the front panel and block diagram windows at the same time.

Labeling Items

Labels are blocks of text that annotate components of front panels and block diagrams. An object first appears in the front panel window with a black or gray rectangle representing a label. If you want to retain the label at this time, enter text from the keyboard. After you enter text into a label, any one of the following actions completes the entry:

- Press <Shift-Return>.
- Press <Enter> on the numeric keypad.
- Click on the enter button in the tool palette.
- Click outside the label.

The label appears on the corresponding block diagram terminal as well as the front panel object.

LabVIEW has two kinds of labels: *owned labels* and *free labels.* Owned labels belong to and move with a particular object and annotate that object only. When you create a control or indicator on the front panel, a blank owned label accompanies it, awaiting input. The label disappears if you do not enter text into it before clicking elsewhere with the mouse.

You can select **Show Label** from the owner's pop-up menu to create or change a label that isn't currently visible. You can hide owned labels, but you cannot copy or delete them independently of their owners.

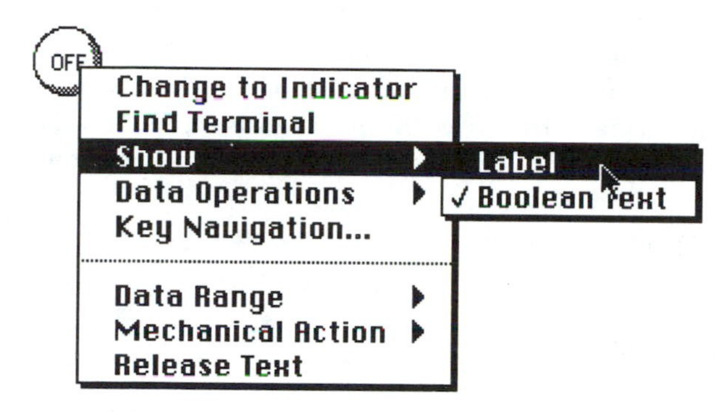

Label the digital control you just created Number 1.

Creating Free Labels. Free labels are not attached to any object, and you can create, move, or dispose of them independently. Use them to annotate your panels and diagrams. You use the Labeling tool to create free labels and to edit both types of label.

To create a free label, select the Labeling tool A from the edit mode palette and click anywhere in empty space. A small, bordered box appears with a text cursor at the left margin ready to accept typed input. Type the text you want to appear in the label. The enter button appears on the tool palette to remind you to end text entry, which you can do by clicking on the enter button, pressing <shift-return>, pressing <enter> on the numeric keypad, or clicking outside the label. If you do not type any text in the label, the label disappears as soon as you click somewhere else.

Create a free label on the front panel that says hippopotamus.

Note

SubVIs do not have labels. They have names, which look just like labels but can't be edited. Function labels, on the other hand, are empty so that you can edit them to reflect the use of the function in the block diagram. For example, you can use the label of an Add function to document what quantities are being added, or why they are being added at that point in the block diagram.

Changing Font, Style, Size, and Color of Text

You can change text attributes in LabVIEW using the options in the **Text** menu. If you select or highlight objects or text, and then make a selection from this menu, the changes apply to everything selected or highlighted. If nothing is selected, the changes apply to the default font and will affect future instances of text.

Change your hippopotamus label so that it uses 18-point font.

Placing Items on the Block Diagram

A user interface isn't much good if there's not a program to support it. You create the actual program by placing functions, subVIs, and structures on the block diagram. To do this, access the **Functions** menu, either by pulling down the menu or by popping up in a blank area of the diagram. Then select the item you want, and it will appear on the diagram. Like front panel items, block diagram objects appear wherever your mouse was if you used the pop-up **Functions** menu.

Drop an **Add** function from the **Arithmetic** palette of the **Functions** menu onto the block diagram.

Editing Techniques

Once you have things in your windows, you will want to be able to move them around, copy them, delete them, etc. Read on to learn how.

Selecting Objects. You must select an item before you can move it. To select an object, click the mouse button while the Positioning tool is on the object. When you select an object, LabVIEW surrounds it with a moving dashed outline called a marquee, shown in the following illustration.

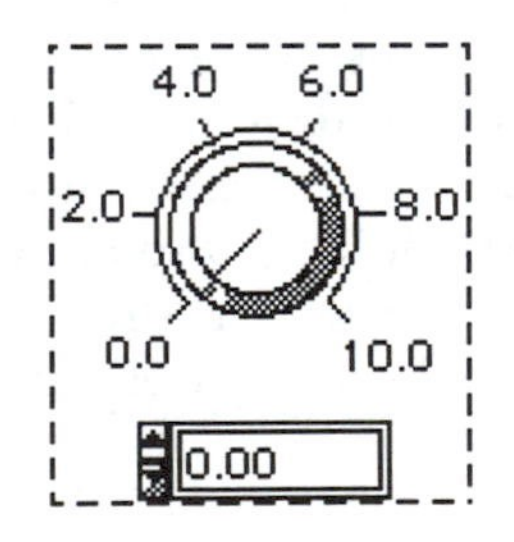

To select more than one object, <shift>-click on each additional object. You can also deselect a selected object by <shift>-clicking on it.

Another way to select single or multiple objects is to drag a selection rectangle around them. To do this, click in an open area with the Positioning tool and drag diagonally until all the objects you want to select lie within or are touched by the selection rectangle that appears. When you release the mouse button, the selection rectangle disappears and a marquee surrounds each selected object. This effect is sometimes referred to as "marching ants." Once you have selected the desired objects, you can move, copy, or delete them at will.

You cannot select a front panel object and a block diagram object at the same time. However, you can select more than one object on the same front panel or block diagram.

Clicking on an unselected object or clicking in an open area deselects everything currently selected. <Shift>-clicking on an object selects or deselects it without affecting other selected objects.

Now select the digital control you created earlier.

Moving Objects. You can move an object by clicking on it with the Positioning tool, holding down the mouse button, and dragging it to the desired location. If you hold down the shift key and then drag an object, LabVIEW restricts the direction of movement horizontally or vertically, depending on which direction you first move the object. You can move selected objects small, precise increments by pressing the appropriate arrow key.

Move your digital control to the other side of the screen.

Duplicating Objects. You can duplicate LabVIEW objects after you have selected them. From the **Edit** menu, select the **Copy** option, place the cursor where you want the new object, and then select the **Paste** option. You can also clone an object by using the Positioning tool to <Ctrl>-click on the object if you use Windows and <option>-click if you use a Mac. Then drag away while still holding down the mouse. You will drag the copy, displayed as a dotted line, with you while the original stays in place, as shown in the following illustration. You can also duplicate both front panel and block diagram objects from one VI to another.

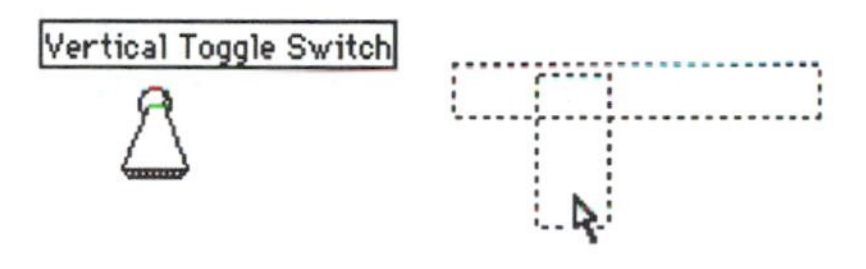

Watch Out!
You cannot duplicate control and indicator terminals on the block diagram—you must copy the items on the front panel.

Copy your digital control using both methods. You should now have three digital controls labeled `Number 1` on your front panel and three corresponding terminals on the block diagram. To find out which one belongs to which, pop up on the control or on the terminal and select **Find Terminal** or **Find Control**.

Deleting Objects. To delete an object, select it and then choose **Clear** from the **Edit** menu or press the <backspace> or <delete> key.

Watch Out!
Block diagram terminals for front panel controls and indicators disappear only when you delete the corresponding front panel controls or indicators.

Although you can delete most objects, you cannot delete control or indicator components such as labels and digital displays. However, you can hide these components by selecting **Show** from the pop-up menu and then deselecting the **Label** or **Digital Display** options.

Delete one of your digital controls.

Resizing Objects. You can change the size of most objects. When you move the Positioning tool over a resizable object, resizing handles appear at the corners of the object, as in the following illustration.

When you pass the Positioning tool over a resizing handle, the cursor changes to the Resizing tool. Click and drag this cursor until the dashed border outlines the size you want, as shown in the following illustration.

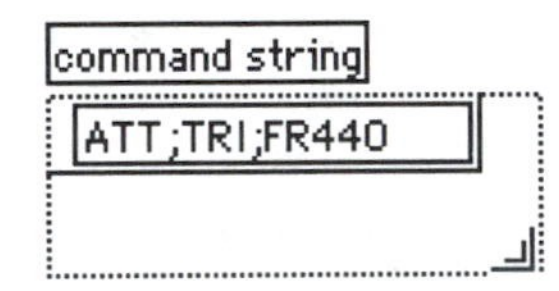

To cancel a resizing operation, continue dragging the frame corner outside the window until the dotted frame disappears. Then release the mouse button. The object maintains its original size.

Resize one of your digital controls.

Moving Object to Front, to Back, Forward, and Backward. Objects can sit on top of and often hide other objects, either because you placed them that way or through some wicked twist of fate. LabVIEW has several commands in the **Edit** menu to move them relative to each other. You may find them very useful for finding "lost" objects in your programs.

- **Move To Front** moves the selected object to the top of a stack of objects.
- **Move Forward** moves the selected object one position higher in the stack.
- **Move To Back** and **Move Backward** work similarly to **Move To Front** and **Move Forward** except that they move items down the stack rather than up.

Coloring Objects. LabVIEW appears on the screen in black and white, shades of gray, or vivid color depending on the capability and settings of your monitor. You can change the color of many LabVIEW objects, but not all of them. For example, block diagram terminals of front panel objects and wires use color codes for the type of data they carry, so you cannot change them. You also cannot change colors in black and white mode.

To change the color of an object or the background screen, pop-up on it with the Coloring tool. The following palette appears in color (assuming you have color enabled on your machine).

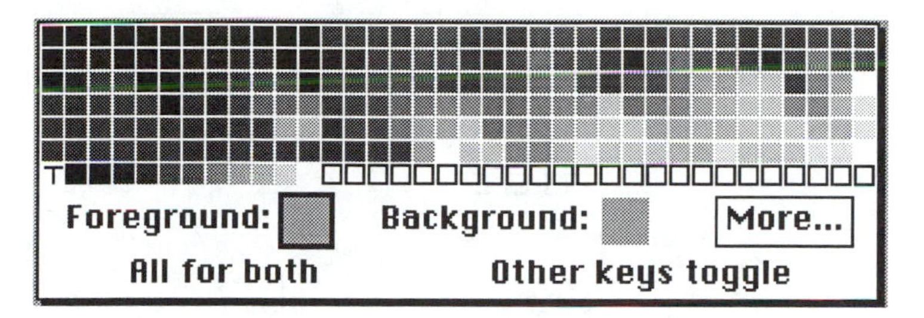

As you move through the palette while depressing the mouse button, the object or background you are coloring redraws with the color the cursor is currently touching. This gives you a preview of the object in the new color. If you release the mouse button on a color, the object retains the selected color. To cancel the coloring operation, move the cursor out of the palette before releasing the mouse button. Selecting the **More** option from the color palette calls up a dialog box with which you can customize colors.

Color one of your digital controls.

Matching Colors. You can also duplicate the color of one object and transfer it to a second object without going through the color palette. <Ctrl>-click under Windows or <option>-click on the Mac with the Coloring tool on the object whose color you want to duplicate. The Coloring tool appears as an eye dropper (we call it the "sucker" tool) and takes on the color of the object you clicked on. Now you can release <Ctrl> or <option> and click on another object with the Coloring tool; that object assumes the color you chose.

Transparency. If you select the box with a "T" in it from the color palette, LabVIEW makes the object transparent. You can layer objects with this feature. For instance, you can place invisible controls on top of indicators, or you can create numeric controls without the standard three-dimensional container. Transparency affects only the appearance of an object. The object responds to mouse and key operations as usual.

Foreground and Background. Some objects have both a foreground and a background that you can color separately. To change between foreground and background in the color palette, you can press <f> for foreground and <b> for background. Pressing <a> for "all" selects both foreground and background. Pressing any other key also toggles the selection between foreground and background. The selection does not toggle until you move the Coloring tool.

EXERCISE 4.1
Editing Practice

1. Select the **Open** option in the **File** menu to open **Editing Exercise.vi**. It is found in EXERCISE.LLB in the EXAMPLES\GENERAL folder or directory. The front panel of the **Editing Exercise** VI contains a number of LabVIEW objects. Your objective is to change the front panel of the VI shown below.

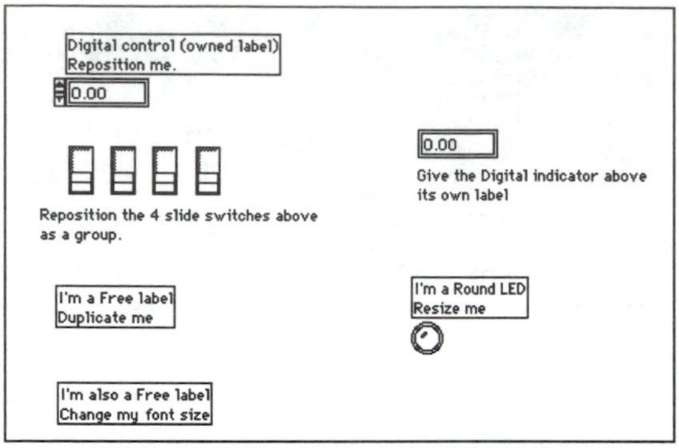

Digital control (owned label)
Reposition me.
0.00

0.00
Give the Digital indicator above its own label

Reposition the 4 slide switches above as a group.

I'm a Free label
Duplicate me

I'm a Round LED
Resize me

I'm also a Free label
Change my font size

2. Make sure you are in edit mode; you will see the editing palette of tools at the top of your window if you are. If you are in run mode, switch to edit mode by clicking on the mode button.
3. Reposition the digital control. Choose the Positioning tool from the tool palette. Click on the digital control and drag it to another location. Notice that the label follows the control—the control *owns* the label. Now, click on a blank space on the panel to deselect the control; then click on the label and drag it to another location. Notice that the control does not follow. An owned label can be positioned anywhere relative to the control, but when the control moves, the label will follow.
4. Reposition the four slide switches as a group. Using the Positioning tool, click in an open area near the four switches, hold down the mouse button, and drag until all the switches lie within the selection rectangle. Click on the selected switches and drag them to a different location.
5. Duplicate the free label. Hold down the <Ctrl> key on a computer running Windows or the <option> key on the Mac, click on the free label, and drag the duplicate to a new location.
6. Change the font size of the free label. Select the text by using the Labeling tool A. You can triple-click on the text, or click on the first letter in the label and drag the cursor to the end of the label. Modify the selected text using the options from the **Text** menu.
7. Create an owned label for the digital indicator. Pop up on the digital indicator by clicking the right mouse button under Windows or by clicking the mouse button while holding down the <Command> key on the Mac, then choose **Show Label** from the pop-up menu. Type Digital Indicator inside the bordered box. Press <Enter> on the numeric keypad, or click the mouse button outside the label to enter the text.
8. Resize the round LED. Place the Positioning tool over a corner of the LED until the tool becomes the resizing cursor. Click and drag the cursor outward to enlarge the LED. If you want to maintain the current ratio of horizontal to vertical size of the LED, hold down the <Shift> key while you resize.
9. Change the color of the round LED. Using the Coloring tool, pop up on the LED. While continuing to depress the mouse button, choose a color from the selection palette. Remember, you use the right mouse button on computers running Windows when you pop up. When you release the mouse button, the object assumes the last color you selected.
10. Your panel should now look something like the one shown.

Digital control (owned label)
Reposition me.
0.00
Reposition the 4 slide switches above as a group.
Digital Indicator Label
0.00
Give the Digital indicator above its own label
I'm a Free label
Duplicate me
I'm a Free label
Duplicate me
I'm also a Free label
Change my font size
I'm a Round LED
Resize me

11. Close the VI by selecting **Close** from the **File** menu. Do not save any changes. Pat yourself on the back—you've mastered LabVIEW's editing techniques!

4.3 Wiring Up

Your neatly arranged, colorful front panel won't do you much good if you don't connect up the wires in your diagram. The following sections detail everything you need to know about wiring techniques.

You use the Wiring tool to connect terminals. The cursor point or "hot spot" of the tool is the tip of the unwound wire segment, as shown.

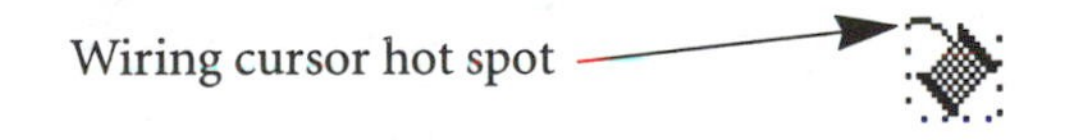

The following figure demonstrates the conventions we will use to explain how to wire. The large rectangle with the boxed number inside represents the mouse. The arrow at the end of this mouse symbol shows where to click, and the number printed on the mouse button indicates how many times to click.

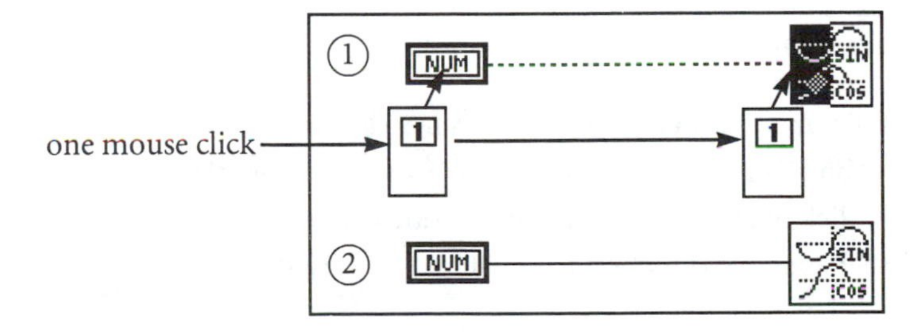

Wiring Conventions

To wire from one terminal to another, click the Wiring tool on the first terminal, move the tool to the second terminal, and then click on the second terminal as shown in the preceding illustration. It does not matter which terminal you click on first. The terminal area blinks when the hot spot of the Wiring tool is correctly positioned on the terminal. Clicking connects a wire to that terminal.

Once you have made the first connection, LabVIEW draws a wire as you move the cursor across the diagram, as if the wire were reeling off the spool. You do not need to hold down the mouse button.

To wire from an existing wire, perform the operation described above, starting or ending the operation on the existing wire. The wire blinks when the Wiring tool is correctly positioned to fasten a new wire to the existing wire.

You can wire directly from a terminal outside a structure to a terminal within the structure using the basic wiring operation. LabVIEW creates a tunnel where the wire crosses the structure boundary, as shown in the following illustration. The first picture shows what the tunnel looks like as you are drawing the wire; the second picture depicts a finished tunnel.

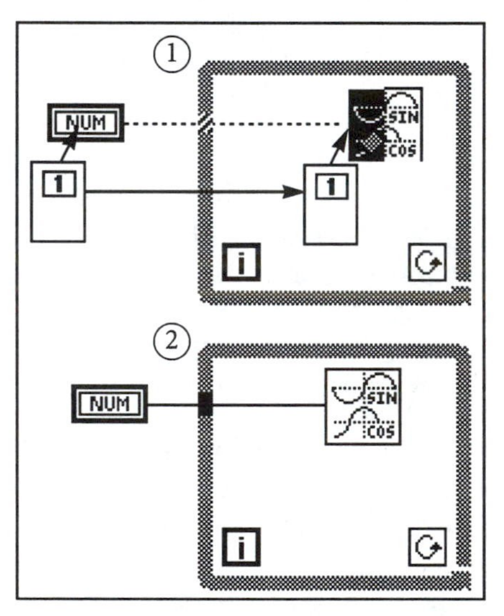

Tunnel Creation

Wiring Tips

The following tips may make wiring a little easier for you:

- You can make a 90-degree turn in your wire, or "elbow" it, once without clicking.
- Click the mouse to tack the wire and change direction.
- Change the direction from which the wire leaves a tack point by pressing the space bar.
- Double-click with the wiring tool to begin or terminate a wire in an open area.
- When wires cross, a small gap appears in the first wire drawn, as if it were underneath the second wire, illustrated below.

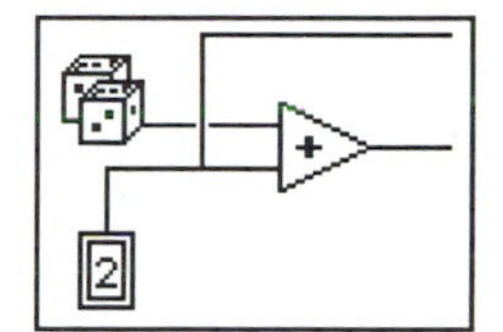

For practice, create a few numeric controls on a front panel and wire them to an input of the **Add** function.

Wire Stretching

You can move wired objects individually or in groups by dragging the selected objects to the new location using the Positioning tool. Wires connected to the selected objects stretch automatically. If you duplicate the selected objects or move them from one diagram into another (for example, from the block diagram into a structure such as a While Loop), LabVIEW leaves behind the connecting wires, unless you select them as well.

Wire stretching occasionally creates wire stubs or loose ends. You must remove these by using the **Remove Bad Wires** command from the **Edit** menu before the VI will execute.

Move the **Add** function with the Positioning tool and watch how the attached wires adjust.

Selecting and Deleting Wires

A wire segment is a single horizontal or vertical piece of wire. The point at which three or four wire segments join is a *junction.* A *bend* in a wire is where two segments join. A wire branch contains all the wire segments from junction to junction, terminal to junction, or terminal to terminal if there are no junctions in between. One mouse click with the Positioning tool on a wire selects a segment. A double-click selects a branch. A triple-click selects an entire wire. Press the delete or backspace key to remove the selected portion of wire.

Select and delete one of your wires; then rewire it.

Moving Wires

You can reposition one or more segments by selecting and dragging them with the Positioning tool. For fine tuning, you can also move selected segments one pixel at a time by pressing the arrow keys on the keyboard. LabVIEW stretches adjacent, unselected segments to accommodate the change. You can select and drag multiple wire segments, even discontinuous segments, simultaneously. When you move a tunnel, LabVIEW normally maintains a wire connection between the tunnel and the wired node.

Move a wire segment using the Positioning tool, then with the arrow keys.

Wiring to Off-Screen Areas

If a block diagram is too large to fit on the screen, you can use the scroll bars to move to an off-screen area and drag whatever objects you need there. Dragging the Wiring tool slightly past the edge of the diagram window while you are wiring automatically scrolls the diagram.

Bad Wires

When you make a wiring mistake, a broken wire—a black dashed line—appears instead of the usual colored wire pattern. Until all bad wires have been vanquished, your run button will appear broken and the VI won't compile. You can remove a bad wire by selecting and deleting it. A better method is to obliterate all bad wires by selecting **Remove Bad Wires** from the **Edit** menu or by using the keyboard shortcut, <Ctrl-B> under Windows and <Cmd-B> on the Mac.

Watch Out!
Sometimes bad wires are mere fragments, hidden under something or so small you can't even see them. A good habit to develop is to Remove Bad Wires whenever you see a broken run arrow. Much of the time, that command will solve your problem.

4.4 Running Your VI

You can run a VI from edit mode or run mode by selecting the **Run** command from the **Operate** menu or by clicking on the run button. If you run from edit mode, LabVIEW compiles the VI, switches to run mode, and runs the VI. When the VI finishes executing, LabVIEW returns you to edit mode. While the VI is executing, the run button changes appearance:

- The VI is running at its top level if the run button looks like this: .
- The VI is executing as a subVI, called by another VI, if the run button looks like this: .

You can run multiple VIs at the same time. After you start the first one, switch to the panel or diagram window of the next one and start it as described above. Notice that if you run a subVI as a top-level VI, all VIs that call it as a subVI are broken until the subVI completes. You cannot run a subVI as a top-level VI and as a subVI at the same time.

EXERCISE 4.2
Building a Thermometer

Now you're going to put together a VI that actually does something! This program will take a voltage reading from a channel on your data acquisition (DAQ) board (or from a simulation if you don't have a board) and display it in a thermometer on the front panel. You should have channel 0 of your board connected to a temperature sensor or similar voltage (preferably between 0 and 1 volt). If you do have a DAQ board, see *Appendix A* for information on how to set it up.

Make sure you save this exercise because you will be adding to it later and you don't want to have to rebuild it!

1. Open a new panel.
2. Drop a thermometer on the panel by selecting it from the **Numeric** palette of the **Controls** menu. Label it **Temperature** by typing inside the owned label box as soon as the thermometer appears on the panel. Rescale the thermometer by dragging over the "10.0" with the Operating tool and entering **100**.

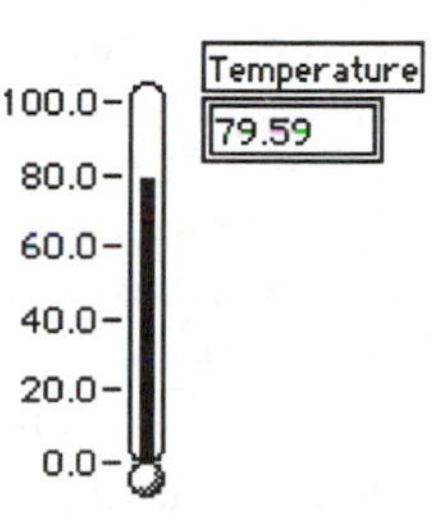

3. Build the block diagram shown in the following illustrtation. You might find it helpful to select **Tile Left and Right** from the **Windows** menu so that you can see both the front panel and the block diagram at the same time. Build (A) if you are using a DAQ board and (B) if you need to use simulated data. *You do not need to build both unless you really want to practice!!*

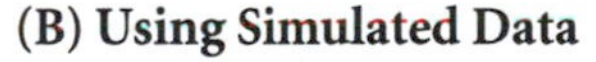

(A) Using a DAQ board **(B) Using Simulated Data**

You will find the **AI Sample Channel** VI, used in (A), under the **Data Acquision** palette of the **Functions** menu. If you do not have a board, use the **Demo Voltage Read** from the **Tutorial** palette, shown in (B). Remember to use the Help window to assist you in wiring to the correct terminals! Notice how, when you place the wiring tool over a node of a function, the corresponding terminal in the Help window will blink to let you see that you are wiring to the right place.

Other block diagram components are described in the following list:

Channel
0 You will need a string constant, located in the **Structs & Constants** palette of the **Functions** menu, to specify the analog input channel you want to use, which may or may not be channel 0. In the **Demo Voltage Read**, this value is ignored. Even though the string constant contains a numeric character, you must use a string data type rather than a numeric data type, or you will not be able to wire it up. When you drop a string constant on the block diagram, it contains no data. You can type a string immediately (if you don't click anywhere else first) and then hit <enter> or click the enter button to enter the data. You can change the value of a constant at any time using the Operating tool or the Labeling tool A.

Device 1 This numeric constant, located in the **Structs & Constants** palette of the **Functions** menu, specifies the device number of your board, which may or may not be 1. In the **Demo Voltage Read**, this value is ignored. When you drop a numeric constant on the block diagram, it contains the highlighted value "0." You can type a new number immediately (if you don't click anywhere else first) and then hit <enter> or click the enter button to enter the data. You can change the value of a constant at any time using the Operating tool or the Labeling tool A.

Under Windows, the device number is assigned in WDAQCONF (it is probably 1 unless you have multiple boards).On a Mac, the device number is the slot number of your DAQ board; you can see this by opening up NI-DAQ in the control panels. This VI assumes you have a temperature sensor wired to channel 0 of your DAQ board (another voltage source is fine, but will probably not provide a very accurate "temperature" unless it reads around 0.8 volts). For more information on device numbers, see *Appendix A* , which tells you how to set up your DAQ boards.

100 This numeric constant simply scales your voltage into a "valid" temperature.

4. Use the Help window, found under the **Help** menu, to show you the wiring pattern of the function terminals. Pay special attention to color coding to avoid broken wires.
5. Run the VI by clicking on the run button. You will see the thermometer display the voltage brought in from you board (or from the simulation). If you can't get your VI to compile, read Chapter 5, *Yet More Foundations*, which explains debugging techniques. Then try again.
6. Save the VI by selecting **Save** from the **File** menu. For now, save it in EXAMPLES\GENERAL\EXERCISE.LLB and name it **Thermometer.vi**. You will be using this VI as a subVI later on in the book. You will learn more about save options in the next chapter.

4.5 Wrap It Up!

LabVIEW has four types of simple controls and indicators: *numeric*, *Boolean*, *string*, and *path*. Each holds a separate data type and has special pop-up options. Control and indicator terminals and wires on the block diagram are color coded according to data type: Numerics are orange, Booleans are blue, strings are pink, and paths are bluish-green.

You place objects on the front panel or block diagram using the **Controls** or **Functions** menu, respectively. You can also access these menus by popping up in an empty section of the panel or diagram.

LabVIEW has special editing techniques fitting to its graphical environment. The Operating tool changes an object's value. The Positioning tool selects, deletes, and moves objects. The Wiring tool creates the wires that connect diagram objects. The Labeling tool creates and changes owned and free labels. Owned labels belong to a particular object and cannot be deleted or moved independently, while free labels have no such restrictions.

To run your VI, click on the run button or select **Run** from the **Operate** menu. If your VI is in edit mode, it will compile (if it can!), execute, and then return to edit mode. If the run button is broken, read the next chapter to learn good debugging techniques.

Additional Exercises

EXERCISE 4.3
Comparison Practice

Build a VI that compares two input numbers. If they are equal, an LED on the front panel turns on.

EXERCISE 4.4
Very Simple Calculator

Build a VI that adds, subtracts, multiplies, and divides two input numbers and displays the result on the front panel. Use this front panel to get started.

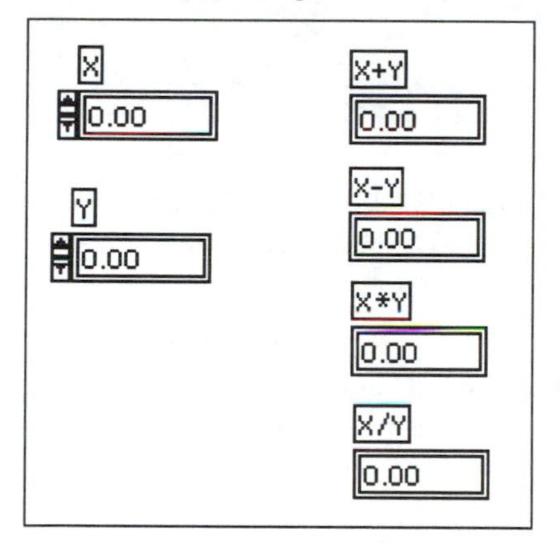

Yet More Foundations

Overview

In this chapter, you will learn more fundamentals of the LabVIEW environment, including LabVIEW's special library files, debugging features, documentation procedures, subVI implementation, and helpful shortcuts to speed your development.

Your Goals

- Learn how to load and save your VIs (and then do it regularly!)
- Learn about LabVIEW's special debugging features and how to make them work for you
- Create your very first subVI
- Know how to document your achievements for all the world to see

Key Terms

VI library
broken VI
single-step mode
execution highlighting
probe
breakpoint
subVI
Icon Editor

5.1 Loading and Saving VIs

You can load a VI by selecting **Open** from the **File** menu, then choosing the VI from the dialog box that appears. As the VI loads, you will see a status window that describes the VIs that are currently being loaded and allows you to cancel the loading process.

You save VIs by selecting **Save** (or a similar option) from the **File** menu. LabVIEW then pops up a file dialog box that resembles dialog boxes in other applications on your operating system, so you can choose where you want to save. If you save VIs as individual files, as you would expect to do, they must conform to the file naming restrictions of your operating system. To avoid these restrictions and save disk space, you can save VIs in a compressed form in a special LabVIEW file called a *VI library*, described momentarily.

LabVIEW references VIs by name. You cannot have more than one VI or subVI with a given name in memory at one time. If you have two VIs with the same name, LabVIEW will load the first VI it finds without loading the other.

Save Options

You can save VIs with one of four save options in the **File** menu.

- Select the **Save** option to save a new VI, specifying a name for the VI and its destination in the disk hierarchy, or to save changes to an existing VI in a previously specified location.
- Select the **Save As...** option to rename the VI in memory and to save a copy of the VI to disk. If you enter a new name for the VI, LabVIEW does not overwrite the disk version of the original VI. If you do not change the name of the VI in the dialog box, LabVIEW prompts you to verify that you want to overwrite the original file.
- When you select the **Save a Copy As...** option, LabVIEW saves a copy of the VI in memory to disk under a different name, which you enter in the dialog box. Unlike **Save As...**, this option does not affect the name of the VI in memory.
- **Save with Options...** brings up a dialog box in which you can choose to save the entire VI hierarchy in memory to disk, the **Save Entire Hierarchy** option. You also have the option to save VIs without block diagrams but make sure you keep an extra copy *with* the block diagram in case you ever need to modify it. To save a specified VI or VIs to a single location without being interrupted by multiple prompts, save **To new location—single prompt**.

Watch Out!
You cannot edit a VI after you save it without a block diagram. Always make a copy of the original VI before you save it without a block diagram.

Filter Rings

At the bottom of your Save or Load dialog box, you will see a filter ring that allows you to see only **VIs, View All** files, or just see those with a **Custom Pattern** that you specify.

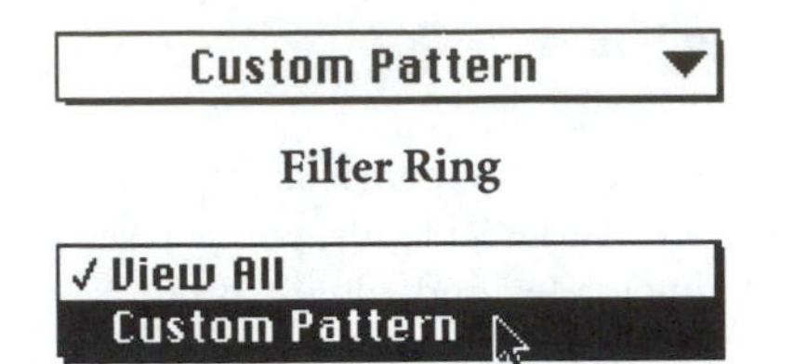

Filter Ring

Filter Ring Options

If you choose **Custom Pattern**, another box appears in which you can specify a pattern. Only files matching that pattern will show up in the dialog box. Notice that an asterisk automatically appears inside the box—it is the "wild card character" and is considered a match with any character or characters. A dialog box showing the custom pattern below would show any file with a .llb extension.

Custom Pattern *.llb

Revert

You can use the **Revert...** option in the **File** menu to return to the last saved version of the VI you are working on. A dialog box will appear to confirm whether to discard any changes in the VI.

5.2 VI Libraries

VI Libraries have the same load, save, and open capabilities as folders and directories. However, your operating system sees them as single files. You can access their contents only from LabVIEW. VI libraries contain only compressed versions of VIs. They can't contain other VI libraries, folders, or directories. Compression and decompression take minimal extra time, and the disk space they save is always substantial. VIs saved in a library are not bound by the operating system naming restrictions (although the library name must still conform). VI names can contain up to 31 characters, including spaces and the .vi extension, and are not case sensitive. Libraries are also easier to port between different platforms.

Watch Out!
If a VI library gets corrupted, every VI in it is lost, so make plenty of backup copies of your work!

In file dialog boxes, a VI library appears as a folder or directory icon with an extension of .llb appended to the name as shown.

BLACKJAK.LLB
demo.llb

You create a VI library by clicking on the **New...** button of the **Save**, **Save As...**, or **Save a Copy As...** dialog box, shown in the following illustration.

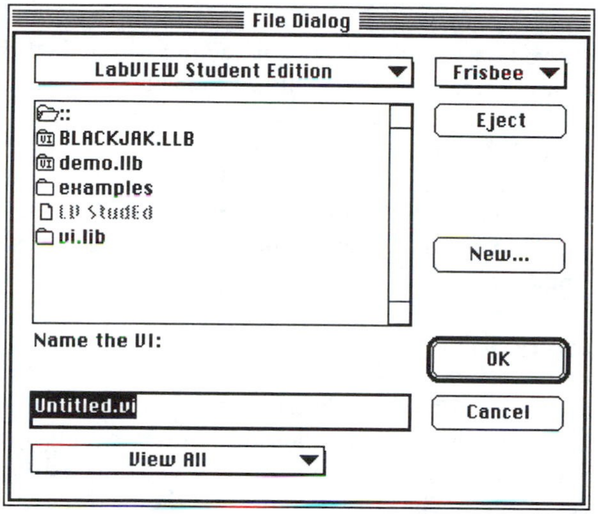

Enter the name of the new library in the dialog box that appears, shown in the next figure, and append a .llb extension. Then click on the VI Library button. If you do not include the .llb extension, LabVIEW adds it.

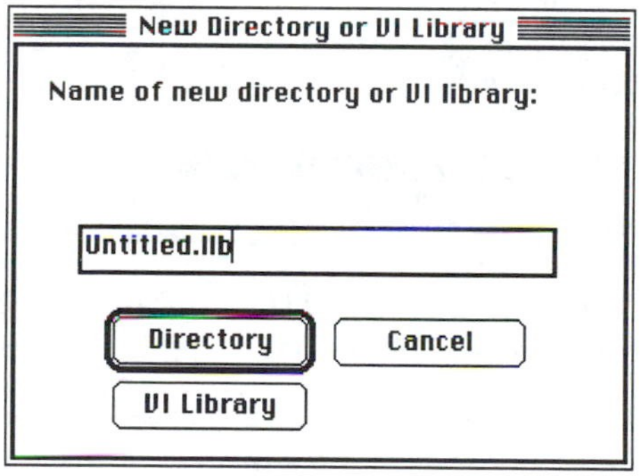

You can remove a file from a VI library using buttons in the VI library dialog box displayed by the **Edit VI Library...** option of the **File** menu.

Accessing Your VI Library from the Functions Menu

The VIs distributed with LabVIEW in the vi.lib folder or directory are organized in VI libraries. These libraries appear as palettes in the **Functions** menu.

Watch Out!
Do not save your own VIs in the vi.lib folder or directory. Placing VIs in vi.lib risks creating a conflict during future installations. Place VIs in .lib folders or directories adjacent to, but not inside, the vi.lib folder or directory instead.

When you start LabVIEW, the application looks in the LabVIEW folder or directory for folders or directories ending in .lib LabVIEW builds palette menus for all VI libraries, folders, or directories that are inside of .lib folders or directories. If you want your VIs to appear in the **Functions** menu, create your own folder or directory ending in .lib, and place folders, directories, or VI libraries inside. When you restart LabVIEW, it creates palettes for each of these folders, directories, and VI libraries and adds them to the **Functions** menu. LabVIEW will not display an individual VI on a palette if it is saved directly into the .lib folder or directory.

Edit VI Library Dialog

You can use the Edit VI Library dialog box to edit the contents of a VI library. The Edit VI Library dialog box, shown in the following illustration, initially displays a list of the files in the VI library. As you move through the list, the creation date and last modification date for the selected file are shown at the bottom of the dialog box.

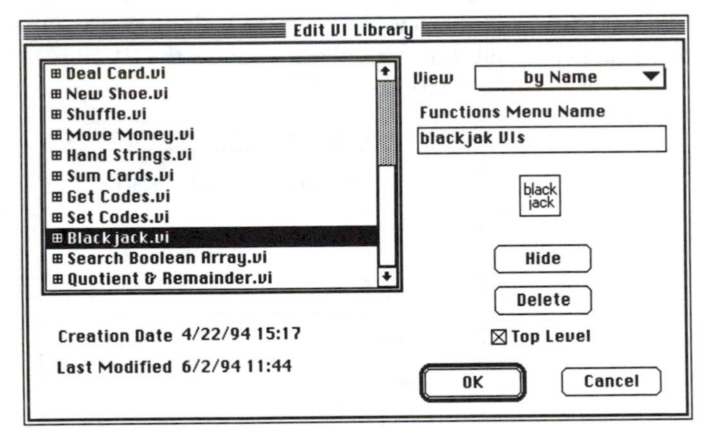

file marked for deletion

This icon indicates that the file is marked for deletion. You select and deselect files for deletion with the Delete/Restore buttons. Marked files are only deleted if you select the OK button.

Icons at the left side of the list indicate the status of each file. The icon shown indicates that the file can be visible in a **Functions** menu palette. You can add a file to the palette or remove it by using the Hide/Show button. It is a good idea to make sure VIs shown in custom palettes have icons!

You can access the **Function Palette** menu, used to place VI icons in the order you want in the **Functions** menu, by selecting it from the **View** menu. The VI list is replaced by an icon grid showing the arrangement of the icons you've selected for visibility in the palette menu. The menu name used for the palette is determined by the **Functions Menu Name** field. To move the VIs around in the palette, simply select and drag them.

If you mark a VI **Top Level,** it will load automatically when you double-click on the icon of its VI library. You can have more than one top-level VI. A top-level VI's name will also appear in a separate section at the top of the Load dialog, so it's easier for you to determine which VIs are main VIs and which are subVIs.

5.3 Debugging Techniques

LabVIEW has many built-in debugging features to help you develop your VIs. This section explains how to use these conveniences to your best advantage.

Fixing a Broken VI

A *broken VI* is a VI that cannot compile or run. The run button appears as a broken arrow to indicate that the VI is broken. A VI is usually broken while you are creating or editing it, until you finish wiring all the icons in the diagram. If it is still broken after everything is wired, first try selecting **Remove Bad Wires** from the **Edit** menu. Often, this command fixes a broken VI by cleaning up little wire fragments that you can't see.

To find out why a VI is broken, click on the broken run button or select **Show Error List** from the **Windows** menu. An information box titled "Error List" appears listing all errors for the VI. You can choose to see the error list for other open VIs using a menu ring at the top of the window. To find out more about a particular error, click on it. The Error List window will display more in-depth information. To locate a particular error in your VI, double-click on the error in the list or highlight it and press the FIND button. LabVIEW brings the relevant window to the front and highlights the object causing the error.

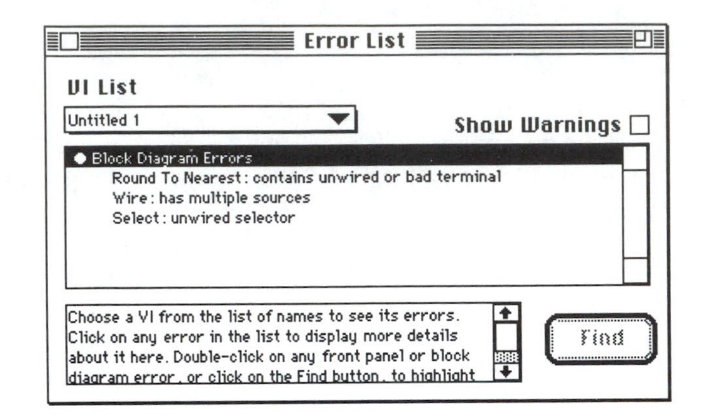

Warnings

You can choose to **Show Warnings** in the Error List window if you want extra debugging help. A warning is something that's not illegal and won't cause a broken run arrow, but does not make sense to LabVIEW, such as a control terminal that is not wired to anything. If you have **Show Warnings** checked and have any outstanding warnings, you will see the warning button in the edit mode palette. You can click on the warning button to see the Error List window, which will describe the warning.

Single-Stepping through a VI

For debugging purposes, you may want to execute a block diagram node by node.

To enable *single-step mode*, click on the step mode button . The symbol changes from a flat line to a square wave . Click on the step mode button at any time to return to normal execution.

While the VI is running in single-step mode, the single-step button appears . Click on this button to execute the next node. When the VI finishes executing, the button disappears.

Execution Highlighting

For debugging purposes, you may want to view an animation of the execution of the VI block diagram. To enable this mode, click on the *execution highlighting* button , which is located in the tool palette if your VI is in run mode. It is NOT accessible from edit mode!

You commonly use execution highlighting in conjunction with single-step mode to gain an understanding of how data flows through nodes. As data passes from one node to another, the movement of data is marked by bubbles moving along the wires. You will notice that highlighting greatly reduces the performance of a VI. The following illustration shows a VI running with execution highlighting enabled.

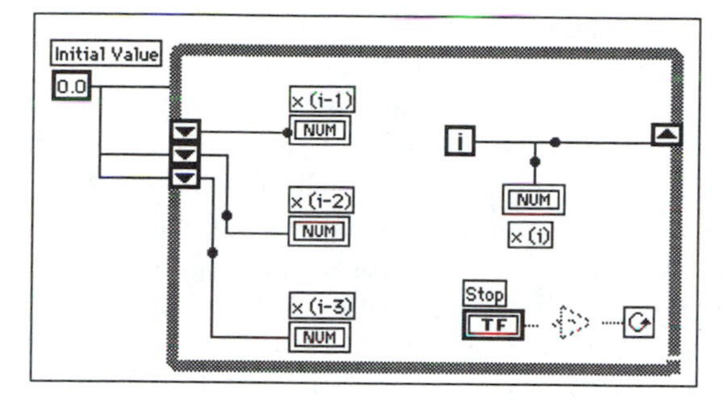

Execution Highlighting

Using the Probe

Use the *probe* to check intermediate values in a VI that executes but produces questionable or unexpected results. For instance, you may have a complicated diagram with a series of operations, any one of which may be returning incorrect data. You could wire an indicator to an output wire to display the intermediate results, or you can use the probe, a simpler technique. To access the probe, you must first be in run mode. Then pop up on the wire and select **Probe**. The probe display, which is a floating window, appears and flashes the values carried by the wire. Each wire can have one probe. You can use the probe in conjunction with execution highlighting and single-step mode to view values more easily.

You cannot change data with the probe, and the probe has no effect on VI execution.

If you lose track of which probe goes with which wire, you can pop up on a probe or a wire and select **Find Wire** or **Find Probe**, respectively, to highlight the corresponding object.

You can also probe using any conventional indicator by selecting **Controls** from the wire's pop-up menu, then choosing the desired indicator with which to probe.

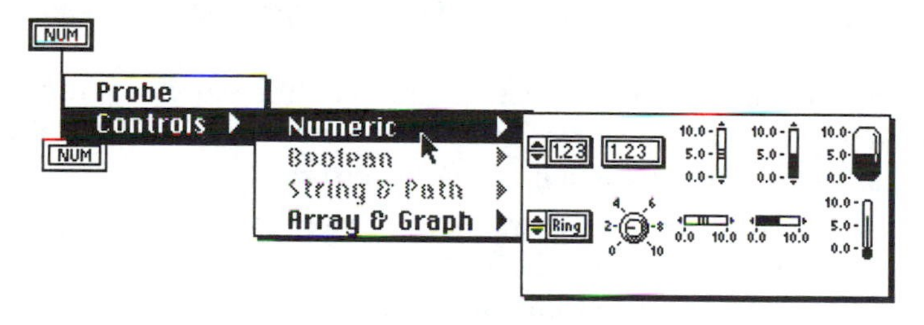

Using Controls to Probe

Setting Breakpoints

Don't panic—*breakpoints* do not "break" a VI; they only suspend its execution. Breakpoints are handy if you want to inspect the inputs to a subVI before it executes. To set a breakpoint, click the breakpoint button on your subVI [icon], located in the tool palette when your subVI is in run mode, which changes to a button with an exclamation point [icon]. When that subVI is called, execution of the higher-level VI will halt so that you can inspect the subVI inputs. Click on the breakpoint button at any time to remove the breakpoint and return to normal call mode.

You can also enable and disable breakpoints with the **VI Setup...** option, found in the icon pane pop-up menu in the subVI's panel. The two methods are interchangeable, and both cause the breakpoint to occur at all calls to the subVI. If a subVI is called from two locations in a block diagram, the breakpoint suspends execution at both calls.

If you want a breakpoint to suspend execution only at a particular call to the subVI, set the breakpoint using the **SubVI Node Setup...** option from the subVI node pop-up menu on the block diagram of the calling VI. **VI Setup** and **SubVI Node Setup** are discussed in Chapter 10, *Miscellaneous Concepts.*

When a subVI encounters a breakpoint, it comes to the foreground and remains in a suspended state. At this time, the values in the subVI controls are the inputs passed by the calling VI; you can change them if you wish. The subVI's indicators will display either default values or values from the last execution of the subVI during which its panel was open.

In the suspended state, two buttons appear on the subVI's run mode palette. Click on the retry button to run the subVI as if it were the top-level VI. You can also use the **Run** command in the **Operate** menu. When the subVI completes, the indicators display results from that execution of the subVI. However, you can change the indicator values if you want to return different values to the calling VI. You can rerun the subVI at the breakpoint as many times as you wish. While rerunning, you can return to the suspended state immediately by using the stop button or Stop command.

Click on the resume button when you are ready to return the subVI's indicator values to the calling VI. If you click the resume button without clicking the retry button first, the VI does not execute before returning.

You cannot set a breakpoint at a function or structure node, although you can encapsulate the function or structure in a subVI and suspend execution at the subVI call.

Debugging Techniques for Executable VIs—When Your Results Aren't Right

If your program executes but does not produce the expected results, use these suggestions as a reference to locate the problem. Don't worry if a few don't make sense to you—some of them refer to structures you haven't learned about yet:

- *Check wire paths* to ensure that the wires connect to the proper terminals. Triple-clicking on the wire with the Positioning tool highlights the entire path. A wire that appears to emanate from one terminal may in fact emanate from another, so look closely to see where the end of the wire connects to the node.
- *Use the Help window* (from the **Windows** menu) to make sure that functions are wired correctly.
- *Verify that the default value is what you expect* if functions or subVIs have unwired inputs.
- *Use breakpoints, execution highlighting, and single-stepping* to determine if the VI is executing as you planned. Make sure you disable these modes when you do not want them to interfere with performance.

- *Use the probe* described in the *Using the Probe* section of this chapter to observe intermediate data values. Also check the error output of functions and subVIs, especially those performing I/O.
- *Observe the behavior of the VI or subVI with various input values.* For numeric controls, you can enter the values NaN (Not A Number) and ±Inf (Infinity) in addition to normal values.
- *Make sure execution highlighting is turned off* in subVIs if the VI runs more slowly than expected. Also, close subVI windows when you are not using them.
- *Check for For Loops that may inadvertently execute zero iterations* and produce empty arrays.
- *Verify that you initialized shift registers properly*, unless you specifically intend them to save data from one execution of the loop to another.
- *Check the order of cluster elements* at the source and destination points. Although LabVIEW detects data type and cluster size mismatches at edit time, LabVIEW does not detect mismatches of elements of the same type. Use the **Cluster Order...** option on the cluster shell pop-up menu to check cluster order.
- *Check the node execution order.* Nodes that are not connected by a wire can execute in any order. The spatial arrangement of these nodes does not control the order. That is, unconnected nodes do not execute from left to right or top to bottom on the diagram like statements do in textual languages.

EXERCISE 5.1
Debugging Challenge

In this exercise, you will troubleshoot and fix a broken VI. Then you will practice using other debugging features, including execution highlighting, single-step mode, and the probe.

1. Open up the VI called **Debug Exercise.vi**, located in EXAMPLES\GENERAL\EXERCISE.LLB.
2. Switch to the block diagram. Notice that the run arrow is broken. You must find out why and rectify the situation so the VI will run.

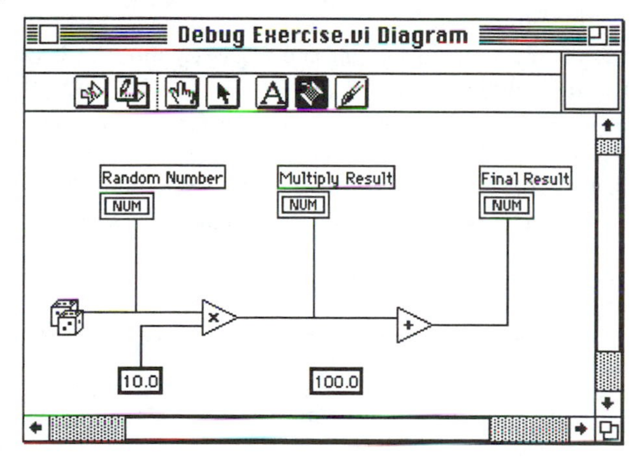

3. Click on the broken run arrow. An Error List dialog box appears describing the errors in the VI.

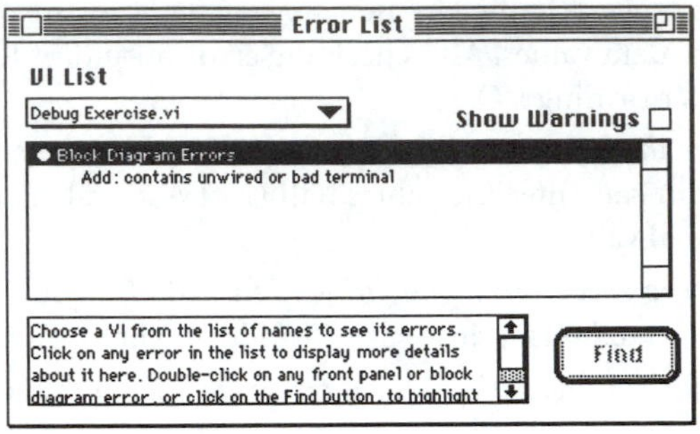

4. Click on the "Add: contains unwired or bad terminal" error. The Error List window will give you a more detailed description of the error. Now double-click on the "Add: contains unwired or bad terminal" error, or click the Find button. LabVIEW highlights the offending function in the block diagram for you, to help you locate the mistake.
5. Now draw in the missing wire. The run button should appear solid. If it doesn't, try to **Remove Bad Wires**. Hint: If you can't find the missing wire, think about how many inputs the **Add** function should have.
6. Switch back to the front panel, make sure you're in run mode, and run the VI a few times.
7. Tile the front panel and block diagram so you can see both at the same time. Enable the execution highlighting and enter single-step mode by pressing the appropriate buttons, located in the tool palette when your VI is in run mode.
8. Run the VI by clicking on the run button. The single-step button will appear—you must click this button each time you wish to execute a node. Notice that the data appears on the front panel as you step through the program. First, the VI generates a random number and then multiplies it by 10.0. Finally, the VI adds 100.0 to the multiplication result. Each of these front panel indicators is updated as new data reaches its block diagram terminals, a perfect example of data flow programming.
9. Now enable the probe by popping up on any wire segment and selecting **Probe**.
10. Step through the VI again and note how the probe displays the data carried by its corresponding wire.
11. Turn off execution highlighting and single step mode by clicking on their respective buttons. You're almost done...
12. Save your VI in your very own VI library. First choose **Save** from the **File** menu. Select **New**, and then choose **VI Library** from the resulting dialog box to create a new VI library called `MYWORK.LLB`. Remember, you can't create a library inside another library!
13. Close the VI by selecting **Close** from the **File** menu. You're finished!

Note
Save all of your subsequent exercises in MYWORK.LLB so you can find them easily!

5.4 Useful Tips

You will find the following list of tips and techniques helpful for developing VIs in LabVIEW:

- Frequently used menu options have equivalent command key shortcuts. For example, to save a VI, you can choose **Save** from the **File** menu, or you can press the Control key on Windows or the Command key on Mac equivalent. These Ctrl or Cmd key equivalents are shown next to their menu items in LabVIEW. Some frequently used Command or Ctrl key shortcuts are:

  <Cmd> or <Ctrl> <r>–Run a VI

  <Cmd> or <Ctrl> <f>–Toggle between the Panel window and the Diagram window

  <Cmd> or <Ctrl> <h>–Toggle the Help window on and off

  <Cmd> or <Ctrl> <b>–Remove all bad wires

  <Cmd> or <Ctrl> <w>–Close the active window
- To rotate through the tools in the tool palette, press <Tab>.
- To change the direction of a wire while wiring, press the space bar.
- To nudge selected objects in the Panel window and the Diagram window, press the arrow keys. Pressing the arrow keys nudges a selected object one pixel in the direction of the arrow. This tip also works for selected wire segments.
- To increment or decrement faster, press <Shift> while you click the increment or decrement button on digital controls.
- To add items quickly to ring controls, press <Shift-Enter> after typing the item name. Pressing <Shift-Enter> accepts the item and positions the cursor to add the next item.
- To duplicate an object, select the object using the Positioning tool, hold down <Ctrl> on a computer running Windows or <option> on the Mac, and drag the mouse. You will drag a copy of the item with you.
- To limit an object to horizontal or vertical motion only, hold down <Shift> and drag the object with the Positioning tool.
- To pick a color from an object, first select the Coloring tool. Place the tool over the object and press and hold down <Ctrl> on a computer running Windows or <option> on the Mac . The tool changes to the Sucker tool. Pick up the object color by clicking on the object. Release the <Ctrl> or <option> key, and then color other objects by clicking on them using the Coloring tool.
- One common error for new LabVIEW users is wiring together two controls or wiring two controls to an indicator. This mistake returns the error message "Signal: has multiple sources." Pop up on the offending control and select **Change to Indicator** to fix this problem.
- To delete a wire as you are wiring, click the right mouse button under Windows or wire off the screen and click on a Mac.

5.5 Creating SubVIs

Much of LabVIEW's power and convenience stems from its modularity. You can build the parts of your program one complete module at a time by creating subVIs. To use a VI as a subVI, you must do two things: design an icon and assign the connector.

Designing the Icon

Every subVI must have an icon to represent it in the block diagram of a calling VI. You can create the icon by selecting **Edit Icon** from the pop-up menu of the icon pane in the upper right-hand corner of the front panel. The Icon Editor window, shown in the following illustration, will appear. Use its tools to design the icon of your choice.

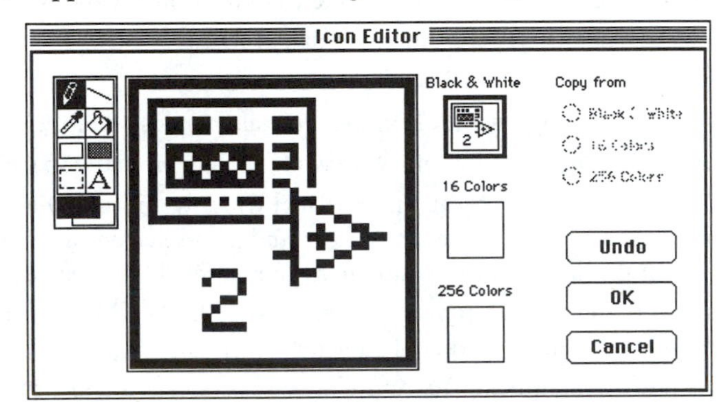

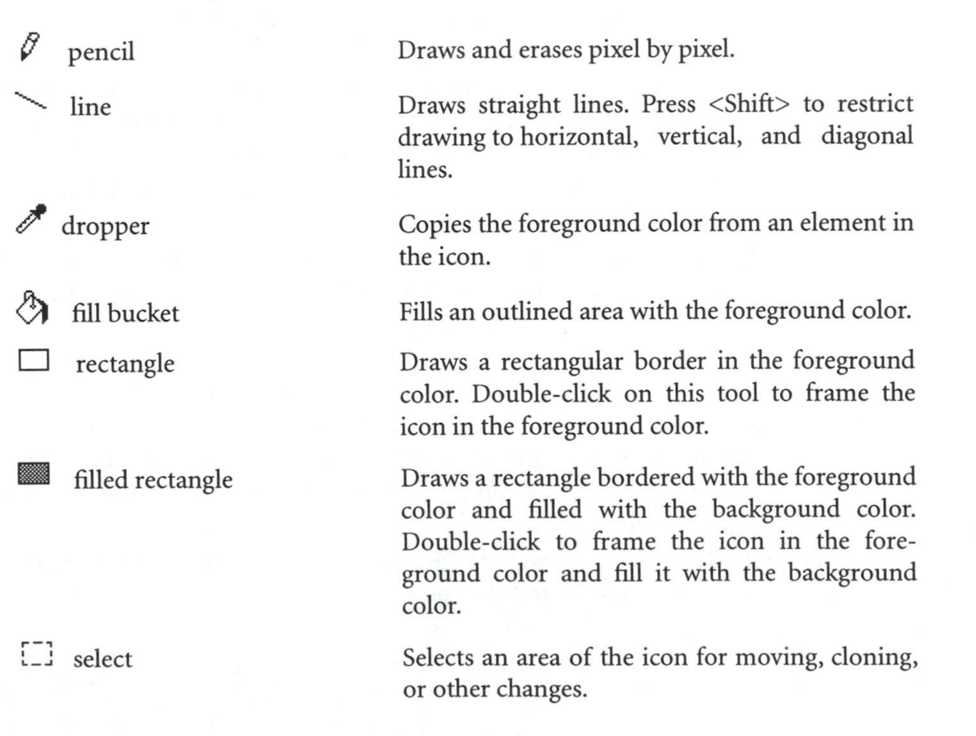

pencil	Draws and erases pixel by pixel.
line	Draws straight lines. Press <Shift> to restrict drawing to horizontal, vertical, and diagonal lines.
dropper	Copies the foreground color from an element in the icon.
fill bucket	Fills an outlined area with the foreground color.
rectangle	Draws a rectangular border in the foreground color. Double-click on this tool to frame the icon in the foreground color.
filled rectangle	Draws a rectangle bordered with the foreground color and filled with the background color. Double-click to frame the icon in the foreground color and fill it with the background color.
select	Selects an area of the icon for moving, cloning, or other changes.

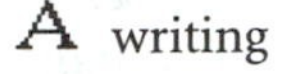 writing	Enters text into the icon design.
foreground/background	Displays the current foreground and background colors. Click on each to get a palette from which you can choose new colors.

The buttons at the right of the editing screen perform the following functions:

- **Undo** Cancels the last operation you performed
- **OK** Saves your drawing as the VI icon and returns to the Panel window
- **Cancel** Returns to the Panel window without saving any changes

Creating the Connector

You have to define connector terminals if you want to pass data in and out of a subVI, just like you must define parameters for a subroutine in a conventional language. To do this, pop up in the icon pane and select **Show Connector**. LabVIEW chooses a connector based on the number of controls and indicators on the front panel. If you want a different one, choose it from the **Patterns** menu, obtained by popping up on the connector. You can rotate and flip your connector if it doesn't have a convenient orientation.

Follow these steps to assign the terminal to a control or indicator:

1. Click (with the left mouse button under Windows) on a terminal in the connector. The cursor automatically changes to the Wiring tool, and the terminal turns dark as shown.

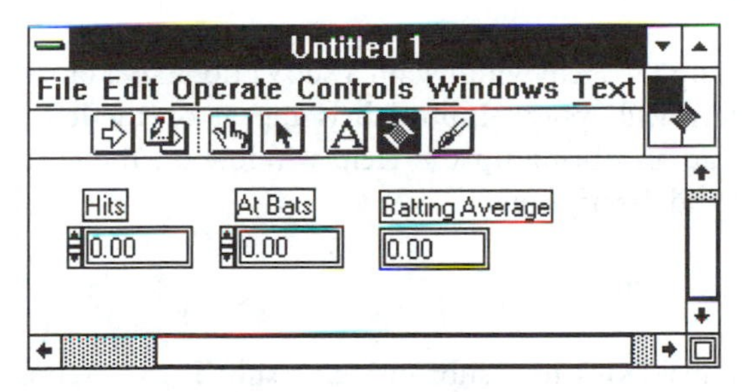

2. Click on the control or indicator you want that terminal to represent. A moving dashed line frames the control or indicator.

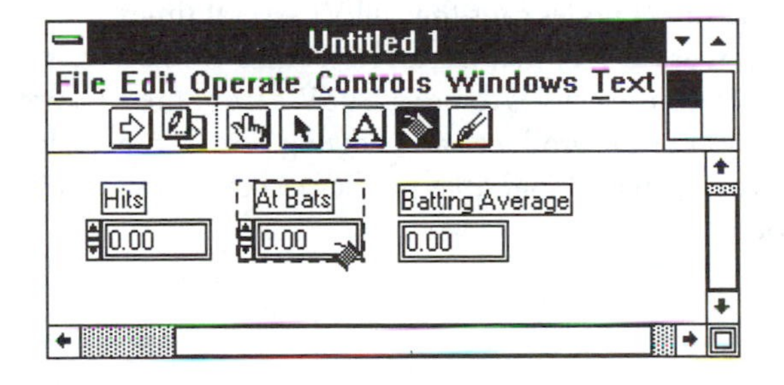

3. Click in an open area on the front panel. The dashed line disappears and the selected terminal dims, indicating that you have assigned the control or indicator to that terminal. The assigned terminal is shown in the following figure.

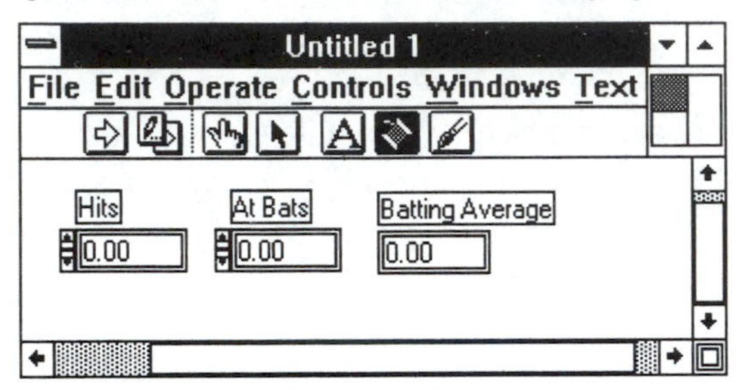

If the terminal is white, you have not made the connection correctly. Repeat the previous steps if necessary.

Note
You can also reverse the order of the first two steps.

If you make a mistake, you can **Disconnect** a particular terminal or **Disconnect All** by selecting the appropriate action from the connector's pop-up menu.

Online SubVI Help

If you bring up the Help window on a subVI node in a block diagram, its description and wiring pattern will appear. Input labels appear on the left, while outputs appear on the right. You can also bring up the Help window on the current VI's icon pane to see its parameters and description.

Using a VI as a SubVI

A subVI is analogous to a subroutine. A subVI node (comprised of icon/connector) is analogous to a subroutine call. The subVI node is not the subVI itself, just as a subroutine call statement in a program is not the subroutine itself. A block diagram that contains several identical subVI nodes calls that subVI several times.

You can use any VI that has an icon and a connector as a subVI in the block diagram of another VI. You select VIs to use as subVIs from the **VI...** option of the **Functions** menu. Choosing this option produces a file dialog box from which you can select any VI in the system. If you open a VI that does not have a connector, you cannot wire inputs or outputs to that node.

5.6 Documentation

It's important to document your VIs so that others can understand them and so you don't forget why or how you did something. This section discusses how you can document your work in LabVIEW.

Creating Descriptions for Individual Objects

If you want to enter a description of a LabVIEW object, such as a control or indicator, choose **Description...** from the **Data Operations** submenu of the object's pop-up menu. Enter the description in the resulting dialog box, shown in the next illustration, and click OK to save it. LabVIEW displays this description whenever you subsequently choose **Description...** from the object's pop-up menu.

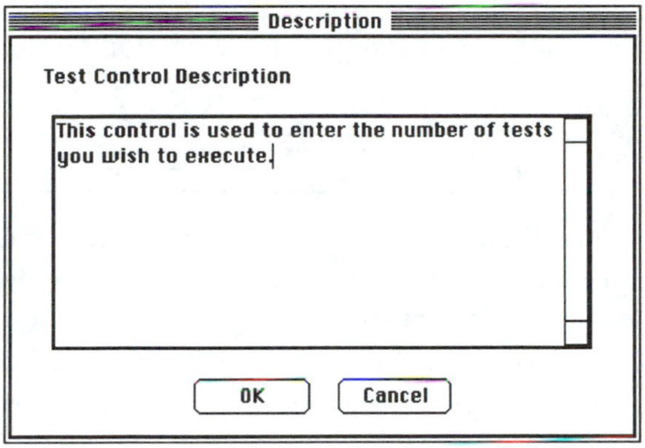

Description

Test Control Description

This control is used to enter the number of tests you wish to execute.

OK Cancel

Note
You can view an object's description in Run mode, but you cannot edit it from there.

Documenting VIs with the Get Info... Option

Selecting **Get Info...** from the **File** menu displays the information dialog box for the current VI.

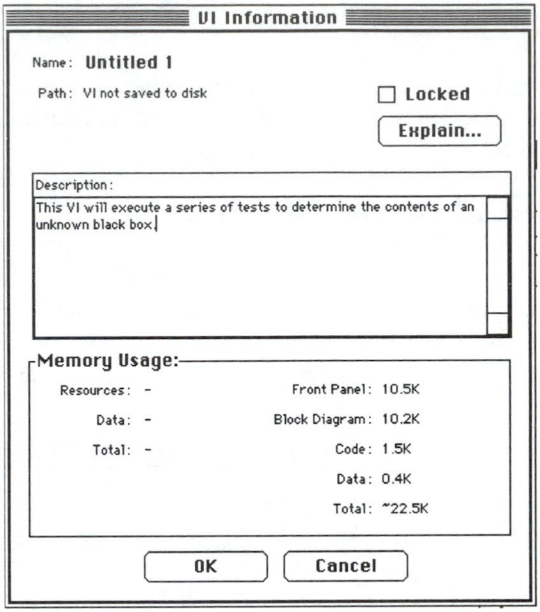

You can use the information dialog box to perform the following functions:

- Enter a description of the VI. The description window has a scrollbar so you can edit or view lengthy descriptions.
- Lock or unlock the VI. You can execute but not edit a locked VI.
- See a list of changes made to the VI since you last saved it by pressing the **Explain** button.
- View the path of the VI.
- See how much memory the VI uses. The Size portion of the information box displays the disk and system memory used by the VI. (This figure applies only to the amount of memory the VI is using and does not reflect the memory used by any of its subVIs.)

Note
Choosing Description from an object's Data Operations menu documents the individual object, while entering text in the Get Info dialog box of a VI documents the entire VI.

EXERCISE 5.2
Creating SubVIs—Practice Makes Perfect

You will turn the **Thermometer** VI you created in the last chapter into a subVI so you can use it in the diagram of another VI.

1. Open **Thermometer.vi** by selecting **Open** from the **File** menu. If you saved it in EXAMPLES\GENERAL\EXERCISE.LLB like we told you to, it should be easy to find.

2. To give you some practice with save options and libraries, resave **Thermometer.vi** in MYWORK.LLB by selecting **Save As...** and then choosing MYWORK.LLB as the destination. Keep the same name for the VI. In fact, you can delete the copy in EXERCISE.LLB if you want to.
3. Create an icon for the VI. Pop up on the icon pane in the front panel (it doesn't work in the diagram) and select **Edit Icon...** from the menu. You will be placed in the icon editor. Use the tools described in section 5.5, *Creating SubVIs*, to create the icon, then click the OK button to return to the main VI. Your icon should appear in the icon pane as shown.

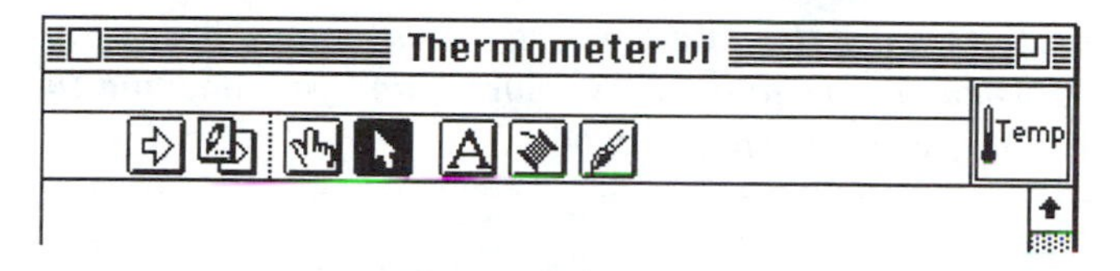

4. Create the connector by popping up in the icon pane and selecting **Show Connector**. Since you have only one indicator on your front panel, your connector should have only one terminal and should appear as a white box, as shown.

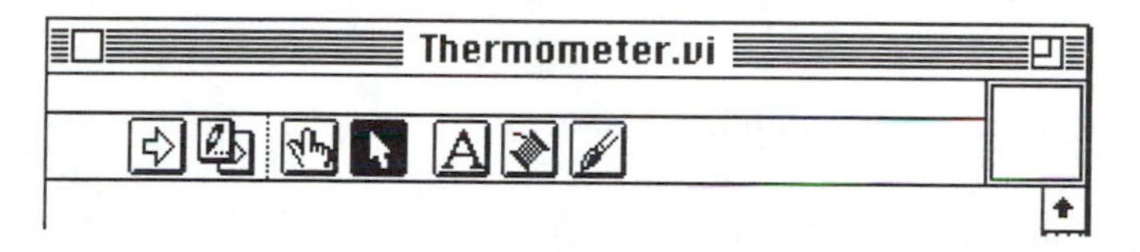

5. Assign the terminal to the thermometer indicator. Using the Wiring tool, click on the terminal in the connector. The terminal will turn black. Then click on the thermometer indicator. A moving dashed line will frame the indicator. Finally, click in an open area on the panel. The dashed line will disappear and the selected terminal will turn from black to grey, indicating that you have assigned the indicator to that terminal. Pop up and choose **Show Icon** to return the icon to the pane.
6. Document the thermometer indicator by selecting **Description...** from its **Data Operations** pop-up menu. Type in the description shown and click OK when you're finished.

Temperature Description

Displays the simulated temperature (Deg F) measurement.

OK Cancel

7. Document **Thermometer.vi** by selecting **Get Info...** from the **File** menu and typing in a description. Click OK to return to the main VI.

8. Now bring up the Help window by choosing **Show Help Window** from the **Windows** menu. When you place the cursor over the icon pane, you will see the VI's description and wiring pattern in the Help window.
9. Save the changes by selecting **Save** from the **File** menu. Good job! You will use this VI as a subVI in the next chapter!

Watch Out!
Frequently, people create an icon for a VI and forget about the connector. If you haven't created the connector, you will be unable to wire inputs and outputs to your VI when you try to use it as a subVI, and you may find the source of your wiring inability very difficult to locate!

5.7 Wrap It Up!

LabVIEW offers several options that let you save your VIs. You will probably want to save them in *VI libraries*, which are special LabVIEW files containing groups of VIs. Your operating system sees VI libraries as single files; only LabVIEW can access the individual VIs inside.

You should take advantage of LabVIEW's many useful debugging features if your VIs do not work right away. You can *single-step* through your diagram node by node, animate the diagram using *execution highlighting*, and suspend subVIs when they are called so you can look at input and output values by setting a *breakpoint*. You can also use the *probe* to display the value a wire is carrying at any time. Each of these features allows you to take a closer look at your VI and its data.

SubVIs are the LabVIEW equivalent of subroutines. You must define an icon and a connector if you wish to use a VI as a subVI. To import a subVI into a calling VI's block diagram, choose **VI...** from the **Functions** menu and then select the subVI you want from the dialog box. SubVIs represent one of LabVIEW's most powerful features. You will find it very easy to develop and debug reusable, low-level subVIs and then call them from higher-level VIs.

As with all programming, it is a good idea to document your work in LabVIEW. You can document a VI by entering a description under **Get Info...** from the **File** menu. This description is also visible in the Help window if you pass the cursor over the VI's icon. You can document individual front panel objects by selecting **Description...** from the **Data Operations** menu and then entering your text in the resulting dialog box.

5.8 Additional Exercises

EXERCISE 5.3
Find the Average

Create a subVI that averages three input numbers and outputs the result. Remember to create both the icon and connector for the subVI.

EXERCISE 5.4
Divide by Zero

Build a VI that generates a random number between zero and ten, then divides it by an input number and displays the result on the front panel. If the input number is zero, the VI lights an LED to flag a "divide by zero" error.

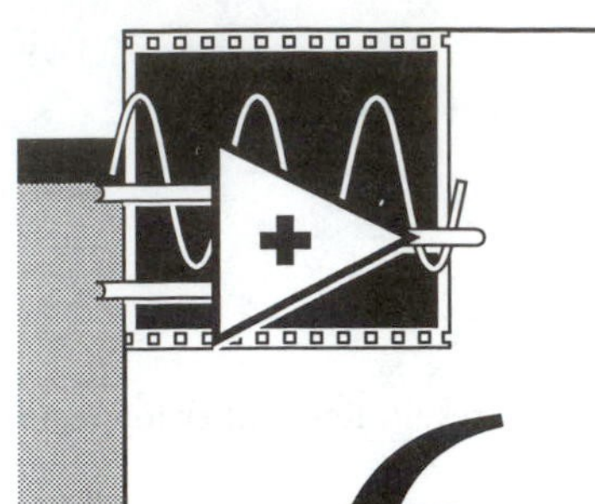

6

Structures

Overview

Structures, an important type of node, govern execution flow in a VI, just as control structures do in a standard programming language. This chapter introduces you to the four structures in LabVIEW: the While Loop, the For Loop, the Case structure, and the Sequence structure. You will also learn how to implement lengthy formulas using the Formula Node, how to pop up a dialog box containing your very own message, and how to control the timing of your programs.

Your Goals

- Know the uses of the While Loop and the For Loop and understand the differences between them
- Recognize the necessity of shift registers in graphical programming
- Understand the two types of Case structures—numeric and Boolean
- Learn how to regulate execution order using Sequence structures
- Use the formula node to implement long mathematical formulas
- Make LabVIEW pop up a dialog box that says anything you want
- Understand how to use LabVIEW's timing functions

Key Terms

For Loop	Case structure
While Loop	selector terminal
iteration terminal	dialog box
conditional terminal	Sequence structure
count terminal	sequence local
shift register	Formula Node

6.1 Two Loops

You use the *For Loop* and *While Loop* to control repetitive operations in a VI. A For Loop executes a specified number times; a While Loop executes until a specified condition is no longer true. You can find the loops under the **Structs & Constants** palette of the **Functions** menu.

The For Loop

A *For Loop*, shown in the following illustration, executes the code inside its borders, called its *subdiagram*, a total of *count* times, where the count equals the value contained in the *count terminal*. You can set the count by wiring a value from outside the loop to the count terminal.

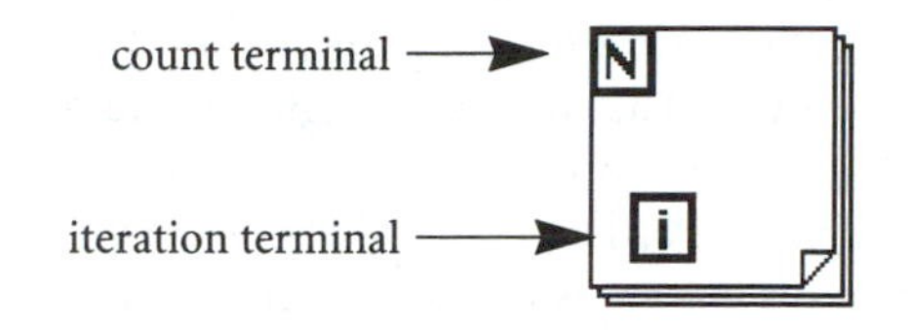

A For Loop

The *iteration terminal* contains the current number of completed loop iterations; 0 during the first iteration, 1 during the second, and so on up to N–1. If you wire 0 to the count terminal, the loop does not execute.

The For Loop is equivalent to the following pseudocode:

```
for i = 0 to N-1
        Execute subdiagram
```

The While Loop

The *While Loop*, shown in the following illustration, executes the subdiagram inside its borders until the Boolean value wired to its *conditional terminal* is FALSE. LabVIEW checks the conditional terminal value at the end of each iteration. If the value is TRUE, another iteration occurs. The default value of the conditional terminal is FALSE, so if you leave it unwired, the loop iterates only once.

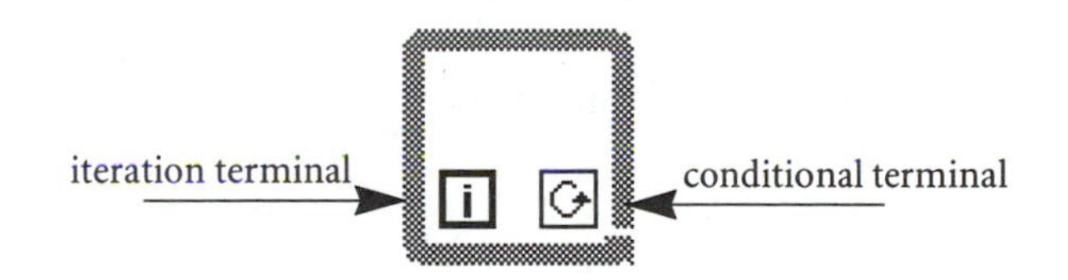

A While Loop

The While Loop's *iteration terminal* behaves exactly like the one in the For Loop.

The While Loop is equivalent to the following pseudocode:

```
Do
        Execute subdiagram (which sets condition)
While condition is TRUE
```

Placing Objects inside Structures

When you first select a structure from the **Structs & Constants** palette of the **Functions** menu, the cursor appears as a miniature of the structure you've selected; for example, (For Loop) or (While Loop). You can then click (and hold down the mouse button) where you want the upper left corner of your structure to be, and drag to define the borders of your structure. When you release the mouse button, the structure will appear containing all you captured in the borders.

Once you have the structure on the diagram, you can place other objects inside either by dragging them in, or by using the pop-up **Functions** menu to create them inside. To make it clear to you that you are dragging something *into* a structure, the *structure's* border will highlight as the object moves inside. When you drag an object *out* of a structure, the *block diagram's* border will highlight as the object moves outside.

If you move an existing structure so that it overlaps another object, the overlapped object will be visible above the edge of the structure. If you drag an existing structure completely over another object, that object will display a thick shadow to warn you that the object is under rather than inside the structure. Both of these situations are shown in the following illustration.

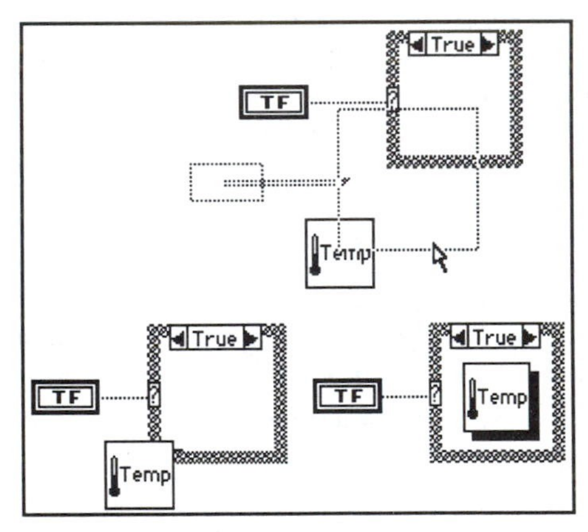

Terminals inside Loops and Other Behavioral Issues

Since LabVIEW operates under dataflow principles, inputs to a loop must pass their data before the loop executes. *Loop outputs pass data out only after the loop completes all iterations.*

As well, *you must place a terminal inside a loop if you want that terminal checked or updated on each loop iteration.* For example, the left While Loop in the following illustration checks its Boolean control each time it loops. When the loop reads a FALSE value, it terminates.

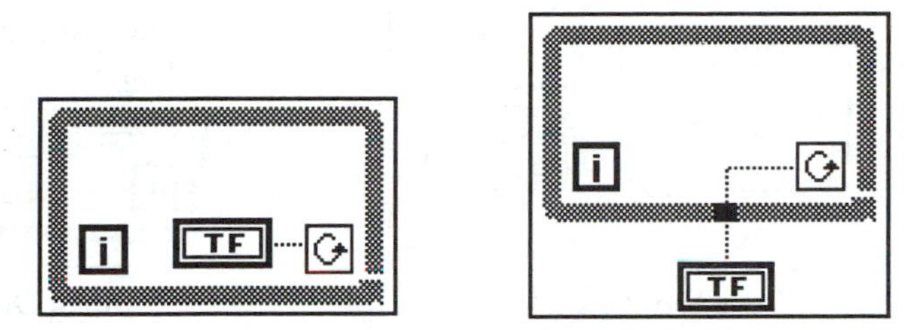

If you place the terminal of the Boolean control outside the While Loop, as shown in the right loop above, you create an infinite loop or a loop that executes only once, depending on the Boolean's initial value. True to dataflow, LabVIEW reads the value of the Boolean before it enters the loop, not within the loop or after completion.

Watch Out!
Remember, the first time through a For Loop or a While Loop, the iteration count is zero! If you want to register how many times the loop has actually executed, you must add one to the count!

EXERCISE 6.1
Counting with Loops

In this exercise, you get to build a For Loop that counts from 0 to 99 and displays the count interactively on the front panel. You will then build a While Loop that counts until you stop it with a Boolean switch. Just for fun (and also to illustrate an important point), you should observe the effect of putting the count indicator outside the While Loop.

1. Create a new panel by selecting **New** from the **File** menu.
2. Build the front panel and block diagram shown. The For Loop is located in the **Structs & Constants** palette of the **Functions** menu. You might use the **Tile Left and Right** command from the **Windows** menu so that you can see both the front panel and the block diagram at the same time.

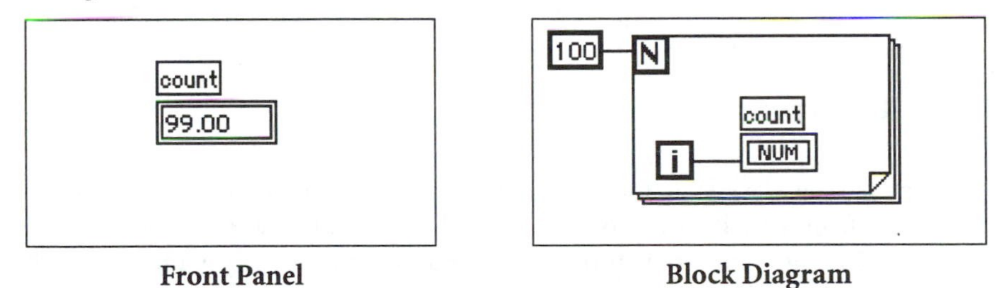

Front Panel **Block Diagram**

3. Run the VI. Notice that the digital indicator counts from 0 to 99, NOT 1 to 100!
4. You can save the VI if you want, but we're through with it. Open up another new window so you can try out the While Loop.

5. Build the VI shown in the following illustration. Remember, Booleans appear on the front panel in their default FALSE position.

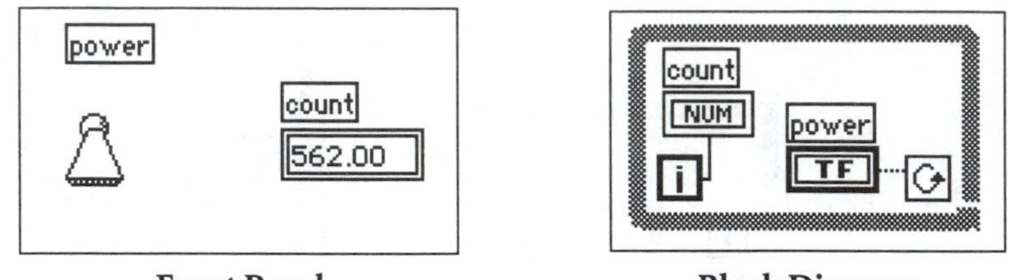

Front Panel **Block Diagram**

6. Flip the switch up to its TRUE position and run the VI. When you want to stop, click on the switch to flip it down to FALSE.
7. With the switch still in the FALSE position, run the VI again. Notice that the While Loop executes once, but only once. Remember, the loop checks the conditional terminal at the *end* of an iteration, so it always executes at least once, even if nothing is wired to it.
8. Now go to the block diagram and move the Loop Count indicator outside the loop as shown.

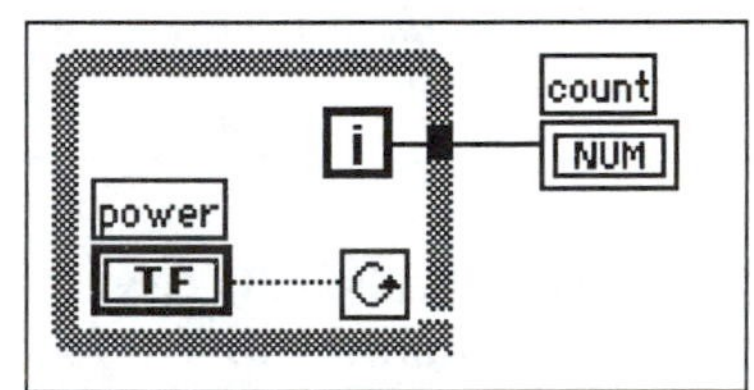

9. Make sure the switch is TRUE and run the VI. Notice that the indicator updates only after the loop has finished executing; it contains the final value of the iteration count, which is passed out after the loop completes. You will learn more about passing data out of loops in Chapter 7, *Arrays and Clusters*. Until then, do not try to pass data out of a For Loop like you just did in a While Loop, or you will get bad wires and you won't understand why.
10. Save and close the VI. Place it in `EXAMPLES\GENERAL\MYWORK.LLB` and call it **Loop Count.vi**. You finished another one!

6.2 Shift Registers

Shift registers (available for While Loops and For Loops) are local variables that transfer values from one iteration of a loop to the next. They are unique to and necessary for LabVIEW's graphical structure. You create a shift register by popping up on the left or right loop border and selecting **Add Shift Register** from the pop-up menu.

The shift register is comprised of a pair of terminals directly opposite each other on the vertical sides of the loop border. The *right* terminal stores the data upon the completion of an iteration. That data is "shifted" at the end of the iteration and appears in the *left* terminal at the beginning of the next iteration, as shown in the following illustration. A shift register can hold any data type—numeric, Boolean, string, array, and so on. The shift register automatically adapts to the data type of the first object that is wired to one of its terminals.

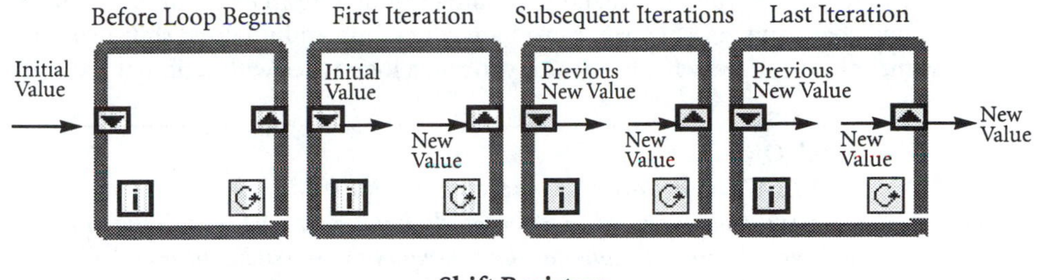

Shift Registers

You can configure the shift register to remember values from several previous iterations, as shown in the following illustration You will find this feature useful when you are averaging data points taken in different iterations. Create additional terminals to access values from previous iterations by popping up on the *left* terminal and choosing **Add Element** from the pop-up menu.

If you're still a little confused, don't worry. Shift registers are a completely new and different concept. Stepping through the next exercise should demonstrate them more clearly for you.

Watch Out!
Make sure to wire directly to the shift register so that you don't accidentally create an unrelated tunnel.

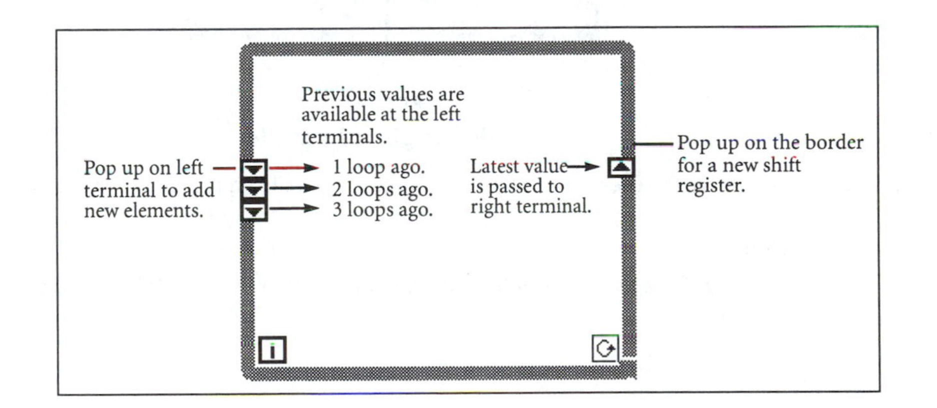

Initializing Shift Registers

You should always initialize your shift registers unless you have a specific reason and make a conscious decision not to. To initialize the shift register with a specific value, wire that value to the left terminal of the shift register from outside the loop, as shown in the following illustration. If you do not initialize, the initial value will be the default value for the shift register data type the first time you run your program. For example, if the shift register data type is Boolean, the initial value will be FALSE. Similarly, if the shift register data type is numeric, the initial value will be zero. The second time you run your VI, an uninitialized shift register will contain values left over from the first run! Study the following figure to make sure you understand what initialization does. The two loops in the left column show what happens when you run a loop containing an initialized shift register twice. The right column shows what happens if you run a loop twice with uninitialized shift registers.

Watch Out!

LabVIEW does not discard values stored in the shift register until you close the VI and remove it from memory. In other words, if you run a VI containing uninitialized shift registers, the initial values for the subsequent run will be the ones left from the previous run. The resulting problems can be very difficult to spot!

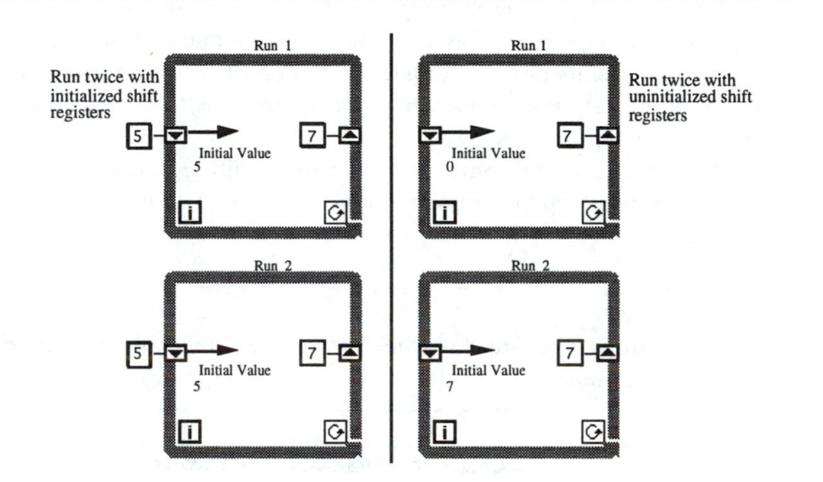

EXERCISE 6.2
Shift Register Example

You will observe the use of shift registers to access values from previous iterations of a loop.

Front Panel

1. Open **Shift Register Example.vi**, located in EXAMPLES\GENERAL\EXERCISE.LLB.

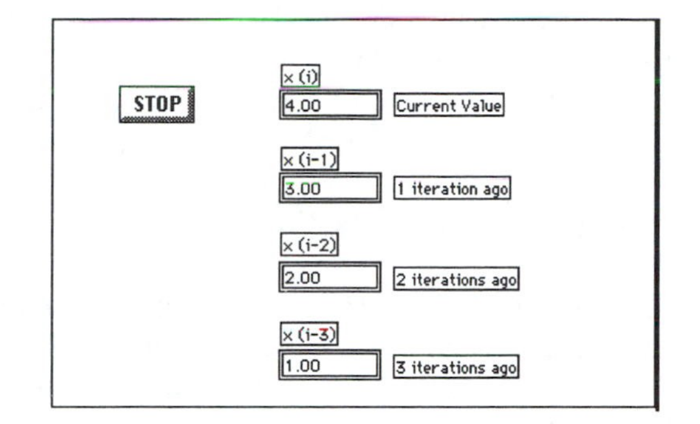

The front panel has four digital indicators. The $x(i)$ indicator will display the current value of the loop count (it is wired to i). The $x(i–1)$ indicator will display the value of the loop count one iteration ago. The $x(i–2)$ indicator will display the value two iterations ago, and so on.

The zero wired to the left terminals initializes the elements of the shift register to zero. At the beginning of the next iteration, the old $x(i)$ value that will shift to the top left terminal to become $x(i–1)$. $x(i–1)$ shifts down into x(i–2), and so on.

2. Open the Diagram window by choosing **Show Diagram** from the **Windows** menu.

Block Diagram

1. After examining the block diagram show both the panel and the diagram at the same time by choosing **Tile Left and Right** from the **Windows** menu.

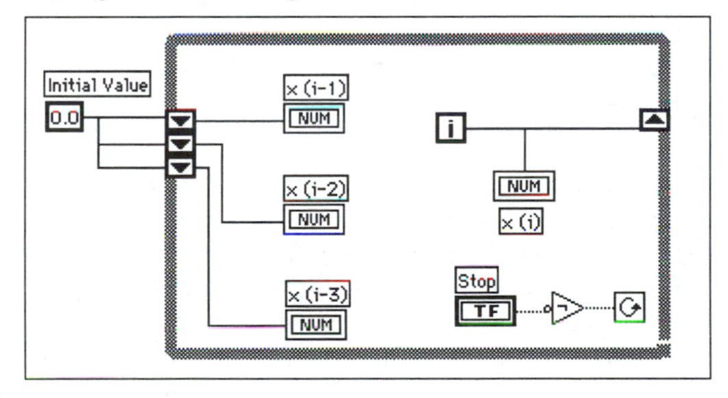

2. Enable the execution highlighting by clicking on the execution highlighting button. The button will turn to .
3. Run the VI and carefully watch the bubbles. If the bubbles are moving too fast, put the VI in single-step mode by clicking on the step mode button. The button will change to . Click on the step button to execute the VI.

   Notice that in each iteration of the While Loop, the VI "funnels" the previous values through the left terminals of the shift register using a first in, first out (FIFO) algorithm. Each iteration of the loop increments the count terminal wired to the right terminal, $x(i)$, of the shift register. This value shifts to the left terminal, $x(i–1)$, at the beginning of the next iteration. The values at the left terminal funnel downward through the terminals. In this example, the VI retains only the last three values. To retain more values, add more elements to the left terminal of the shift register.

4. Close the VI. Do not save any changes. Another job well done.

Why You Need Shift Registers

Observe the following example. In loop (A), you are creating a running sum of the iteration count. Each time through the loop, the new sum is saved in the shift register. At the end of the loop, the total sum of 45 is passed out to the numeric indicator. In loop (B), you have no shift registers, so you do not save values between iterations. Instead, you add zero to the current "*i*" each time, and only the last value of 9 will be passed out of the loop.

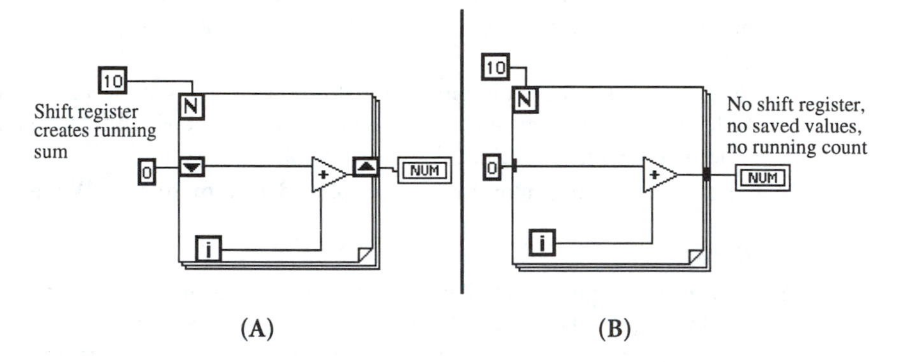

(A) (B)

6.3 Case Structures

A *Case structure* is LabVIEW's method of executing conditional text, sort of like an "if-then-else" statement. You can find it in the **Structs & Constants** palette of the **Functions** menu. The Case structure, shown in the following illustration, has two or more subdiagrams, or cases, exactly one of which executes, depending on the value of the Boolean or numeric scalar you wire to the *selector terminal.* If a Boolean value is wired to the selector, the structure has two cases, FALSE and TRUE. If a numeric is wired to the selector, the structure can have from 0 to 2^{15}–1 cases. Initially only the 0 and 1 cases are available, but you can easily add more. When you first place it on the panel, the Case structure appears in its Boolean form; it assumes numeric values as soon as you wire a numeric data type to its selector terminal.

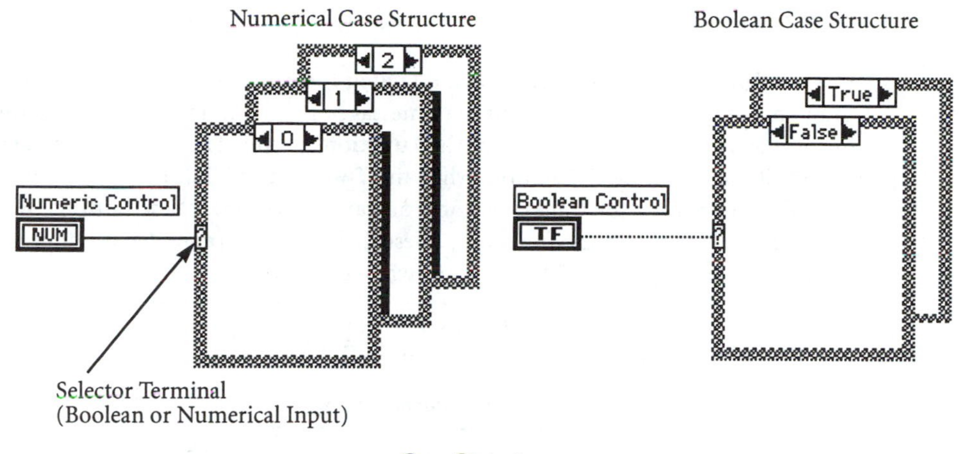

Case Structures

Case structures can have multiple subdiagrams, but you can only see one case at a time (unlike it appears in the previous illustration), sort of like a stacked deck of cards. Clicking on the decrement ◄ (left) or increment ► (right) button at the top of the structure displays the previous or next subdiagram, respectively.

If you wire a floating-point number to the selector, LabVIEW rounds that number to the nearest integer value. LabVIEW coerces negative numbers to 0, and changes any value higher than the highest-numbered case to equal the number of the highest case.

You can position the selector anywhere along the left border, but you have to wire something to it. The selector automatically adjusts to the data type. If you change the value wired to the selector from a numeric to a Boolean, cases 0 and 1 change to FALSE and TRUE. If other cases exist (2 through *n*), LabVIEW does not discard them, in case the change in data types is accidental. However, you must delete these extra cases before the structure can execute.

Wiring Inputs and Outputs

The data at all input terminals (tunnels and selection terminal) is available to all cases. Cases are not required to use input data or to supply output data, but *if any one case supplies output data, all must do so.* If you do not wire data to an output tunnel from every case, the tunnel turns white and the broken Run button appears.

Adding Cases

If you pop up on the Case structure border, the resulting menu gives you the options to **Add Case After** and **Add Case Before** the current case. These options create a new case that appears either after or before the currently shown case. You can also choose to copy the currently shown case by selecting **Duplicate Case**. You can delete the current case by selecting **Remove Case**.

Dialog Boxes

The **One Button Dialog** and **Two Button Dialog** functions, shown in the following illustration, bring up a *dialog box* containing a message of your choice. You can find these functions in the **Time & Dialog** palette of the **Functions** menu. The **One Button Dialog** stays open until you click the OK button, while the **Two Button Dialog** box remains until you click either the OK or the Cancel button. You can also rename these buttons by inputting "button name" strings to the functions. These dialog boxes are *modal*; in other words, you can't click on any other LabVIEW window while they are open.

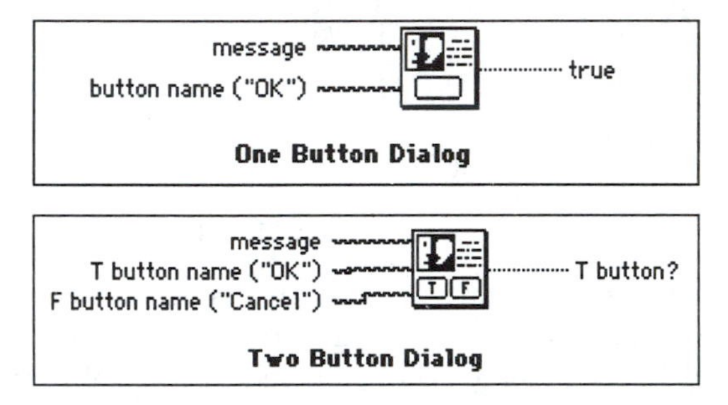

EXERCISE 6.3
Square Roots

You will build a VI that returns the square root of a positive input number. If the input number is negative, the VI returns an error.

Front Panel

1. Open a new panel.
2. Build the front panel shown in the following illustration.

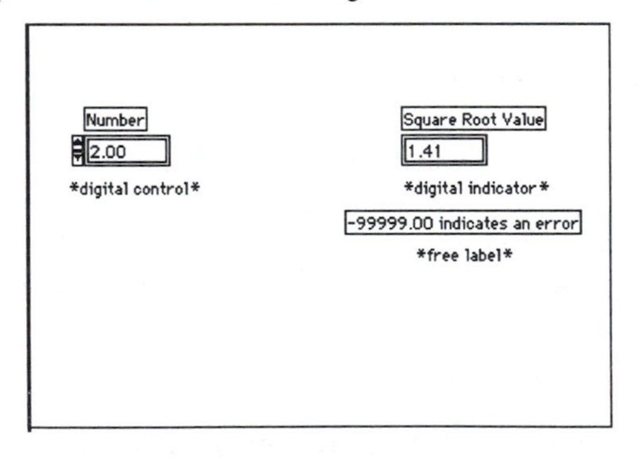

The Number digital control supplies the input number. The Square Root Value indicator will display the square root of the number.

Block Diagram

1. Open the Diagram window.

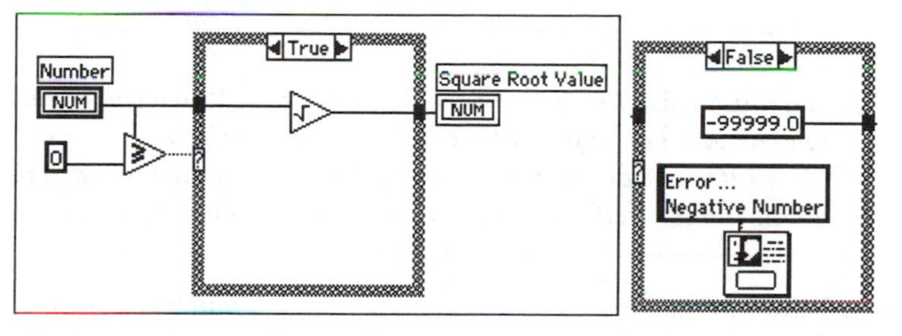

2. Place the Case structure (**Structs & Constants** menu) in the Diagram window. Enlarge the case structure by dragging one corner with the Positioning tool.

   By default, the Case structure selector terminal is Boolean. It would automatically change to numeric if you wired something with a numeric data type to the terminal. In this case, however, the **Greater or Equal?** function returns a Boolean value.

   You can display only one case at a time. To change cases, click on the arrows in the top border of the Case structure.

3. Select the other diagram objects and wire them as shown in the preceding illustration. *Make sure to use the Help window to display terminal inputs and outputs!*

   Greater or Equal? function (**Comparison** menu). In this exercise, checks whether the number input is negative. The function returns a TRUE if the number input is greater than or equal to 0.

   Square Root function (**Arithmetic** menu). In this exercise, returns the square root of the input number.

   -99999.0 0 **Numeric Constants** (**Structs & Constants** menu).

   One Button Dialog function (**Time & Dialog** menu). In this exercise, displays a dialog box that contains the message "Error...Negative Number."

   Error... Negative Number **String Constant** (**Structs & Constants** menu). Enter text inside the box with the Operating tool. (You will study strings in detail in Chapter 9, *Strings and File I/O.*)

   In this exercise, the VI will execute either the TRUE case or the FALSE case. If the number is greater than or equal to zero, the VI will execute the TRUE case. The TRUE case returns the square root of the number. The FALSE case outputs a –99999.00 and displays a dialog box with the message "Error...Negative Number."

Note
You must define the output tunnel for each case. When you create an output tunnel in one case, tunnels appear at the same position in the other cases. Unwired tunnels look like white squares. Be sure to wire to the output tunnel for each unwired case, clicking on the tunnel itself each time, or you might accidentally create another tunnel.

---

4. Return to the front panel and run the VI. Try a number greater than zero and one less than zero.
5. Save and close the VI. Name it **Square Root.vi** and place it in MYWORK.LLB.

VI Logic

```
if (Number >= 0) then
            Square Root Value = SQRT (Number)
      else
            Square Root Value = -99999.00
            Display Message "Error...Negative Number"
end if
```

6.4 Sequence Structures

Determining the execution order of a program by arranging its elements in a certain sequence is called *control flow.* BASIC, C, and most other programming languages have inherent control flow because statements execute in the order in which they appear in the program. LabVIEW uses the *Sequence structure* to obtain control flow within a dataflow framework. A Sequence structure executes frame 0, followed by frame 1, then frame 2, until the last frame executes. Only when the last frame completes does data leave the structure.

The Sequence structure, shown in the following illustration, looks like a frame of film and can be found in the **Structs & Constants** palette of the **Functions** menu. Like the Case structure, only one frame is visible at a time—you must click the arrows at the top of the structure to see other frames. You create new frames by popping up on the structure border and selecting **Add Frame After** or **Add Frame Before.**

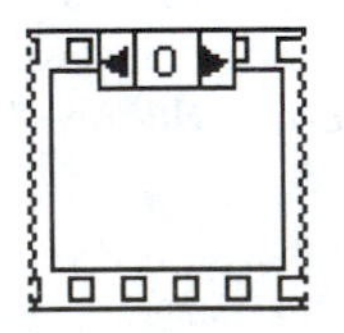

Sequence Structure

You use the Sequence structure to control the order of execution of nodes that are not data dependent. Within each frame, as in the rest of the block diagram, data dependency determines the execution order of nodes.

Output tunnels of Sequence structures can have only one data source, unlike Case structures, whose outputs have one data source per case. The output can emit from any frame, but keep in mind that data leaves the structure only when it completes execution entirely, not when the individual frames finish. Data at input tunnels is available to all frames.

Sequence Locals

To pass data from one frame to any subsequent frame, you must use a terminal called a *sequence local.* To obtain a sequence local, choose **Add Sequence Local** from the *structure border* pop-up menu. This option is not available if you pop up too close to another sequence local or over the subdiagram display window. You can drag the terminal to any unoccupied location on the border. Use the **Remove** command from the sequence local pop-up menu to remove a terminal.

When it first arrives on the diagram, a sequence local terminal is just a small yellow box. When you wire source data to the sequence local, an outward-pointing arrow appears in the terminal of the frame containing the data source. The terminals in subsequent frames contains an inward-pointing arrow, indicating that the terminal is a data source for that frame. In frames before the source frame, you cannot use the sequence local (after all, it hasn't been assigned a value), and it appears as a dimmed rectangle. The following figure shows the sequence local terminal in its several forms.

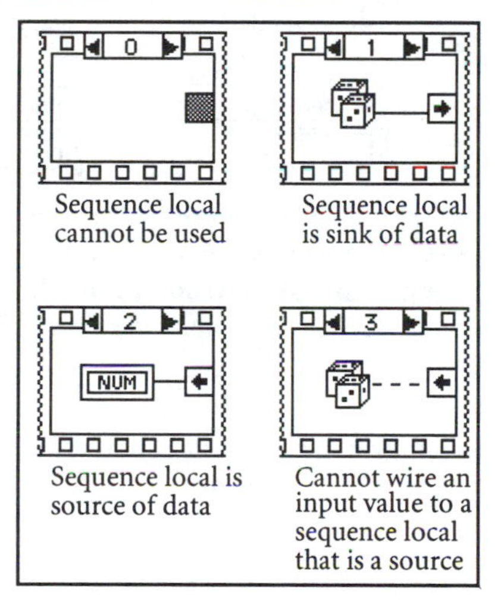

Sequence Local Terminals

Timing

Sometimes you will find it useful to control or monitor the timing of your VI. **Wait (ms)** and **Tick Count (ms)**, located in the **Time & Dialog** palette of the **Functions** menu, accomplish these tasks. **Wait (ms)** causes your VI to wait a specified number of milliseconds before it continues execution—handy if you have a loop you want to execute once a second or so. **Tick Count (ms)** returns the value of the internal clock in milliseconds; it is commonly used to calculate elapsed time, as in the next exercise. Be warned that the internal clock doesn't have great resolution—one tick of the clock is about 55 ms on Windows computers and 17 ms on Macs, an operating system limitation that LabVIEW can't work around. These functions are shown in the following illustration.

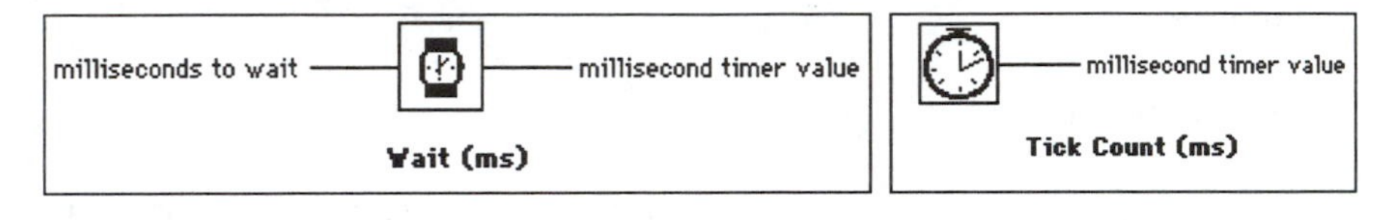

EXERCISE 6.4
Matching Numbers

You will build a VI that computes the time it takes to match a number you specify with a randomly generated number.

Front Panel

1. Open a new front panel.
2. Build the front panel shown.

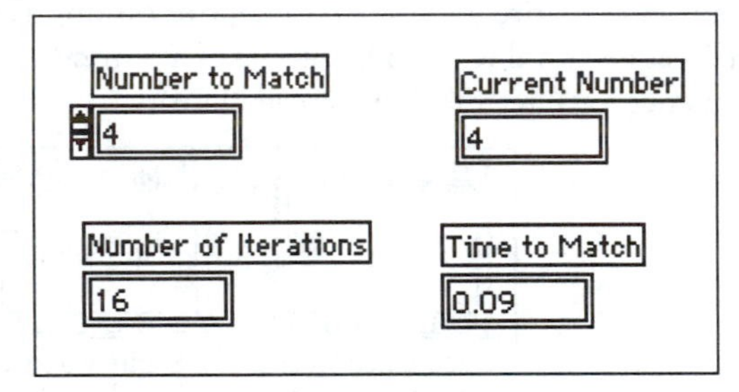

3. Change the precision of Number to Match, Current Number, and Number of Iterations to zero by selecting **Format & Precision...** from their pop-up menus. Enter "0" for Digits of Precision so that no digits are displayed to the right of the decimal point.

Block Diagram

1. Open the Diagram window and build the block diagram shown. You will have to build three separate frames of the sequence structure.

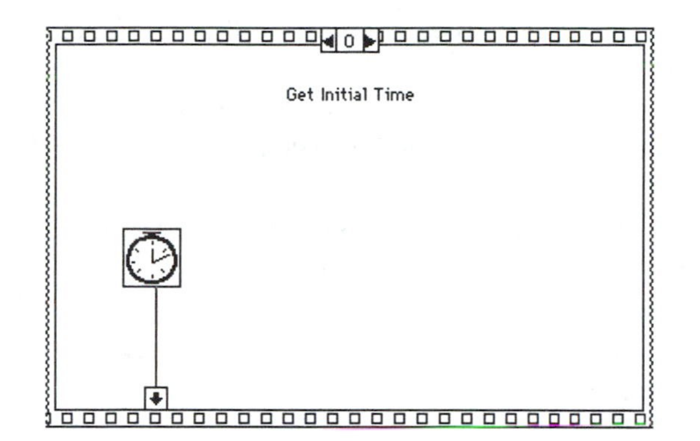

Frame 0

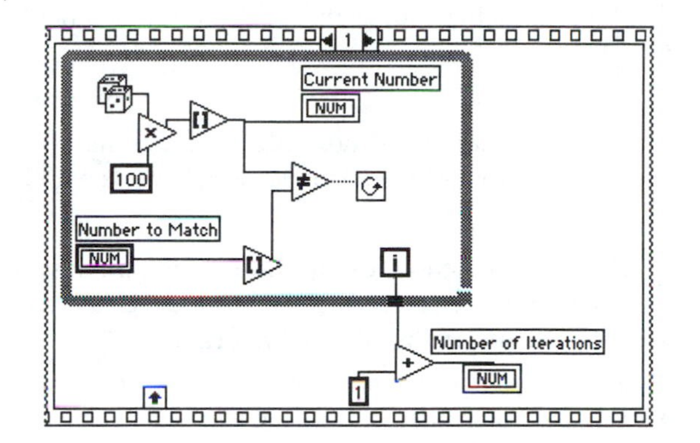

Frame 1

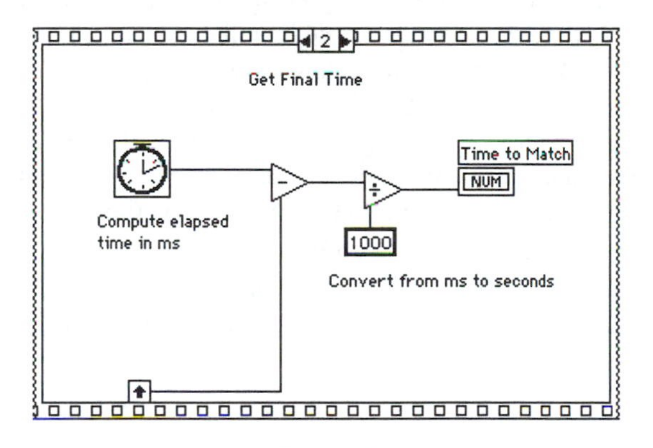

Frame 2

2. Place the Sequence structure (**Structs & Constants** menu) in the Diagram window. Enlarge the structure by dragging one corner with the Positioning tool. To create a new frame, pop up on the frame border and choose **Add Frame After** from the pop-up menu.
3. Create the sequence local by popping up on the bottom border of Frame 0 and choosing **Add Sequence Local** from the pop-up menu. The sequence local will appear as an empty

square. The arrow inside the square will appear automatically when you wire to the sequence local.

4. Build the diagram. Some new functions are listed below. Make sure to use the Help window to display terminal inputs and outputs when you wire!

   Tick Count (ms) function (**Time & Dialog** menu). Returns the number of milliseconds that have elapsed since power on.

   Random Number (0–1) function (**Arithmetic** menu). Returns a random number between 0 and 1.

   Multiply function (**Arithmetic** menu). In this exercise, multiplies the random number by 100. In other words, the function returns a random number between 0.0 and 100.0.

   Round to Nearest function (**Arithmetic** menu). In this exercise, rounds the random number between 0 and 100 to the nearest whole number.

   Not Equal? function (**Comparison** menu). In this exercise, compares the random number to the number specified in the front panel and returns a TRUE if the numbers are not equal; otherwise, this function returns a FALSE.

   In Frame 0, the **Tick Count (ms)** function returns the current time in milliseconds. This value is wired to the sequence local, so it will be available in subsequent frames. In Frame 1, the VI executes the While Loop as long as the number specified does not match the number that the **Random Number (0–1)** function returns. In Frame 2, the **Tick Count (ms)** function returns a new time in milliseconds. The VI subtracts the old time (passed from Frame 0 through the Sequence local) from the new time to compute the time elapsed.

5. Return to the front panel and turn on execution highlighting by clicking on . Remember, you must be in run mode to access this button. Execution highlighting slows the VI enough to see the current generated number on the front panel.

6. Enter a number inside the Number to Match control and run the VI. To speed things up, you can turn off execution highlighting by clicking on .

7. Use the **Save** command to save the VI in `MYWORK.LLB` as **Time to Match.vi**, and close the VI.

6.5 The Formula Node

The *Formula Node* is a resizable box that you use to enter algebraic formulas directly into the block diagram. You will find this feature extremely useful when you have a complicated formula to solve. For example, consider the equation $y = x^2 + x + 1$. If you implement this equation using regular LabVIEW arithmetic functions, the block diagram looks like the one shown below.

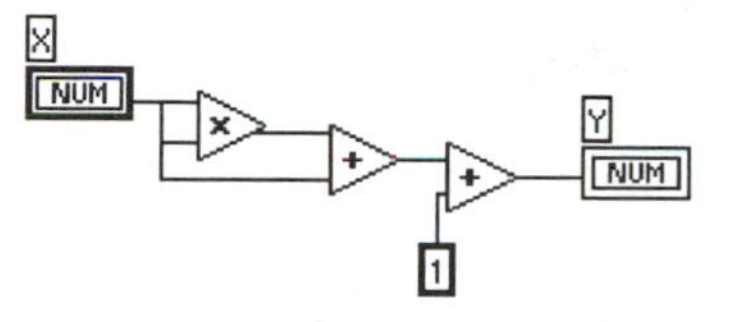

You can implement the same equation using a Formula Node, as shown in the following illustration.

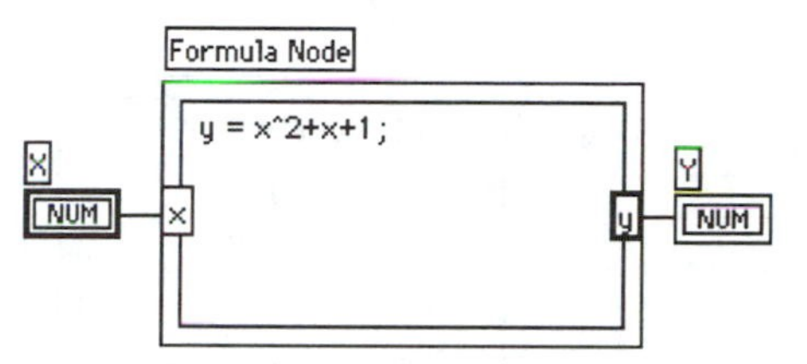

With the Formula Node, you can directly enter a complicated formula or formulas, in lieu of creating block diagram subsections. Simply enter the formula or formulas inside the box. You create the input and output terminals of the Formula Node by popping up on the border of the node and choosing **Add Input** or **Add Output** from the pop-up menu. Then enter variable names into the input and output boxes. Names are limited to two characters and are case sensitive. *Each formula statement must terminate with a semicolon (;).*

You place the Formula Node on the block diagram by selecting it from the **Structs & Constants** palette of the **Functions** menu.

The following operators and functions are available inside the Formula Node.

```
Formula Node operators, lowest precedence first:
=                           assignment
? :                         conditional
||  &&                      logical
==  !=  >  <  >=  <=        relational
+  -  *  /  ^               arithmetic
+  -  !                     unary

Formula Node functions:
abs  acos  acosh  asin  asinh  atan  atanh  ceil
cos  cosh  cot  csc  exp  expm1  floor  getexp  getman
int  intrz  ln  lnp1  log  log2  max  min  mod  rand
rem  sec  sign  sin  sinc  sinh  sqrt  tan  tanh
```

For more information, refer to the Formula Node in the Function Reference section of this book.

The following example shows how you can perform conditional branching inside a Formula Node. Consider the following code fragment, similar to Exercise 6.3, that computes the square root of x if x is positive, and assigns the result to y. If x is negative, the code assigns –99 to y.

```
if (x >= 0) then
        y = sqrt(x)
else
        y = -99
end if
```

You can implement the code fragment using a Formula Node, as shown in the next illustration.

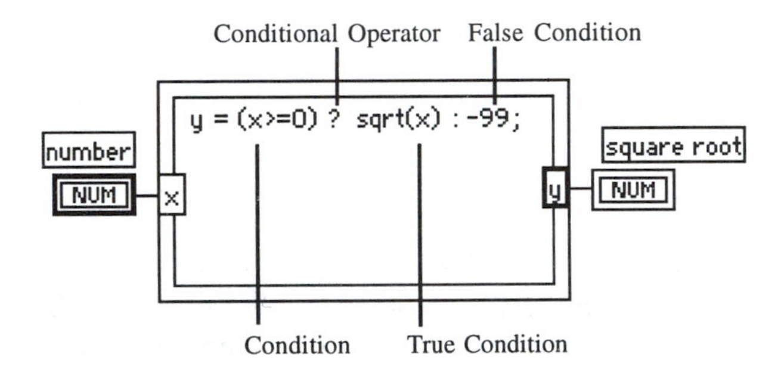

EXERCISE 6.5
Formula Fun

You will build a VI that uses the Formula Node to evaluate the equation $y = \sin(x)$ and graph the results.

Front Panel

1. Open a new panel. Select **Waveform Graph** from the **Array & Graph** palette of the **Controls** menu. Label it Graph. The graph indicator will display the plot of the equation $y = \sin(x)$

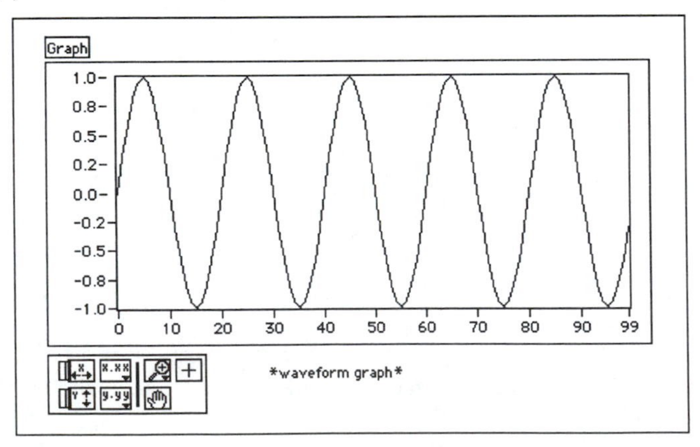

Block Diagram

1. Build the block diagram shown.

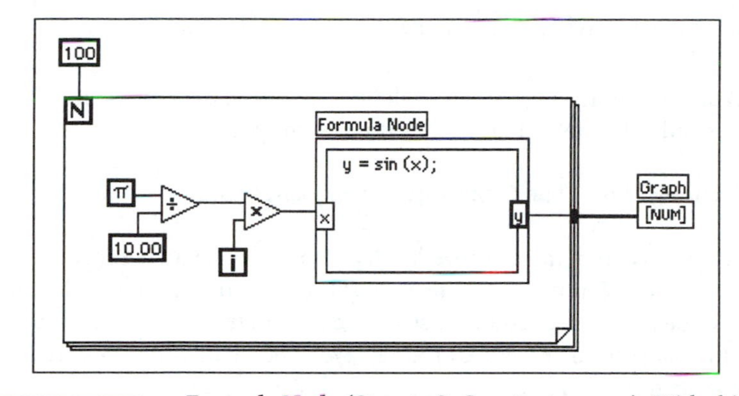

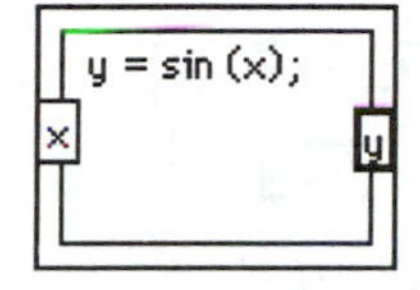

Formula Node (**Structs & Constants** menu). With this node, you can directly enter formula(s). Create the input terminal by popping up on the border and choosing **Add Input** from the pop-up menu. You create the output terminal by choosing **Add Output** from the pop-up menu.

When you create an input or output terminal, you must give it a variable name. The variable name must exactly match the one you use in the formula. Remember, variable names are limited to two characters and are case sensitive.

Note
Notice that a semicolon (;) terminates the formula statement.

π constant (**Structs & Constants** menu).

During each iteration, the VI multiplies the iteration terminal value by $\pi/10$. The multiplication result is wired to the Formula Node, which computes the sine of the result. The VI then stores the result in an array at the For Loop border (you will learn all about arrays in Chapter 7, *Arrays and Clusters*). After the For Loop finishes executing, the VI plots the array.

2. Return to the front panel and run the VI.
3. Save the VI in `MYWORK.LLB` and name it **Formula Node Exercise.vi**. Close the VI.

VI Logic

```
for i = 0 to 99
      x = i* (π/10)
      y = sin (x)
      array [i] = y
next i
Graph (array)
```

6.6 Problems in Wiring Structures

Reading the following sections will help you avoid creating faulty connections with structures and will help you diagnose the problems you do have.

Assigning More Than One Value to a Sequence Local

You can assign a value to the local variable of a Sequence structure in *only one* frame, although you can use the value in all subsequent frames. The following illustration shows the value *pi* (π) assigned to the sequence local in frame 0. If you try to assign another value to this same local variable in frame 1, you get a bad wire. This error is a variation of the multiple sources error.

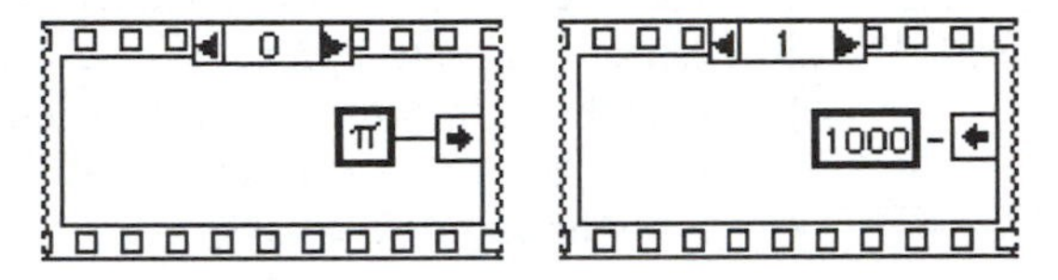

Failing to Wire a Tunnel in All Cases of a Case Structure

Wiring from a Case structure to an object outside the structure results in a bad tunnel if you do not connect output data from all cases to the object, as shown in part 1 of the following illustration. This is a variation of the no source error because at least one case would not provide a data value if it executed. Wiring to the tunnel in all cases, as shown in part 2 of this example, corrects the problem. This is not a multiple sources violation because only one case executes and produces only one output value per each execution of the Case structure.

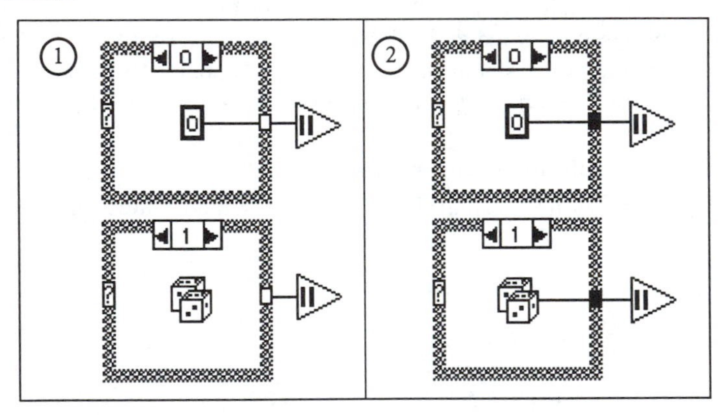

Overlapping Tunnels

Because LabVIEW creates tunnels as you wire, tunnels sometimes overlap each other. Overlapping tunnels, shown below, do not affect the execution of the diagram, but they can make editing difficult. You should avoid creating overlapping tunnels. If they occur, drag one tunnel away to expose the other.

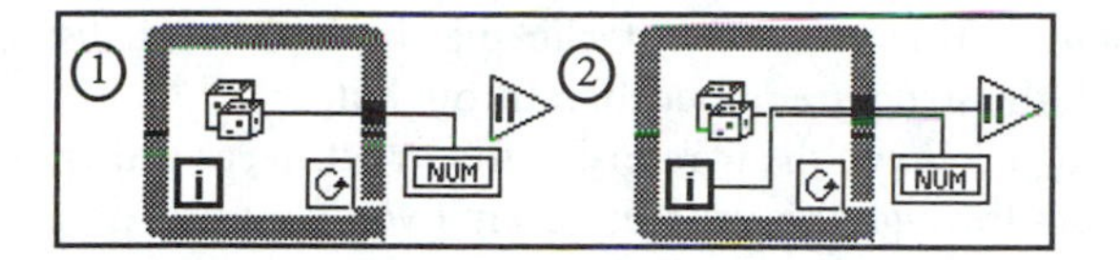

It can be difficult to tell which tunnel is on top. You can make mistakes if you try to wire to one of them while they overlap.

If you need to wire from an object inside a structure to an object outside when one wire segment to or from the structure border already exists, do not wire through the structure again. Instead, begin the second wire at the tunnel.

Wiring from Multiple Frames of a Sequence Structure

The following illustration shows another variation of the multiple sources error. Two Sequence structure frames attempt to assign values to the same tunnel. The tunnel turns white to signal this error.

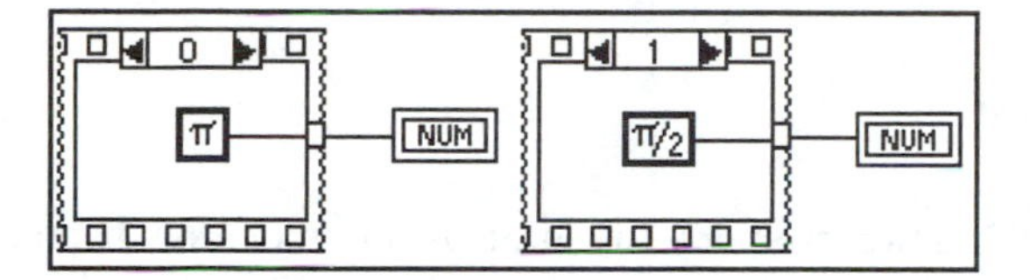

Wiring underneath Rather Than through a Structure

To wire through a structure you must click either in the interior or on the border of the structure, as shown.

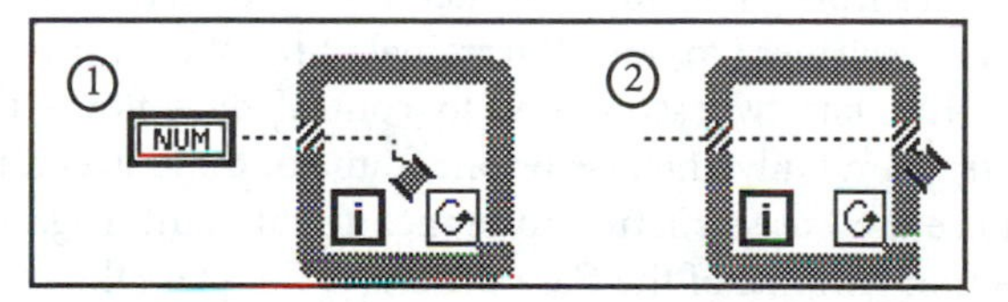

If you do not click in the interior or on the border of the structure, the wire passes underneath the structure, as shown in the following illustration.

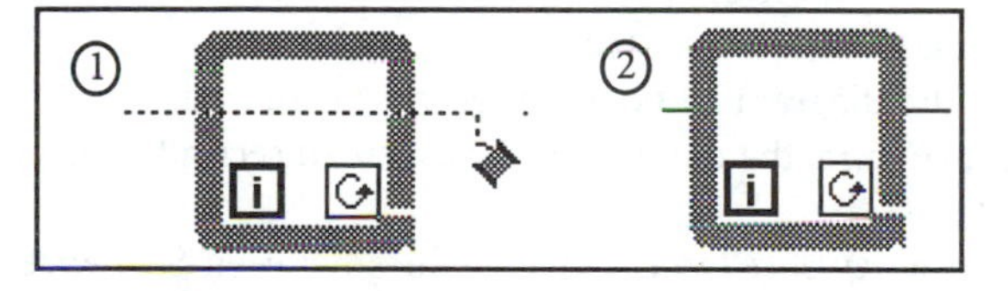

When the Wiring tool crosses the left border of the structure, a highlighted tunnel appears to indicate that LabVIEW will create a tunnel at that location as soon as you click the mouse button. If you continue to drag the tool through the structure without clicking the mouse until the tool touches the right border of the structure, a second highlighted tunnel appears on the right border. If you continue to drag the wiring tool past the right border of the structure without clicking, both tunnels disappear, and the wire passes underneath the structure rather than through it.

If you tack down the wire inside the structure, as shown in the following illustration, the wire goes through the structure even if you continue dragging the wiring tool past the right border.

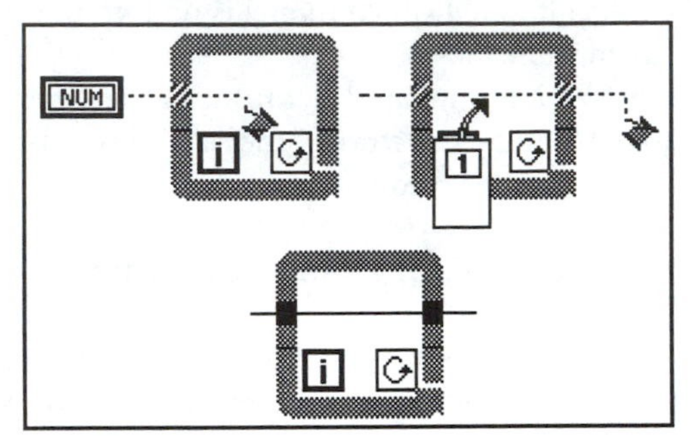

6.7 Wrap It Up!

LabVIEW has two structures to repeat execution of a subdiagram—the *While Loop* and the *For Loop*. Both structures are resizable boxes. You place the subdiagram to be repeated inside the border of the loop structure. The While Loop executes as long as the value at the *conditional terminal* is TRUE. The For Loop executes a specified number of times.

Shift registers, available for While Loops and For Loops, are local variables that transfer values from one iteration to the beginning of the next. You can configure shift registers to access values from many previous iterations. For each iteration you want to recall, you must add a new element to the left terminal of the shift register.

LabVIEW has two structures to control data flow—the *Case structure* and the *Sequence structure*. Only one case or one frame of these structures is visible at a time.

You use the Case structure to branch to different diagrams depending on the input to the selection terminal of the Case structure. You place the subdiagrams inside the border of each case of the Case structure. The case selection can be either Boolean (2 cases) or numeric (2^{15}–1 cases). LabVIEW automatically determines the selector terminal type when you wire a Boolean or numeric control to it.

You use the Sequence structure to execute diagram functions in a specific order. The portion of the diagram to be executed first is placed in the first frame (Frame 0) of the Sequence structure, the diagram to be executed second is placed in the second frame, and so on.

You use *sequence locals* to pass values between Sequence structure frames. The data passed in a sequence local is available only in frames subsequent to the frame in which you created the sequence local, NOT in those frames that precede the frame.

With the *Formula Node*, you can directly enter formulas in the block diagram. This feature is extremely useful when a function equation is complicated. Remember that variable names are case sensitive and that each formula statement must end with a semicolon (;).

The **Time & Dialog** palette of the **Functions** menu provides functions to pop up *dialog boxes* and control or monitor VI timing. The **One Button Dialog** and **Two Button Dialog** functions let you pop up a dialog box containing the message of your choice. The **Wait (ms)** function pauses your VI for the specified number of milliseconds; **Tick Count (ms)** returns to you the value of the internal clock.

6.8 Additional Exercises

EXERCISE 6.6
Equations

Build a VI that uses the Formula Node to calculate the following equations.

$$y_1 = x^3 + x^2 + 5$$
$$y_2 = m \times x + b$$

Use only one Formula Node for both equations. (Remember to put a semicolon [;] after each equation in the node.) Name the VI **Equations.vi**.

EXERCISE 6.7
Calculator

Build a VI that functions like a calculator. The front panel should have digital controls to input two numbers, and a digital indicator to display the result of the operation (add, subtract, multiply, or divide) that the VI performs on the two numbers. Use a slide control to specify the operation to be performed. Name the VI **Calculator.vi**.

Hint: You will want to use Text Labels (obtained from the pop-up menu) on the slide control to specify the function (add, subtract, multiply, and divide). Slides with text labels behave very much like text ring controls. If you want to add another text marker to the slide, pop up on the text display and select **Add Item After** or **Add Item Before** and then type in that marker's text. You can use the Labeling tool to change text.

EXERCISE 6.8
Combination For/While Loop

Using only a While Loop, build a combination For Loop/While Loop that stops either when it reaches "N" (specified inside a front panel control), or when a user pushes a stop button. Name the VI **Combo For/While Loop.vi**.

Hints: Don't forget to that a While Loop only executes while the conditional terminal reads a TRUE value. Also, remember that while executing a loop, LabVIEW does not update indicators or read controls that are outside of the loop. Your stop button MUST be inside of your loop if you want correct functionality.

EXERCISE 6.9
Dialog Display

Write a VI that reads the value of a front panel switch, then pops up a dialog box indicating if the switch is on or off. Name the VI **Dialog Display.vi**. If you've developed the bad habit of using the continuous run button, now is the time to break it, or you will get yourself stuck in an endless loop! If you do get stuck, use the keyboard shortcut to stop your VI: <Control>-<.> or <Command>-<.>.

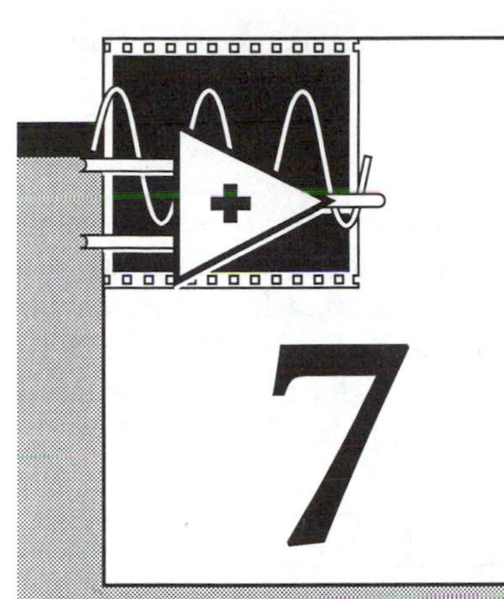

7

Arrays and Clusters

Overview

In this chapter, you will learn about two new data types—arrays and clusters. These composite data types allow you great flexibility in data storage and manipulation. You will also see how you can use built-in functions to manage arrays.

Your Goals

- Understand how to create and use array controls and indicators
- Be able to use auto-indexing to create an array
- Learn about the built-in array manipulation functions
- Grasp the concept of polymorphism
- Understand how clusters differ from arrays
- Assemble and break apart clusters using the **Bundle** and **Unbundle** functions

Key Terms

array
cluster
auto-indexing
polymorphism
bundle
unbundle

7.1 What Are Arrays?

An *array* is a collection of data elements that are all the same type. An array can have one or more dimensions and up to 2^{31} elements per dimension, memory permitting. Arrays in LabVIEW can be any type, except array, chart, or graph. You access each array element by its *index*. The index is in the range 0 to N–1 where N is the number of elements in the array. The *one-dimensional* (1D) array shown in the following illustrates this structure. Notice that the *first* element has index 0, the *second* element has index 1, and so on.

index	0	1	2	3	4	5	6	7	8	9
10-element array	1.2	3.2	8.2	8.0	4.8	5.1	6.0	1.0	2.5	1.7

You will find that waveforms are often stored in arrays, with each point in the waveform comprising an element of the array. If you have waveforms from several channels being read from a DAQ board, they will be stored in a two-dimensional (2D) array, with each column in the 2D array corresponding to one channel's data.

7.2 Creating Array Controls and Indicators

It takes two steps to make more complex data types like arrays and clusters. You create the array control or indicator by combining an *array shell*, shown in the following illustration, with a data object, which can be numeric, Boolean, or string. The data object cannot be another array, a chart, or a graph. You will find the array shell in the **Array & Graph** palette of the **Controls** menu.

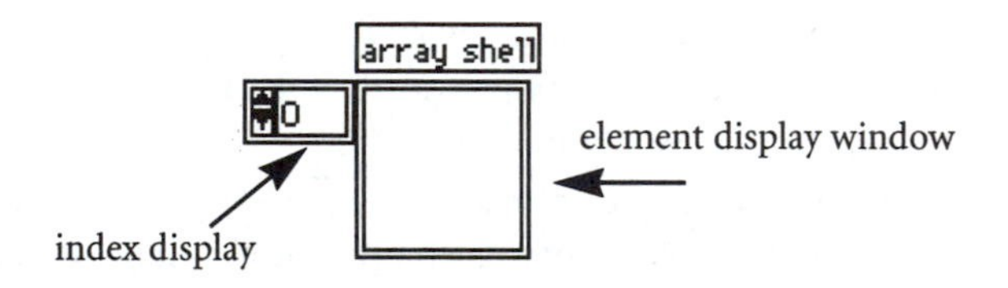

To create an array, you drag a data object into the element display window or deposit the object directly into the window using the window's pop-up menu. The element display window resizes to accommodate its new type, as shown in the following illustration.

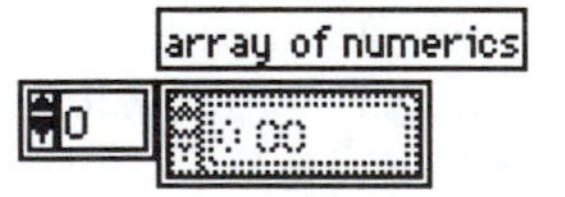

If you want to resize the element display window, use the Positioning tool and make sure it turns into the standard resizing brackets when you place it on the corner of the window (you will probably have to position it slightly inside the box). If you want to show more elements, grab the window corner with the grid cursor of the Positioning tool . and stretch. You will then have multiple elements visible. The element closest to the index display always corresponds to the element number displayed.

When you first drop an array shell on the front panel, its block diagram terminal is black, characteristic of an undefined data type. The terminal also contains brackets, shown in part (A) of the next illustration, which are LabVIEW's way of denoting array structures. When you assign the array a type (by placing a control or indicator in its element display window), then the array's block diagram terminal turns from black to its new type's color and lettering (although it retains its brackets), as in part (B). You will notice that array wires are thicker than wires carrying a single value.

Two-Dimensional Arrays

A 2D array requires two indices to locate an element, a column index and a row index, both of which are zero based. The following figure shows an N-column by M-row array that contains N times M elements.

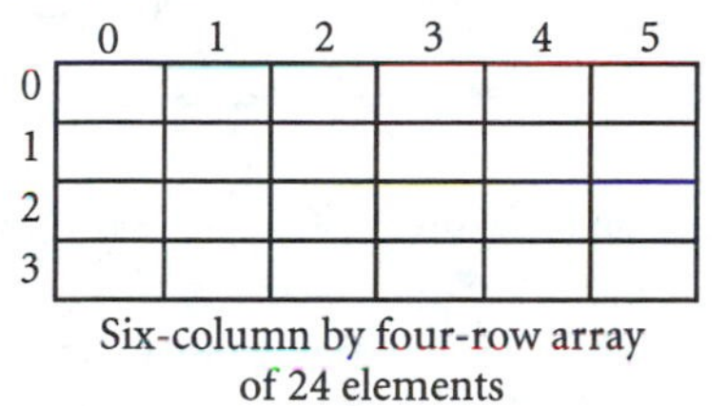

Six-column by four-row array
of 24 elements

You add dimensions to the array control or indicator by popping up on the array control or indicator *index display* and choosing **Add Dimension** from the pop-up menu. The next illustration shows a 2D digital control array.

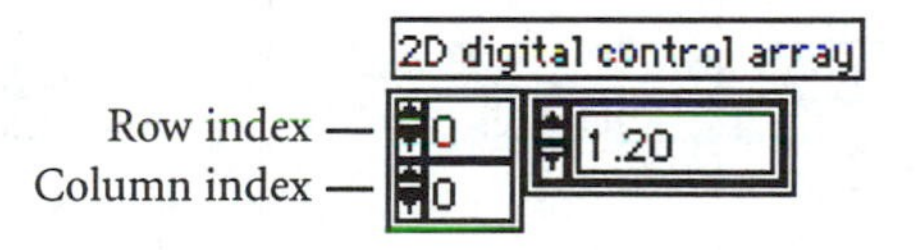

Creating Arrays Using Auto-Indexing

The For Loop and the While Loop can index and accumulate arrays at their boundaries automatically. This capability is called *auto-indexing*. The following figure shows a For Loop auto-indexing an array at its boundary. Each iteration creates the next array element. After the loop completes, the array passes out of the loop to the indicator. Notice that the wire becomes thicker as it changes to an array type at the loop border.

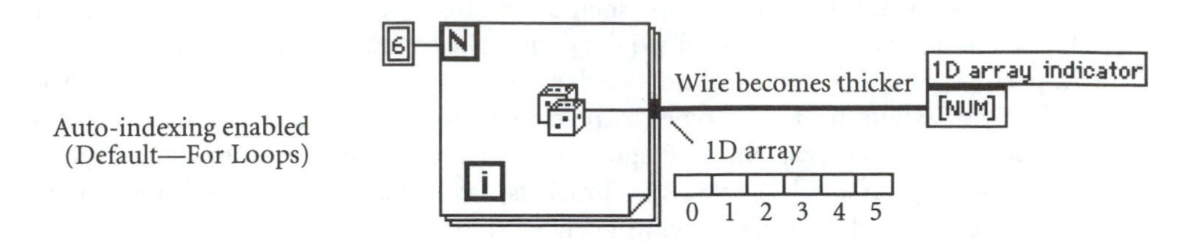

If you need to tunnel a value out of a loop without creating an array, you must disable auto-indexing by popping up on the tunnel (the black square) and choosing **Disable Indexing** from the pop-up menu. In the next illustration, auto-indexing is disabled, and only the last value returned from the **Random Number (0–1)** function passes out of the loop. Notice that the wire remains the same size after it leaves the loop.

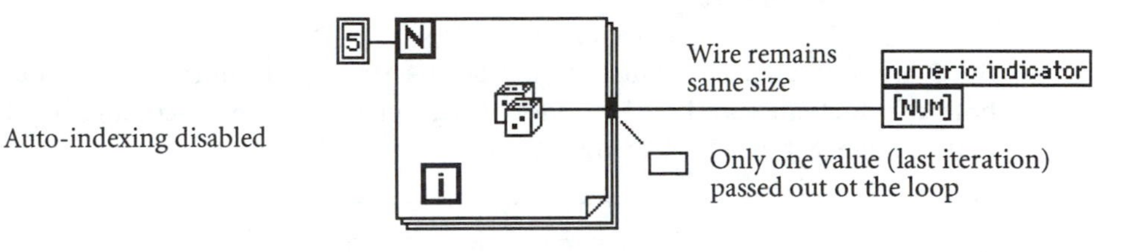

Indexing also applies when you are wiring arrays into loops. If indexing is enabled as in loop (a), the loop will index off one element from the array each time it iterates. If indexing is disabled as in loop (b), the entire array passes into the loop at once.

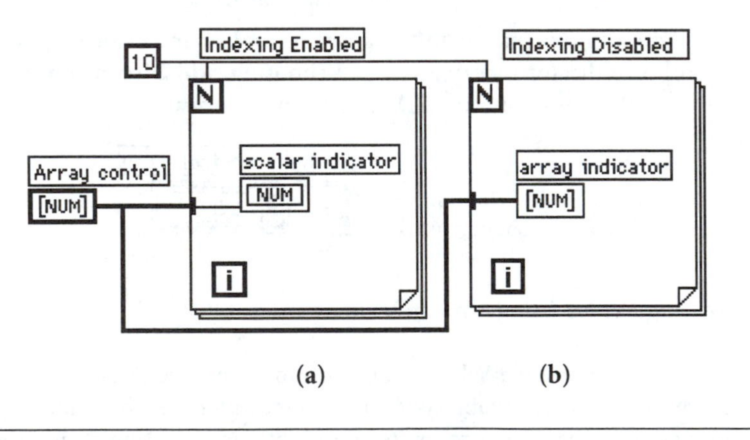

(a) (b)

Watch Out!

Because For Loops are often used to process arrays, LabVIEW enables auto-indexing by default when you wire an array into or out of a loop. By default, LabVIEW does not enable auto-indexing for While Loops. You must pop up on the array tunnel and choose Enable Indexing from the pop-up menu. Pay close attention to the state of your indexing, lest you develop errors that are tricky to spot.

Creating Two-Dimensional Arrays

You can use two For Loops, one inside the other, to create a 2D array. The inner For Loop creates a row, and the outer For Loop "stacks" these rows to fill in columns of a matrix. The following illustration shows two For Loops creating a 2D array of random numbers using auto-indexing.

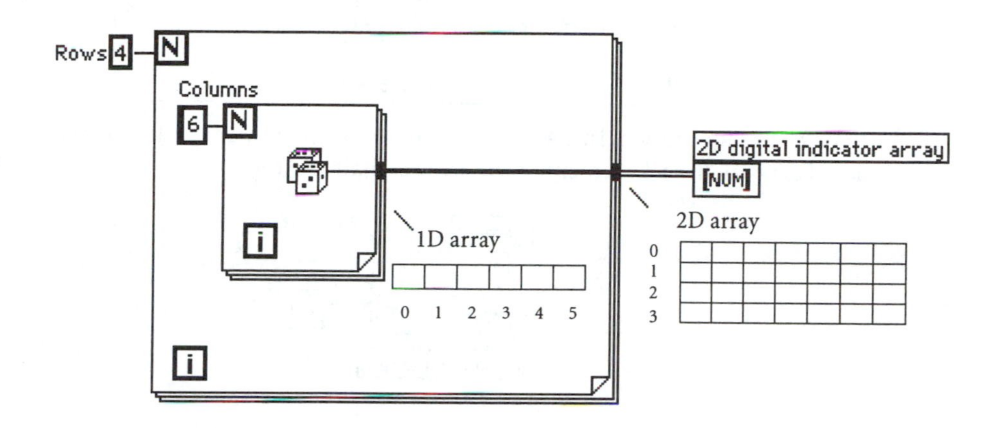

Using Auto-Indexing to Set the For Loop Count

When you enable auto-indexing on an array *entering* a For Loop, LabVIEW automatically sets the count to the array size, thus eliminating the need to wire a value to **N**. If you enable auto-indexing for more than one array, or if you set the count by wiring to **N**, the count becomes the smaller of the two choices. In the following figure, the array size, and not **N**, sets the For Loop count because the array size is the smaller of the two.

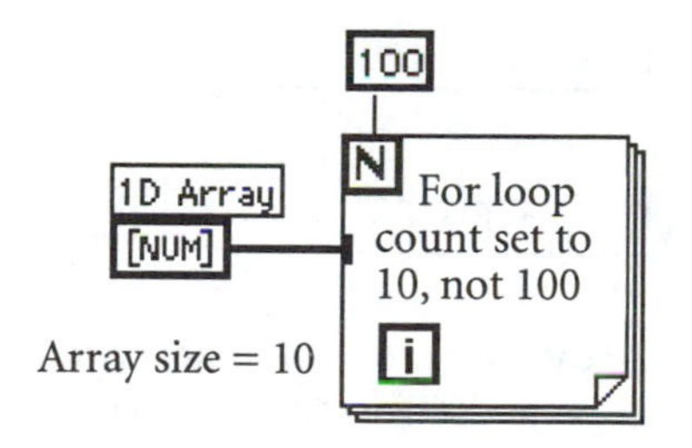

7.3 Functions for Manipulating Arrays

LabVIEW has many functions to manipulate arrays in the **Array & Cluster** palette of the **Functions** menu. Always keep in mind that arrays (and all other LabVIEW structures) are zero indexed—the first element has an index of zero, the second has an index of one, and so on. Some common functions are discussed below:

- **Initialize Array** will create and initialize an n-dimension array with the value of your choice. This function is useful for allocating memory for arrays of a certain size.

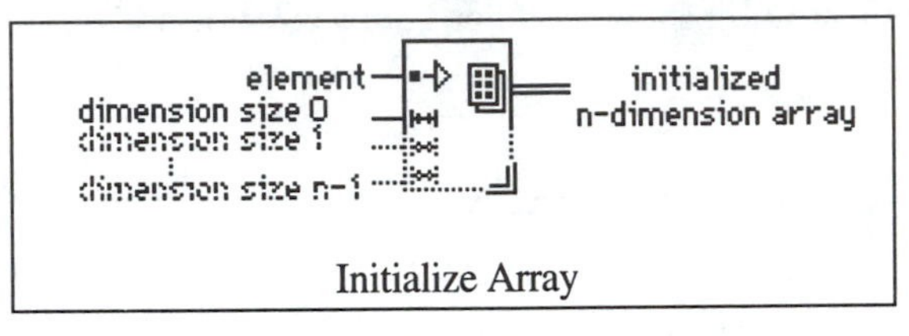

Initialize Array

In this illustration, **Initialize Array** shows how to initialize a 10-element, one dimensional array, with each element of the array containing a zero.

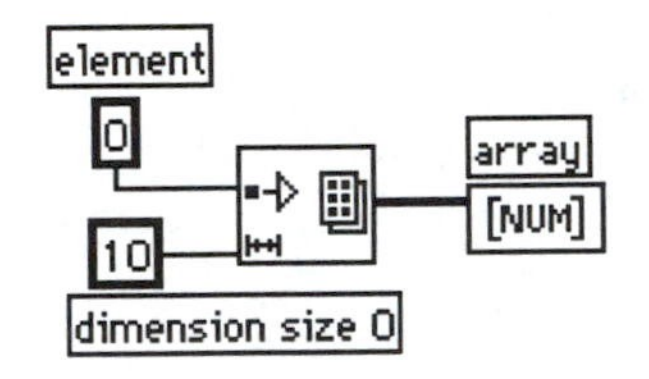

- **Array Size** returns the number of elements in the input array. If the input array is *n*-dimensional, **Array Size** returns an *n*-element, one-dimensional array, with each element containing a dimension size.

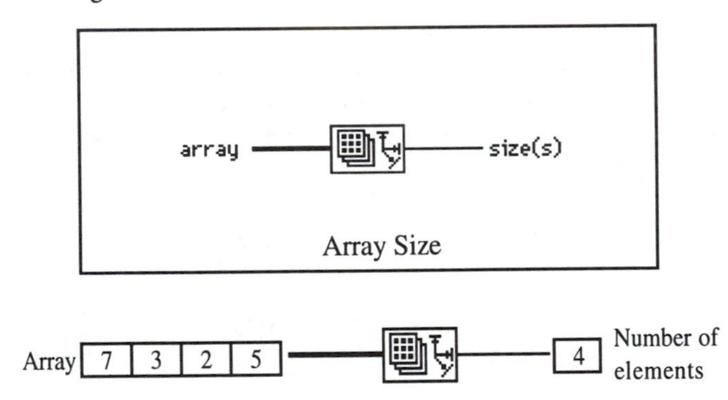

Array Size

- **Build Array** concatenates, or combines, two arrays or adds extra elements to an array. The function looks like [icon] when placed in the diagram window. You can resize this function to increase the number of inputs. **Build Array** has two types of inputs, array and element, so it can accommodate both arrays and single-valued elements.

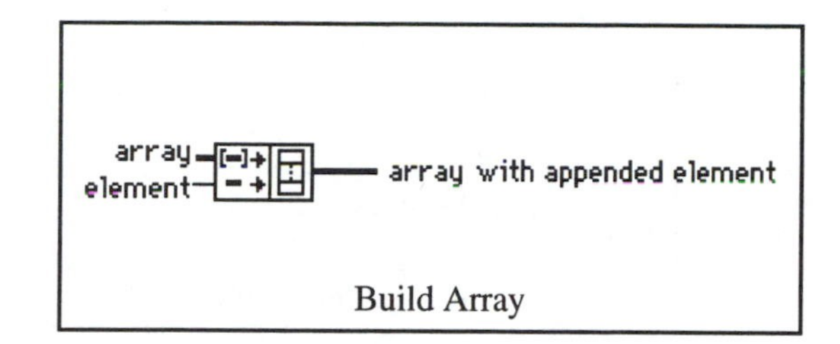

Build Array

For example, the **Build Array** function shown below has been configured to concatenate two arrays and one element into a new array.

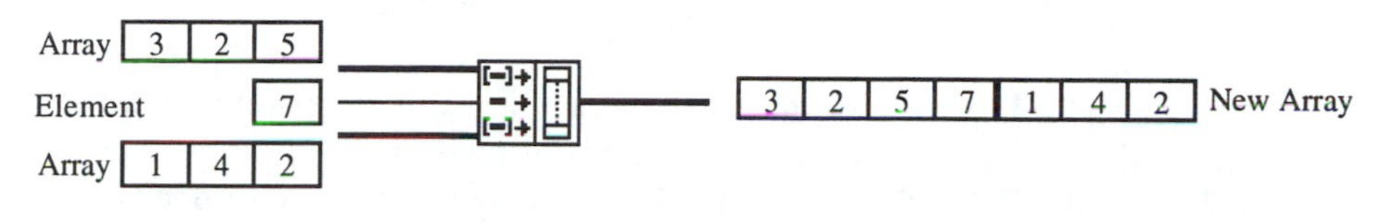

Watch Out!
Pay special attention to the inputs of a Build Array function. Array inputs have brackets, while element inputs do not. They are NOT interchangeable and can cause lots of confusing bad wires if you're not careful.

- **Array Subset** returns a portion of an array starting at **index** and containing **length** elements.

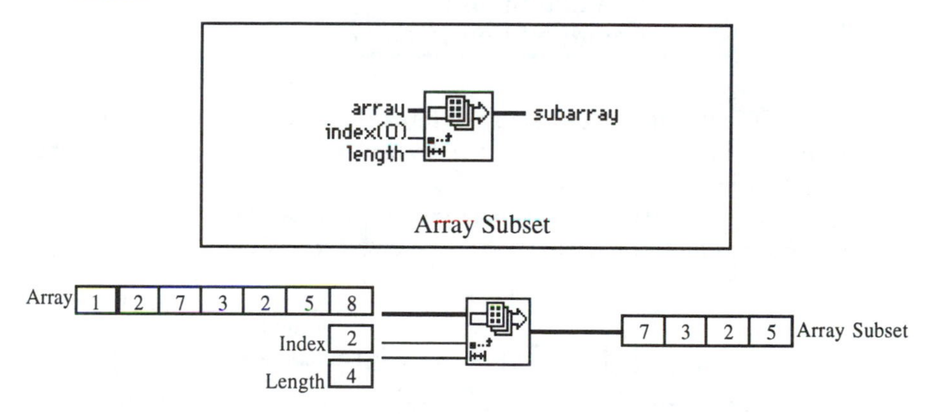

Array Subset

- **Index Array** accesses an element of an array. An example of an **Index Array** function accessing the third element of an array is shown in the following illustration. Notice that the third element's index is two because the index starts at zero; that is, *the first element has an index of zero.*

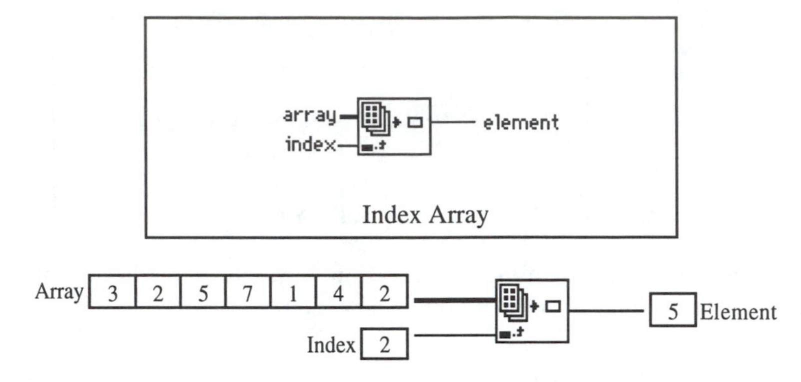

Here, the **Index Array** function is extracting a scalar element from an array. You also can use this function to *slice off* one or more dimensions of a 2D array to create a subarray of the original. To do this, stretch the **Index Array** function to include two index inputs, and select the **Disable Indexing** command on the pop-up menu of the index terminal as shown.

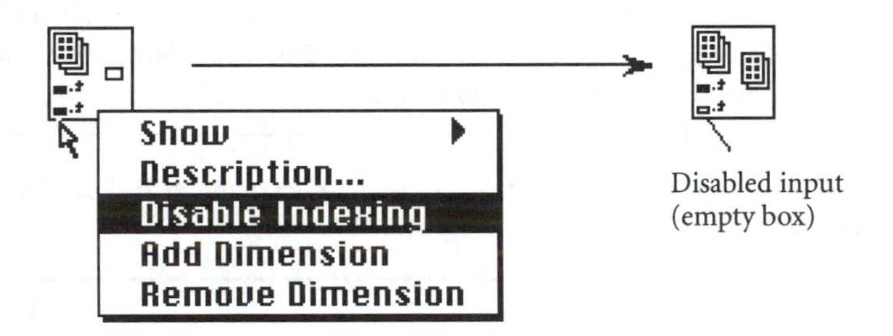

Notice that the index terminal symbol changes from a solid to an empty box when you disable indexing. You can restore a disabled index with the **Enable Indexing** command from the same menu.

You can extract subarrays along any combination of dimensions. The following illustration shows how to extract 1D row or column arrays from a 2D array.

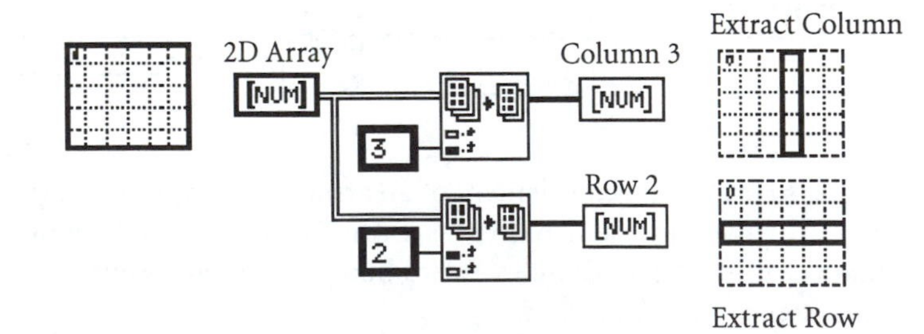

EXERCISE 7.1
Array Acrobatics

You will finish building a VI that concatenates two arrays and then indexes out the element in the middle of the concatenated array.

Front Panel

1. Open **Array Exercise.vi**, located in EXAMPLES\GENERAL\EXERCISE.LLB.

   The front panel contains two input arrays (each showing three elements), two digital controls, and an output array (showing eight elements). The VI will concatenate the arrays and the control values in the following sequence to create the new array.

   Initial Array – Element 1 – Element 2 – Terminal Array

   The front panel already is built. You will finish building the diagram. Note that an array control or indicator will appear greyed out until you or the program assigns it some data.

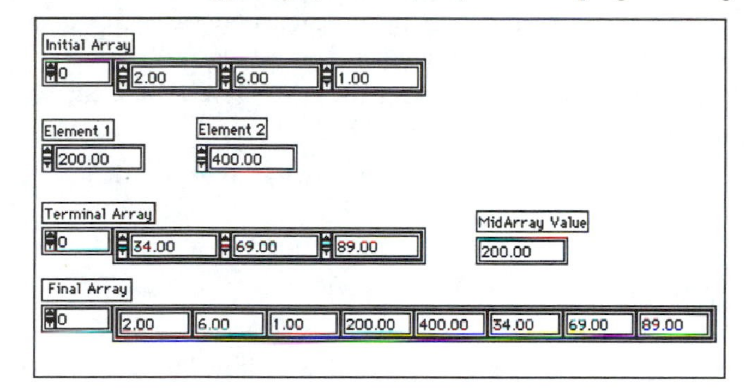

2. Make sure you are in edit mode.

Block Diagram

1. Build the block diagram shown. Use the Help window to find the correct terminals on the functions.

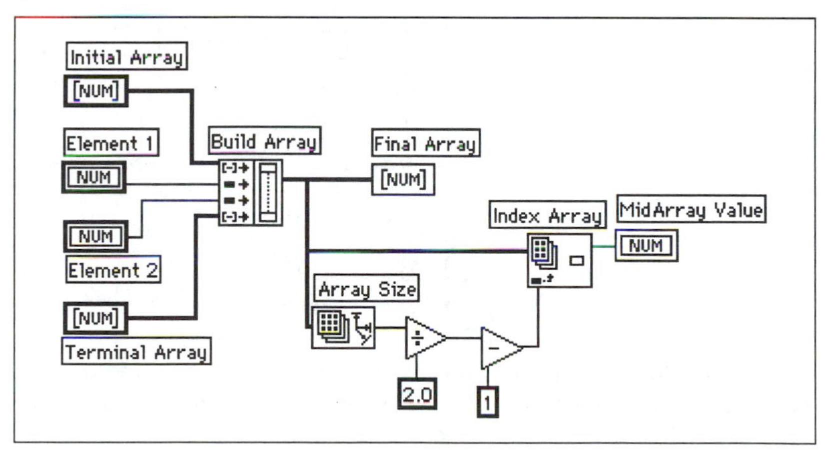

Build Array function (**Array & Cluster** menu). In this exercise, concatenates the input data to create a new array in the following sequence: Initial Array – Element 1 – Element 2 – Terminal Array.

The function looks like when placed in the diagram window. Place the Positioning tool on the lower right corner and resize the function until it includes four inputs.

Change the top input of the **Build Array** function to an array by popping up on it and choosing **Change to Array** from the pop-up menu. Repeat for the bottom input. The symbol indicates an array input.

Array Size function (**Array & Cluster** menu). Returns the number of elements in the concatenated array.

Index Array function (**Array & Cluster** menu). In this exercise, returns the element in the middle of the array.

You will build the array using the **Build Array** function. You then calculate the index for the middle of the array by taking the length of the array, dividing it by two, and subtracting one, to account for the zero-based array index.

2. Return to the front panel and run the VI. Try several different number combinations.
3. Save and close the VI.

7.4 Polymorphism

The LabVIEW arithmetic functions, **Add, Multiply, Divide,** and so on, are polymorphic. *Polymorphism* is just a big word for a simple principle: The inputs to these functions can be of different data types. For example, you can add a scalar to an array or add two arrays together. The following illustration shows some of the polymorphic combinations of the **Add** function.

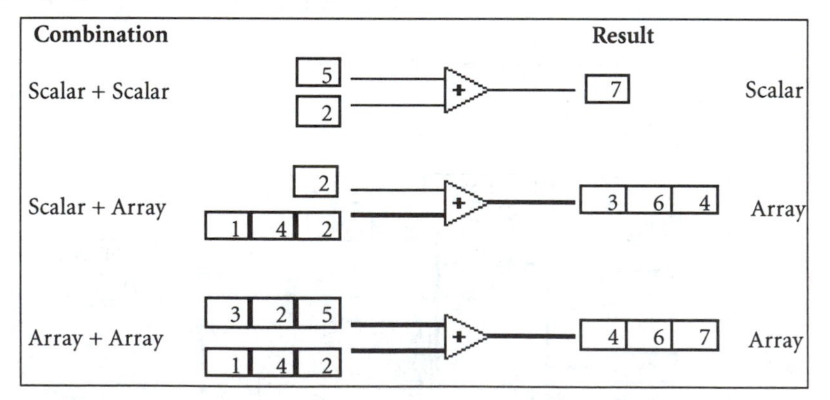

In the first combination, the result is a scalar. In the second combination, the scalar is added to *each* element of the array. In the third combination, each element of one array is added to the corresponding element of the other array.

Note

If you are doing arithmetic on two arrays of different lengths, the resulting array will be the same size as the smaller of the two. In other words, LabVIEW operates on corresponding elements in the two arrays until one of the arrays runs out of elements. The remaining elements in the longer array are ignored.

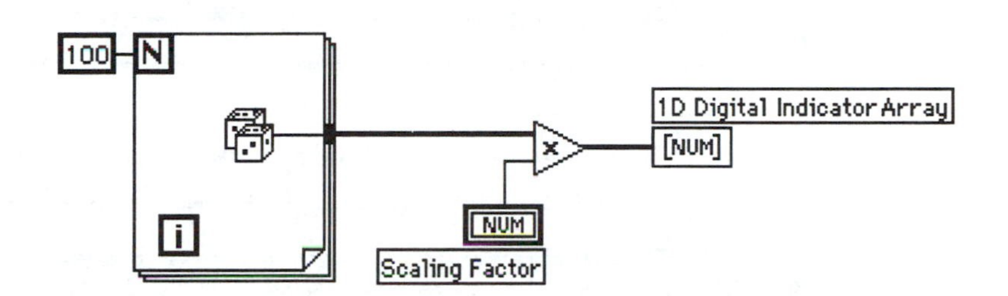

Here, each iteration of the For Loop generates one random number that is stored in the array created at the border of the loop. After the loop finishes execution, the **Multiply** function multiplies each element in the array by the scaling factor. The front panel indicator then displays the array.

The next figure illustrates some of the possible polymorphic combinations of the **Add** function.

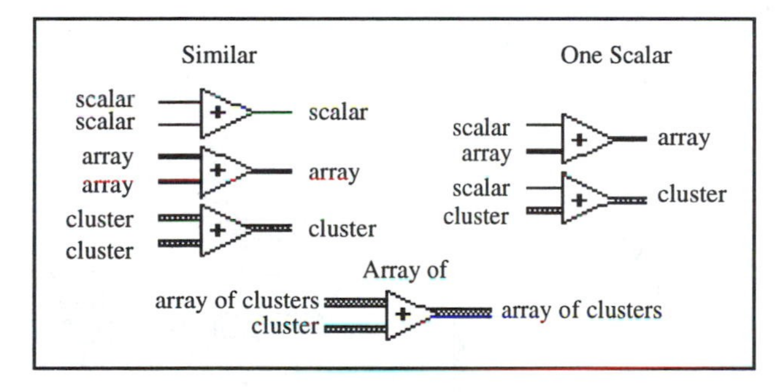

EXERCISE 7.2
Polymorphism

You will build a VI that demonstrates polymorphism on arrays.

Front Panel

1. Open a new panel and re-create the one shown.

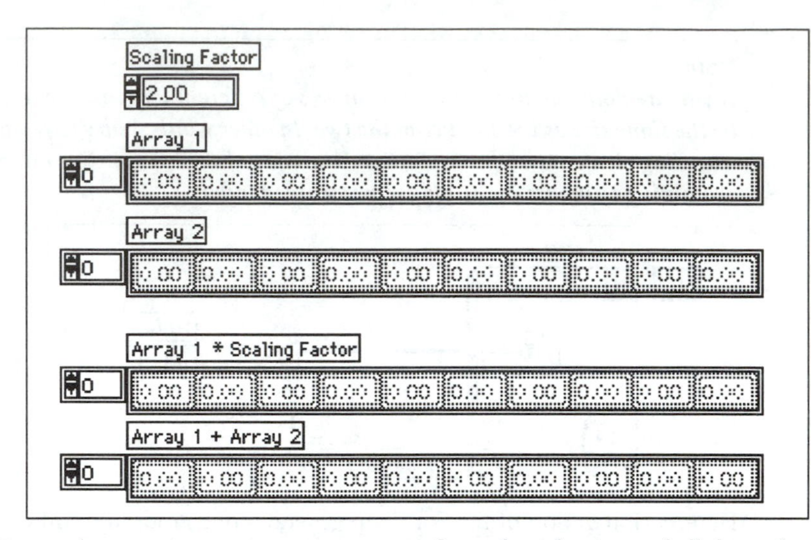

Remember, to create an array, you must first select the Array shell from the **Array & Graph** palette of the **Controls** menu. Then you must put a numeric control or indicator into the shell's data object window, either by selecting it from the pop-up menu in the window or by dragging it inside. To see more than one element in the array, you must grab and drag the corner of the filled element display window with the grid cursor of the Positioning tool.

Block Diagram

1. Build the diagram shown.

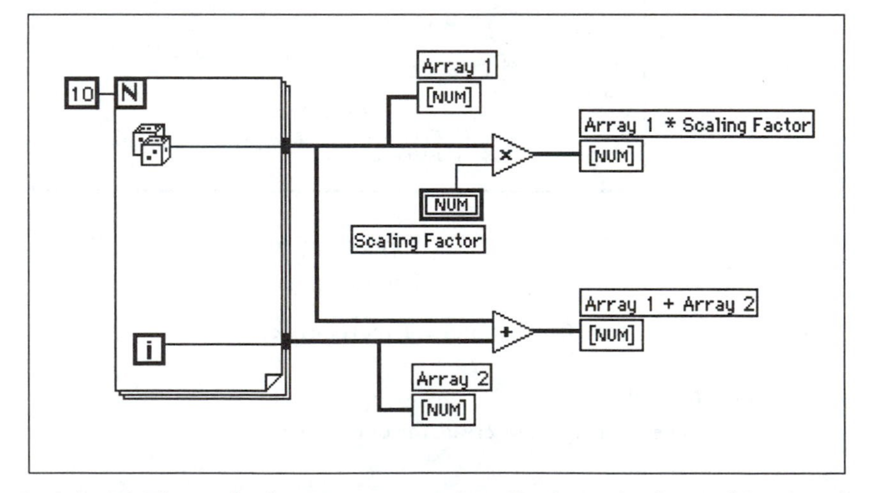

2. Run the VI. The results demonstrate several applications of polymorphism in LabVIEW. For example, Array 1 and Array 2 could be incoming waveforms that you wish to scale.
3. Save the VI as **Polymorphism Example.vi** and place it in MYWORK.LLB. Close the VI.

7.5 Clusters

A *cluster*, like an array, is a data structure that groups data. However, a cluster can group data of different types; it is analogous to a *record* in Pascal or a *struct* in C. A cluster may be thought of as a *bundle* of wires, much like a telephone cable. Each wire in the cable represents a different element of the cluster. Because a cluster constitutes only one "wire" in the block diagram, clusters reduce wire clutter and the number of connector terminals that subVIs need. You will find that the cluster data type appears frequently when you plot your data.

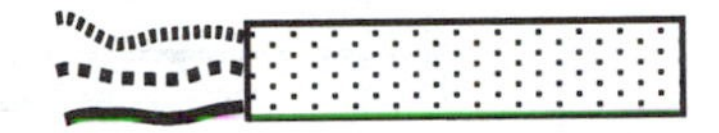

Bundling Data

Although cluster and array elements are both ordered, you access cluster elements by *unbundling* them all at once rather than indexing one at a time. You may think of unbundling a cluster as unwrapping a telephone cable and having access to the different-colored wires. Unlike arrays, clusters have a fixed size, or a fixed number of wires in them.

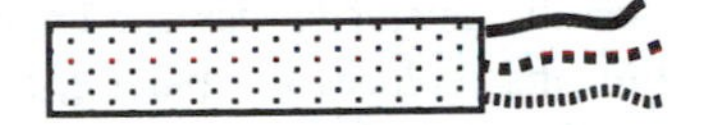

Unbundling Data

You can connect cluster terminals only if they have the same type; in other words, both clusters must have the same number of elements, and matching corresponding elements. Polymorphism applies to clusters as well as arrays, as long as you have compatible data types in your clusters.

7.6 Creating Cluster Controls and Indicators

You create cluster controls and indicators by placing a *cluster* shell (**Array & Graph** palette of the **Controls** menu) in the Panel window. You can then place any objects inside the cluster that you can place in the LabVIEW Panel window. Like arrays, you can deposit objects directly inside the cluster by popping up inside the cluster, or you can drag an object into a cluster. *Objects inside a cluster must be all controls or all indicators.* You cannot combine both controls and indicators inside the same cluster. The cluster assumes the data direction (control or indicator) of the first object you place inside. The following illustration shows a cluster with four controls.

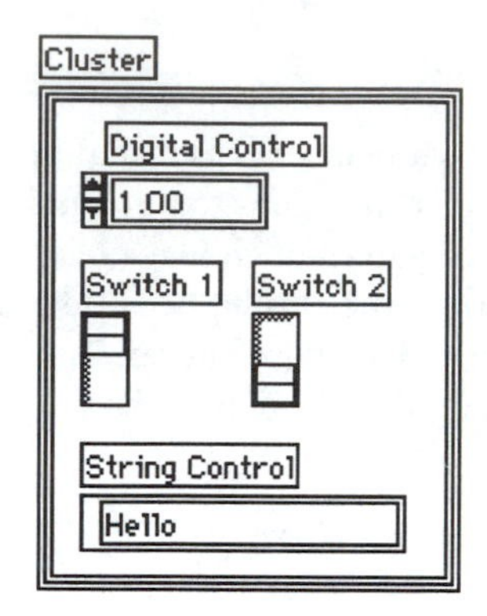

Cluster Order

Cluster elements have a logical order unrelated to their position within the shell. The first object inserted into the cluster is element 0, the second is element 1, and so on. If you delete an element, the order adjusts automatically. You can change the order within the cluster by popping up on the cluster *border* and choosing **Cluster Order** from the pop-up menu. A new set of buttons replaces the palette, and the cluster appearance changes as shown in the next figure. *You must keep track of your cluster order if you want to access individual elements in the cluster;* ***an element is accessed by order, not by name!***

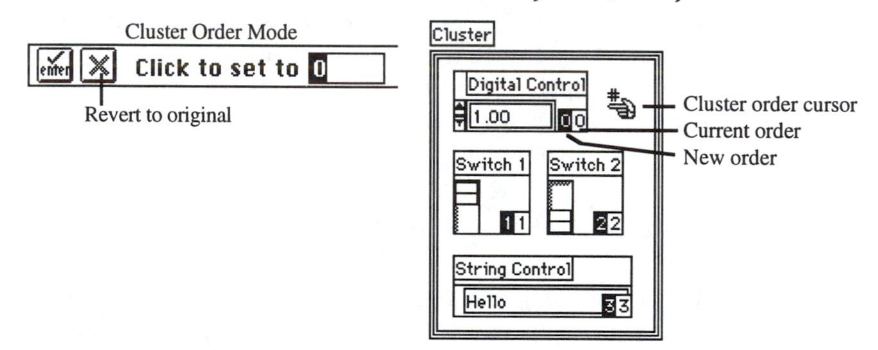

The white boxes on the elements show their current places in the cluster order. The black boxes show the element's new place in the order. Clicking on an element with the cluster order cursor sets the element's place in the cluster order to the number displayed inside the tools palette. You can change this order by typing a new number into that field. You can revert to the old order by clicking on the X button. When you have the order the way you want, you can set it by clicking on the enter button and exiting the cluster order edit mode.

Using Clusters to Pass Data to and from SubVIs

The connector pane of a VI can have a maximum of 20 terminals. Do not pass information to all 20 terminals when calling a VI as a subVI—the wiring can be very tedious and you can easily make a mistake. By bundling a number of controls into a cluster, and passing the cluster to the subVI, you can use a single terminal to pass several values. Similarly, one terminal can pass multiple outputs from the subVI. You can use clusters to work around the 20-terminal limit for the connector. You can also use fewer and therefore larger terminals, resulting in neater wiring.

Bundling Your Data

The **Bundle** function (**Array & Cluster** menu) assembles individual components into a single new cluster or allows you to replace elements in an existing cluster. The function appears as when you place it in the diagram window. You can increase the number of inputs by dragging a corner of the function with the Positioning tool. When you wire to each input terminal, a symbol representing the data type of the wired element appears in the originally empty terminal. The order of the resultant cluster will be the order of inputs to the **Bundle**, top to bottom.

If you just want to create a new cluster, you do not need to wire an input to the center "cluster" input of the **Bundle** function.

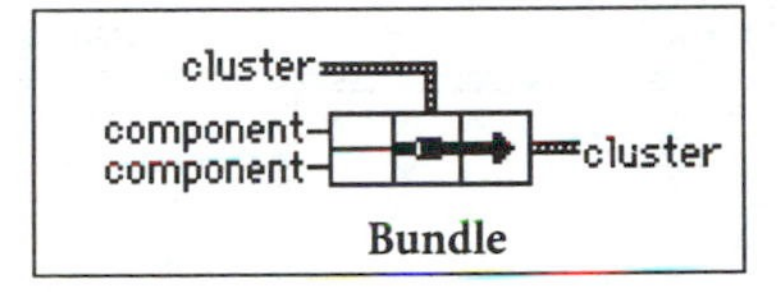

Bundle

Replacing a Cluster Element

If you want to *replace* an element in a cluster, first resize the **Bundle** to contain the same number of input terminals as there are elements in the cluster. Then wire the cluster to the middle terminal of the **Bundle** function and wire the new value(s) of the element(s) you want to replace to the corresponding inputs.

Unbundling Your Clusters

The **Unbundle** function (**Array & Cluster** menu) splits a cluster into each of its individual components. *The output components are arranged from top to bottom in the same order they have in the cluster.* The elements' order in the cluster is the only way you can tell the difference between them. The function appears as when you place it in the diagram window. You can increase the number of outputs by dragging a corner of the function with the Positioning tool. The **Unbundle** function must be resized to contain the same number of inputs as elements in the input cluster, or it will produce bad wires. When you wire the input cluster to the correctly sized **Unbundle**, the previously blank output terminals will assume the symbols of the data types in the cluster.

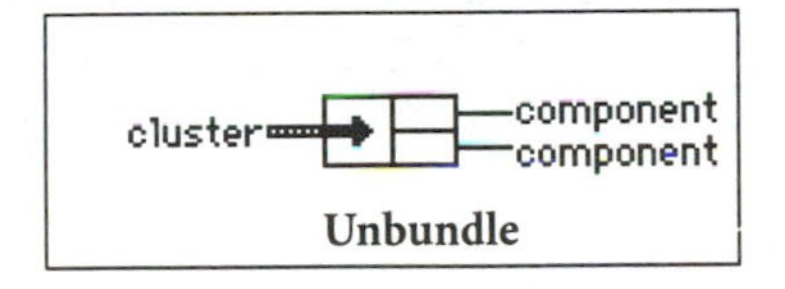

Unbundle

Watch Out!
Cluster order is integral when accessing cluster data. For example, if you have two Booleans, it would be easy to mistakenly access Switch 2 instead of Switch 1, since they are referenced in the Unbundle function by order, NOT by name. Your VI will wire with no errors, but your results will be incorrect.

EXERCISE 7.3
Clusters

In this exercise, you will build a VI that checks whether the value of the Numeric 1 digital control in the input cluster is greater than or equal to zero. If it is not, the VI will take the absolute value of each control. Then the VI multiplies all values by 0.5, whether Numeric 1 is positive or negative, and displays the results in Output Cluster. Use the front panel and block diagram shown to get started. Remember that, like arrays, you create clusters in two steps: First, drop the cluster shell; then place objects inside. Save the VI as **Cluster Comparison.vi**.

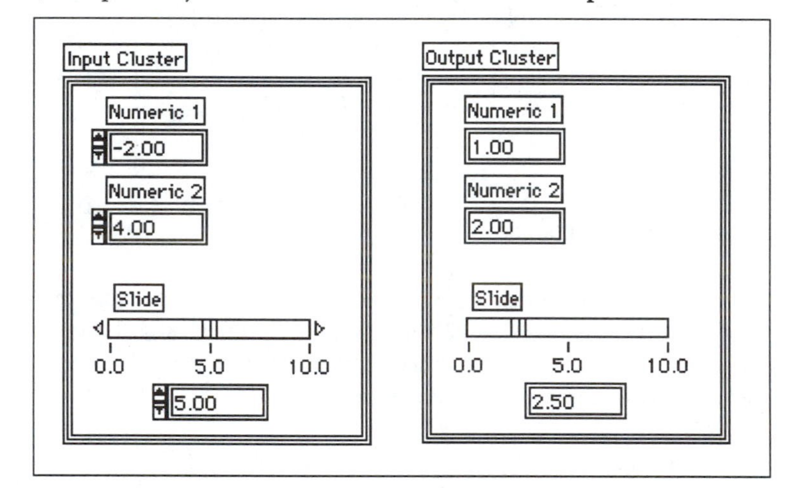

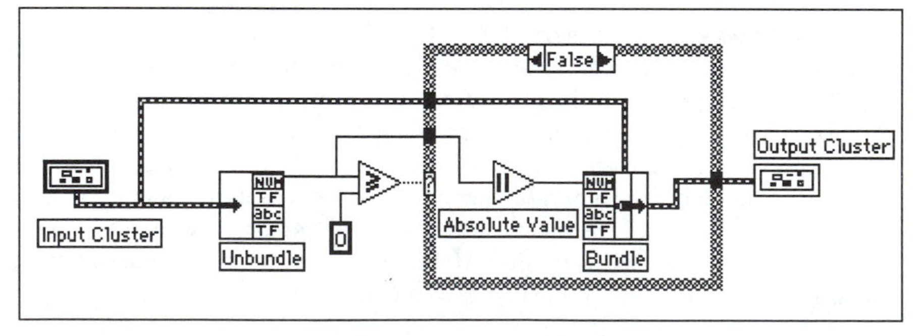

7.7 Wrap It Up!

A collection of data elements of the same type is called an *array.* In LabVIEW, arrays can be of any data type, except chart, graph, or another array. You must create an array using a two-step process: First, you place an array shell (**Array & Graph** palette of the **Controls** menu) in the window, and then you add the desired control or indicator to the shell.

LabVIEW offers many functions to help you manipulate arrays, such as **Build Array** and **Index Array**, in the **Array & Cluster** menu. In this lesson, you used array functions to work with only 1D arrays; however, these functions are smart and will work similarly with multidimensional arrays.

Both the For Loop and the While Loop can accumulate arrays at their borders using *auto-indexing*, a useful feature when you need to create and process arrays. Remember that by default, LabVIEW enables indexing in For Loops and disables indexing for While Loops.

Polymorphism is a fancy name for the ability of a function to adjust to input data of different data types. You studied polymorphic capabilities of arithmetic functions; however, many other functions are also polymorphic.

Clusters also group data, but unlike arrays, they will accept data of different types. You must create them on the panel in a two-step process. Keep in mind that objects inside a cluster must be all controls or all indicators. You cannot combine both controls and indicators within one cluster.

Clusters are useful for reducing the number of wires or terminals associated with a VI. For example, if a VI has many front panel controls and indicators that you need to associate with terminals, it is easier to group them as a cluster and have only one terminal.

The **Unbundle** function (**Array & Cluster** menu) splits a cluster into each of its individual components. You can use the components as ordin2ary terminals in the block diagram. The **Bundle** function (**Array & Cluster** menu) assembles all the individual components into a single cluster or can replace an element in a cluster.

7.8 Additional Exercises

EXERCISE 7.4
Reversal of Fortune ***Challenge***

Build a VI that reverses the order of an array containing 100 random numbers. For example, array[0] becomes array[99], array[1] becomes array[98], and so on. Name the VI **Reverse Random Array.vi**.

EXERCISE 7.5
Taking a Subset

Build a VI that generates an array containing 100 random numbers and displays a portion of the array; for example, from index 10 to index 50. (Hint: Use the **Array Subset** function [**Array & Cluster** menu] to extract the portion of the array.) Name the VI **Subset Random Array.vi**.

EXERCISE 7.6
Dice! ***Challenge***

Build a VI that simulates the roll of a die (possible values 1–6) and keeps track of the number of times that the die rolls each value. Your input is the number of times to roll the die, and the outputs include (for each possible value) the number of times the die fell on that value. Name the VI **Die Roller.vi.**

Hint: You will need to use a shift register in order to keep track of values from one iteration of a loop to the next.

EXERCISE 7.7
Multiplying Array Elements

Build a VI that takes an input 1D array and then multiplies pairs of elements together (starting with elements 0 and 1), and outputs the resulting array. For example, the input array with values 1 23 10 5 7 11 will result in the output array 23 50 77. Name the VI **Array Pair Multiplier.vi.**

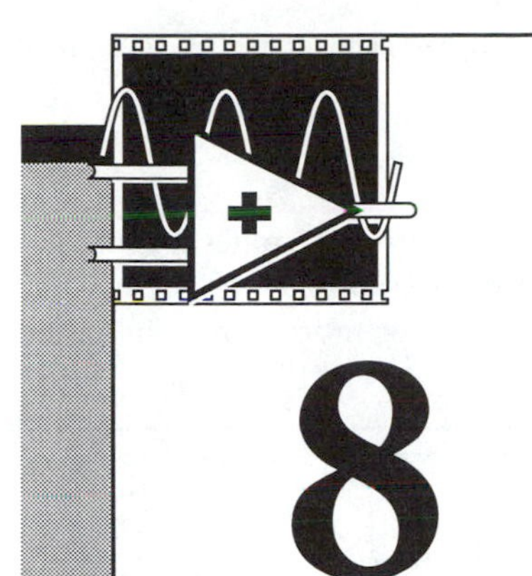

8

Charts and Graphs

Overview

LabVIEW's charts and graphs let you display plots of data in a graphical form. Charts interactively plot data, appending new data to old so you can see the current value in context with previous data. Graphs plot pre-generated arrays of values in a more traditional fashion. In this chapter, you will learn about charts and graphs, their appropriate data types, and several ways to use them.

Your Goals

- Understand the uses for charts and graphs
- Be able to recognize a chart's three modes: strip, scope, and sweep
- Understand mechanical action of Boolean switches
- Recognize the difference in functionality of charts and graphs
- Know the data types accepted by charts and graphs for both single and multiple plots
- Be able to customize the appearance of charts and graphs by changing the scales and using the palette and the legend

Key Terms

waveform chart	legend
waveform graph	palette
XY graph	mechanical action
strip mode	latch action
sweep mode	switch action
scope mode	

8.1 Waveform Charts

LabVIEW has one kind of waveform chart, with three different update modes for interactive data display. The waveform chart, located in the **Array & Graph** palette of the **Controls** menu, is a special numeric indicator that can display one or more plots. The figure below shows an example of a multiple-plot waveform chart.

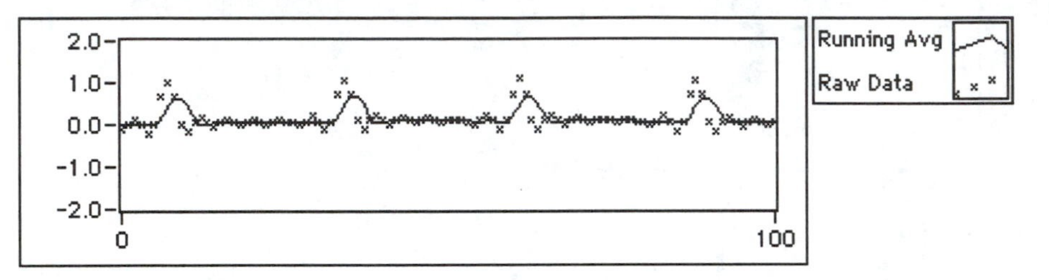

Chart Update Modes

The waveform chart has three update modes—*strip chart*, *scope chart*, and *sweep chart*, shown in the following illustration. The update mode can be selected by popping up on the waveform chart and choosing one of the options from the **Data Operations ▶ Update Mode** menu. (In the run mode, select **Update Mode** from the chart pop-up menu.)

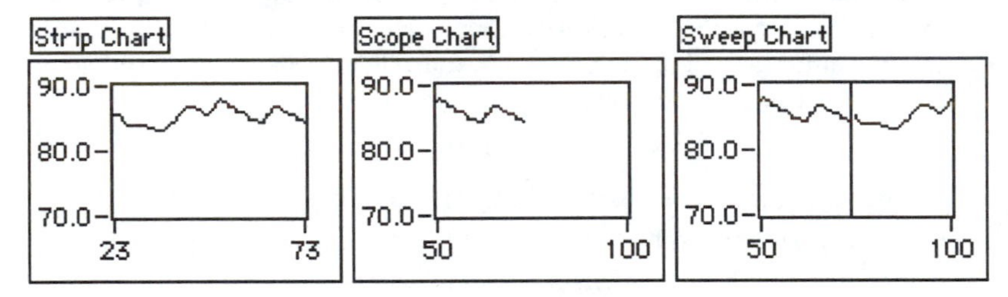

Waveform Chart Update Modes

The strip chart has a scrolling display similar to a paper strip chart. The scope chart and the sweep chart have retracing displays similar to an oscilloscope. On the scope chart, when the plot reaches the right border of the plotting area, the plot erases, and plotting begins again from the left border. The sweep chart acts much like the scope chart, but the display does not blank when the data reaches the right border. Instead, a moving vertical line marks the beginning of new data and moves across the display as new data is added. Because there is less overhead in retracing a plot, the scope chart and the sweep chart operate significantly faster than the strip chart.

Single-Plot Charts

You can directly wire a scalar output to a waveform chart, as shown in the following illustration. In this example, a new temperature value will be plotted on the chart each time the loop iterates.

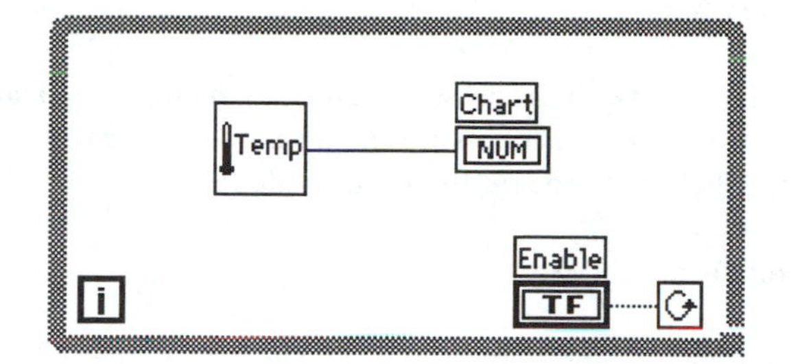

Wiring a Multiple-Plot Chart

Waveform charts can also accommodate more than one plot. You must bundle the data together using the **Bundle** function (**Array & Cluster** menu). In the next figure, the **Bundle** function "bundles" or groups the outputs of the three different VIs that acquire temperature into a cluster so they can be plotted on the waveform chart. Notice the change in the waveform chart terminal icon. To add more plots, simply increase the number of **Bundle** input terminals by resizing using the Positioning tool.

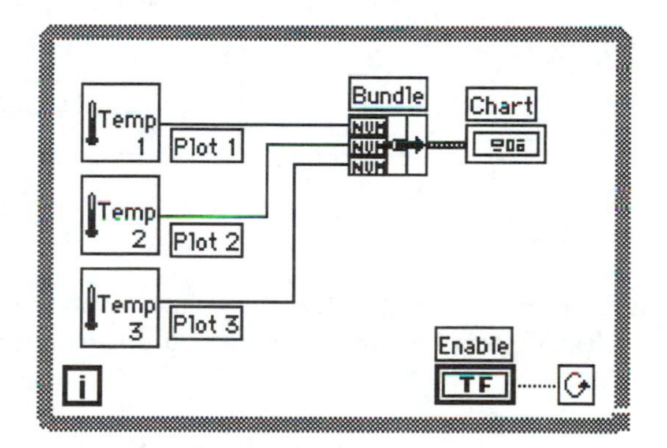

For an online example of charts, their modes, and their expected data types, open and run **Charts.vi** in GRAPH.LLB in the EXAMPLES directory or folder.

Show or Hide the Digital Display

Like many other numeric indicators, charts have the option to show or hide the digital display (pop up to get the **Show** option). The digital display shows the most recent value displayed by the chart.

The Scrollbar

Charts also have scrollbars that you can show or hide. You can display older data that has scrolled off the chart using the scrollbar.

Clearing the Chart

To clear a chart from edit mode, select **Clear Chart** from the **Data Operations** submenu in the chart's pop-up menu. If you are in run mode, **Clear Chart** is a pop-up menu option. You can also click the **Clear** button on the palette.

Stacked and Overlaid Plots

If you have a multiple-plot chart, you can choose whether to display all plots on the same set of scales, called an *overlaid* plot; or you can give each plot its own scale, called a *stacked* plot. You can select **Stack Plots** or **Overlay Plots** from the chart's pop-up menu to toggle the type of display. The following figure illustrates the difference between stacked and overlaid plots.

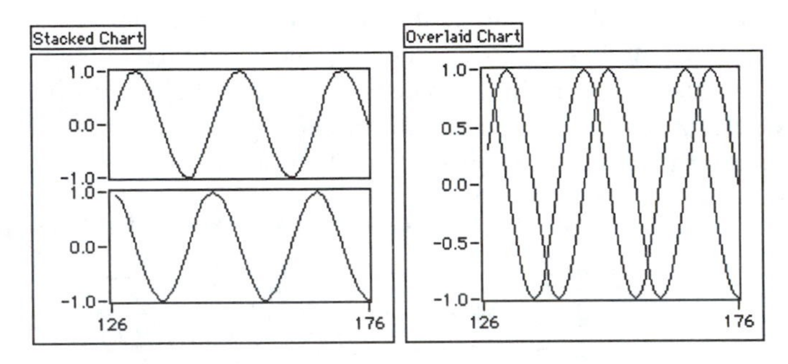

Chart History Length

By default, a chart can store up to 1024 data points. If you want to change the size of this buffer, select **Chart History Length...** from the pop-up menu and specify a new value of up to 100,000 points.

EXERCISE 8.1
Temperature Monitor

You will build a VI to measure temperature and display it on the waveform chart. This VI will measure the temperature using the **Thermometer** VI you built in the previous lesson as a subVI.

Front Panel

1. Open a new panel. You will re-create the panel shown.

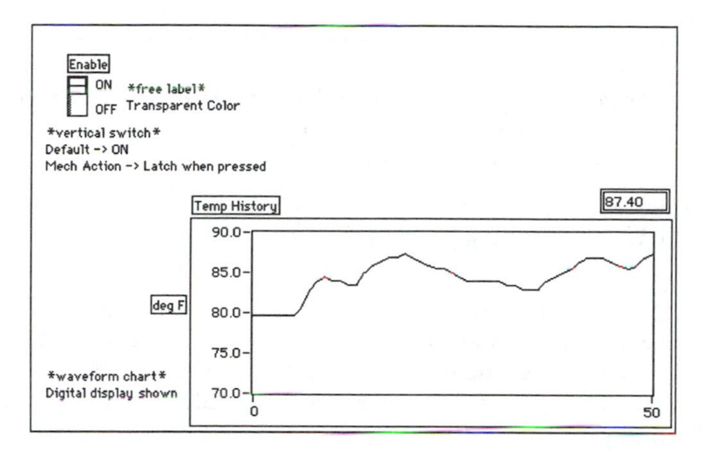

2. Place a vertical switch (**Boolean** menu) in the Panel window. Label the switch Enable. You will use the switch to stop the acquisition.
3. Place a waveform chart (**Array & Graph** menu) in the Panel window. Label the waveform chart Temp History. The waveform chart will display the temperature in real time.
4. The waveform chart has a digital display that shows the latest value. Pop up on the waveform chart and choose **Show ▶ Digital Display** from the pop-up menu.
5. Because the temperature sensor measures room temperature, rescale the waveform chart to correctly display the temperature. Using the Labeling tool, double-click on "10.0" in the waveform chart scale, type 90, and click outside the text area. The click enters the value. You also can press <enter> to input your change to the scale. Change "0.0" to 70 in the same way.

Block Diagram

1. Open the Diagram window and build the diagram shown.

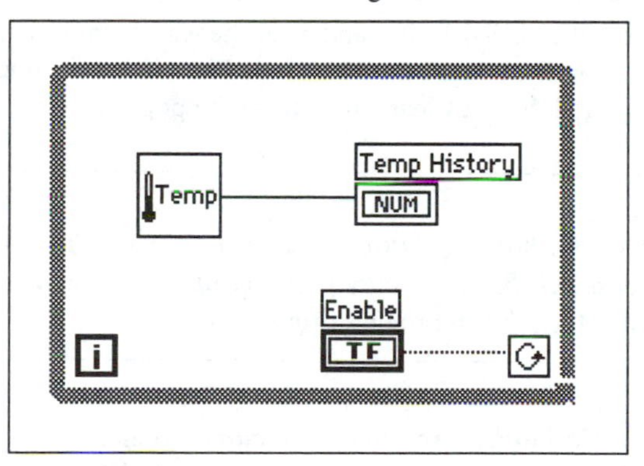

2. Place the **While Loop** (**Structs & Constants** menu) in the Diagram window. Enlarge the loop by dragging one corner with the Positioning tool.

Note
The Positioning tool will change into a frame corner at each corner of the While Loop. Drag one frame corner to resize the loop.

---

3. Place the two terminals inside the While Loop.
4. Import the **Thermometer** subVI.

   Temp **Thermometer** VI. This VI returns one temperature measurement from the temperature sensor. You should have written it in Chapter 4. Load it using the **VI...** option of the pop-up **Functions** menu. It should probably be in `MYWORK.LLB`. If you don't have it, you can use **Digital Thermometer.vi**, located in the **Tutorial** palette of the **Functions** menu.
5. Wire the diagram as shown in the preceding illustration.
6. Return to the front panel and turn on the vertical switch by clicking on it with the Operating tool. Run the VI.

   Remember, the While Loop is an indefinite looping structure. The diagram within its border will execute as long as the specified condition is TRUE. In this example, as long as the switch is on (TRUE), the **Thermometer** VI will take and return a new measurement and display it on the waveform chart.
7. To stop the acquisition, click on the vertical switch. This action causes the loop condition to be FALSE and the loop ends.
8. The waveform chart has a display buffer that retains a number of points after they have scrolled off the display. While in edit mode, show the scroll bar by popping up on the waveform chart and selecting **Show ▶ Scrollbar** from the pop-up menu. You can use the Positioning tool to adjust the scroll bar's size and position.

   To scroll through the waveform chart, click and hold on either arrow in the scroll bar.

   To clear the display buffer and reset the waveform chart, pop up on the waveform chart and choose **Data Operations ▶ Clear Chart** from the pop-up menu. If you have switched to run mode, select **Clear Chart** from the pop-up menu.

Note
The display buffer default size is 1024 points. While in edit mode, you can increase or decrease this buffer size by popping up on the waveform chart and choosing Chart History Length from the pop-up menu.

Using Mechanical Action of Boolean Switches. You may notice that each time you run the VI, you first must turn on the vertical Enable switch and then click on the run button. You can modify the mechanical action of Boolean control to circumvent this inconvenience. LabVIEW offers six possible choices for the mechanical action of a Boolean control.

Switch When Pressed action changes the control value each time you click on the control with the Operating tool. The action is similar to that of a ceiling light switch, and is not affected by how often the VI reads the control.

Switch When Released action changes the control value only after you release the mouse button during a mouse click within the graphical boundary of the control. The action is not affected by how often the VI reads the control. This mode is similar to what happens when you click on a check mark in a dialog box; it becomes highlighted but does not change until you release the mouse button.

Switch Until Released action changes the control value when you click on the control. It retains the new value until you release the mouse button, at which time the control reverts to its original value. The action is similar to that of a door buzzer, and is not affected by how often the VI reads the control.

Latch When Pressed action changes the control value when you click on the control. It retains the new value until the VI reads it once, at which point the control reverts to its default value. (This action happens whether or not you continue to press the mouse button.) This action is similar to that of a circuit breaker and is useful for stopping While Loops or having the VI do something only once for each time you set the control.

Latch When Released action changes the control value only after you release the mouse button. When your VI reads the value once, the control reverts to the old value. This action guarantees at least one new value. As with **Switch When Released**, this mode is similar to the behavior of buttons in a dialog box; the button becomes highlighted when you click on it, and latches a reading when you release the mouse button.

Latch Until Released changes the control value when you click on the control. It retains the value until your VI reads the value once or until you release the mouse button, whichever occurs last.

For example, consider a vertical switch—its default value is off (FALSE).

9. Modify the vertical switch in your VI so that you need not turn the switch to TRUE each time you run the VI.

a. Turn on the vertical switch.

b. Pop up on the switch and choose **Data Operations ▶ Make Current Value Default** from the pop-up menu. This will make the *on* position the default value.

c. Pop up on the switch and choose **Mechanical Action ▶ Latch When Pressed** from the pop-up menu.

10. Run the VI. Click on the vertical switch to stop the acquisition. The switch will move to the *off* position briefly, and then change back to *on* after the While Loop conditional terminal reads one FALSE value.

Adding Timing

When you ran the VI, the While Loop executed as quickly as possible. You may want to take data at certain intervals, however, such as once per second or once per minute.

You can control loop timing using the **Wait (ms)** function (**Time & Dialog** menu). This function ensures that no iteration is shorter than the specified number of milliseconds.

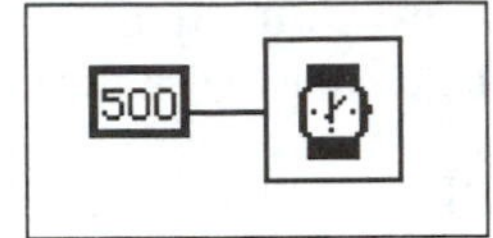

11. Modify the VI to take a temperature measurement about once every half-second by placing the code segment shown above into the While Loop.

    Wait (ms) function (**Time & Dialog** menu) ensures that each iteration waits a half-second (500 ms) before continuing.

    The Numeric Constant wired to the **Wait (ms)** function specifies a wait of 500 ms (one half-second). Thus, the loop executes approximately once every half-second, since the other functions take only a few milliseconds.
12. Run the VI. Try different values for the number of milliseconds.
13. Save and close the VI. Name it **Temperature Monitor.vi** and place it in MYWORK.LLB.

8.2 Chart and Graph Components

Graphs and charts have many powerful features that you can use to customize your plots. This section covers how to configure these options.

Playing with the Scales

Charts and graphs automatically adjust their horizontal and vertical scales to reflect the points plotted on them. You can turn this *autoscaling* feature on or off using the **AutoScale X** and **AutoScale Y** options from the **Data Operations** menu or the **X Scale** or **Y Scale** submenus of the pop-up menu. You can also control these autoscaling features from the palette. LabVIEW defaults to **AutoScaling On**. However, autoscaling may cause the chart or graph to be slower, depending upon the computer and video system you use.

If you don't want to autoscale, you can change the horizontal or vertical scale directly using the Operating or Labeling tool, just as you can with any other LabVIEW control or indicator. LabVIEW sets point density automatically based on the range you choose.

The X and Y scales each have a submenu of options.

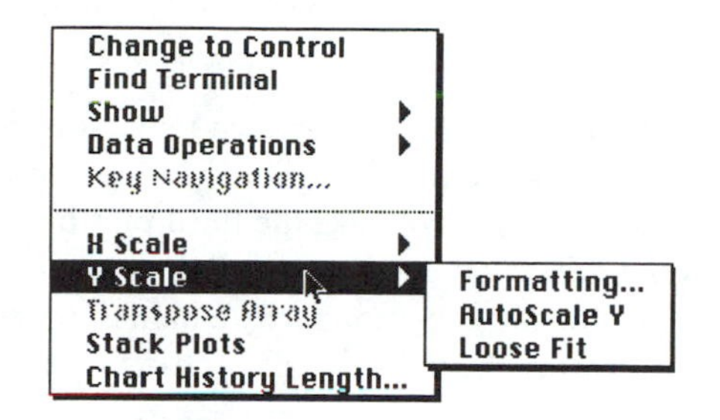

Use **AutoScale** to turn the autoscaling option on or off.

Normally, the scales are set to the exact range of the data when you perform a fit operation. You can use the **Loose Fit** option if you want LabVIEW to round the scale to "nicer" numbers. With a loose fit, the numbers are rounded to a multiple of the increment used for the scale. For example, if the markers increment by 5, then the minimum and maximum values are set to a multiple of 5.

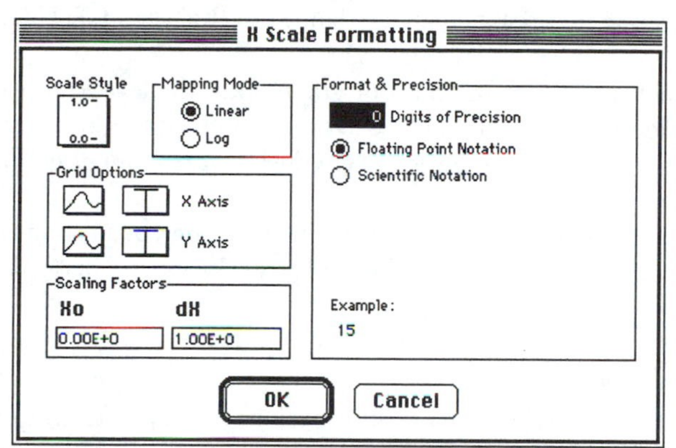

The **Formatting...** option brings up a dialog box, shown in the preceding illustration, that allows you to configure the following things:

- The **Scale Style** menu lets you select major and minor tick marks for the scale, or none at all. Click on it to see your choices. A major tick mark corresponds to a scale label, while a minor tick mark denotes an interior point between labels. This menu also lets you select the markers for a given axis as either visible or invisible.
- The **Mapping Mode** lets you select either a linear or a logarithmic scale for the data display.
- The **Grid Options** lets you select no gridlines, gridlines only at major tick mark locations, or gridlines at both major and minor tick marks. You can also change the color of the gridlines here.
- You can set **Xo**, the X value you want to start at, and **dX**, the increment between X values, in the **Scaling Factors** section.
- The **Format & Precision** section lets you choose the number of digits of precision as well as the notation (**Floating Point** or **Scientific**) display.

The Legend

Charts and graphs use a default style for each new plot unless you have created a custom plot style for it. If you want a multiple-plot chart or graph to have certain characteristics for specific plots (for example, to make the third plot blue), you can set these characteristics using the *legend*, which can be shown or hidden using the **Show** submenu of the chart or graph pop-up menu. You can also specify a name for each plot in the legend. An example of a legend is shown in the following illustration.

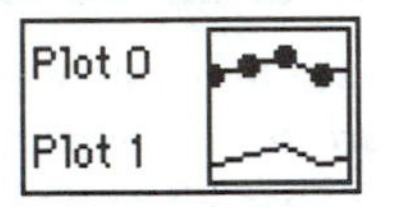

When you select **Legend**, only one plot appears. You can show more plots by dragging down a corner of the legend with the Resizing tool. After you set plot characteristics, the plot retains those settings, regardless of the whether the legend is visible. If the chart or graph receives more plots than are defined in the legend, LabVIEW draws them in default style.

When you move the chart or graph body, the legend moves with it. You can change the position of the legend relative to the graph by dragging only the legend to a new location. *Resize the legend on the left to give labels more room or on the right to give plot samples more room.*

By default, each plot is labeled with a number, beginning with 0. You can modify this label the same way you modify other LabVIEW labels (just start typing with the labeling tool).

Each plot sample has its own pop-up menu to change the plot, line, color, and point styles of the plot.

- The **Point Style** and **Line Style** options display different plot styles.
- The **Interpolation** option determines how LabVIEW draws lines between data points. The **Line** option draws a straight line between plotted points. The **None** option does not draw a line, making it suitable for a scatter plot (in other words, you get points only). The **Step** option links points with a right-angled elbow. You can use this option to create histogram-like plots.
- The **Color** option displays the color palette so you can select the plot color. You can also color the plots on the legend with the Coloring tool. You can change the plot colors while in run mode.

Using the Palette

With the palette, you can access several useful functions while the VI executes. You can clear the chart or graph, scale the X or Y axis, and change the display format of the scale at any time. The palette, which you access from the **Show** menu of the chart or graph pop-up menu, is shown in the following illustration.

If you press the button, LabVIEW will autoscale the X data of the graph. If you press the button, LabVIEW will autoscale the Y data of the graph. If you want the graph to autoscale either of the scales continuously, click on the lock switch to lock autoscaling on.

The button and the button give you run-time control over the format and precision of the X and Y scale markers, respectively.

The remaining three buttons let you control the operation mode for the graph. Normally, you are in standard operate mode, meaning that you can click in the graph to move cursors around. If you press the pan button, then you switch to a mode where you can scroll the visible data by clicking and dragging sections of the graph with the pan cursor. If you click on the zoom button, you get a pop-up menu that lets you choose from several methods of zooming, shown in the following illustration.

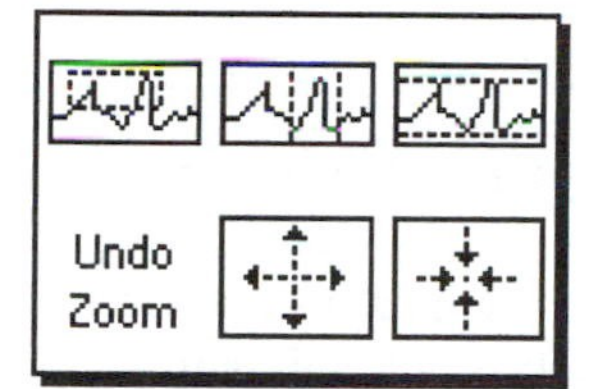

Following is a description of each of these zoom options:

Zoom by Rectangle. In other words, drag the cursor to draw a rectangle around the area you want to zoom in on. When you release the mouse button, your axes will rescale to accommodate the zoom.

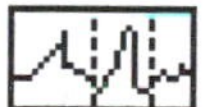

Zoom by Rectangle, with zooming restricted to X data (the Y scale remains unchanged).

Zoom by Rectangle, with zooming restricted to Y data (the X scale remains unchanged).

Undo
Zoom

Undo Last Zoom, and reset the scales to their previous setting.

Zoom In about a Point. If you hold down the mouse on a specific point, the graph will continuously zoom in until you release the mouse button.

Zoom Out about a Point. If you hold down the mouse on a specific point, the graph will continuously zoom out until you release the mouse button.

Note
For the last two modes, and , shift-clicking will zoom in the other direction.

8.3 Graphs

Unlike charts, which plot interactively, graphs plot pregenerated arrays of data all at once. LabVIEW provides two types of graph for greater flexibility: *waveform graphs* and *XY graphs*. Both types look identical on the front panel of your VI, but have very different functionality. An example of a graph with several graph options enabled is shown in the following picture.

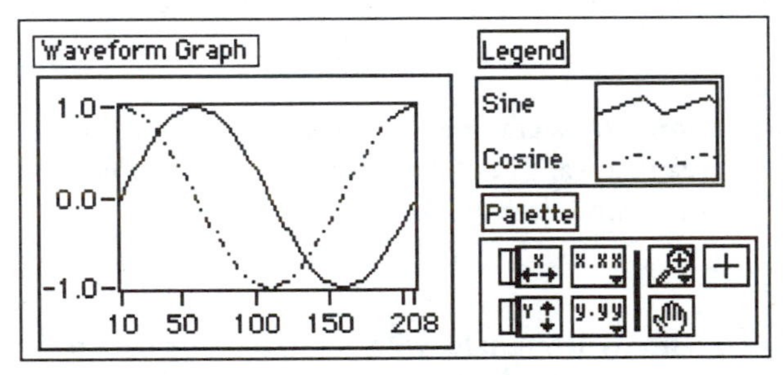

You obtain both types of graph indicators from the **Array & Graph** palette of the pop-up **Controls** menu. The *waveform graph* plots only single-valued functions with uniformly spaced points, such as acquired time-varying waveforms. The waveform graph is ideal for plotting arrays of data in which the points are evenly distributed. The *XY graph* is a general-purpose, Cartesian graph ideal for plotting multivalued functions such as circular shapes or waveforms with varying timebases. The two types of graph take different types of input, so you must be careful not to confuse them.

The legend, palette, and other graph options are described in Section 8.2, *Chart and Graph Components.*

Single-Plot Waveform Graphs

For basic single-plot graphs, an array of Y values can pass directly to a waveform graph, as shown in the following illustration. This method assumes the initial X value and the delta X value are 0 and 1, respectively. Once wired, the graph terminal in the block diagram now appears as an array indicator.

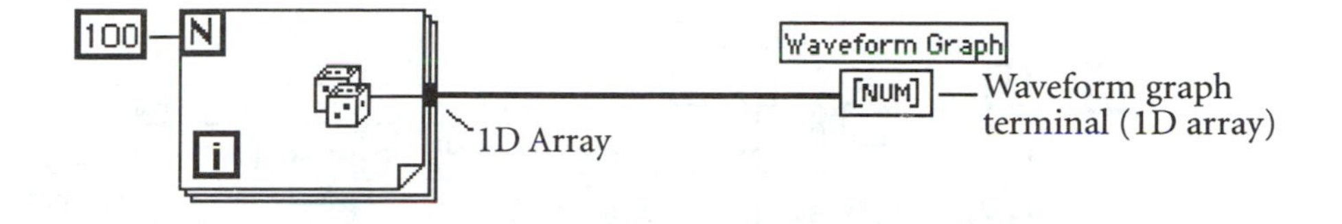

If you want the flexibility to change the timebase for the graph (for example, you start sampling at a time other than "Xo=0," or your samples are spaced more than "delta X=1" apart), you can wire a cluster of data consisting of the initial X value, the delta X value, and a data array to the waveform graph. Notice that the graph terminal appears as a cluster indicator in the following illustration.

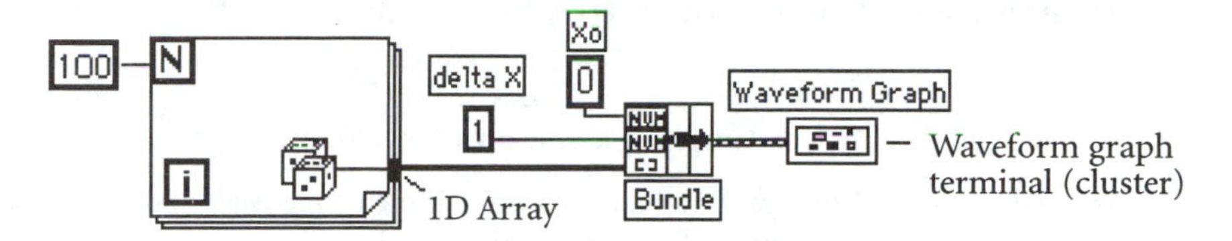

Multiple-Plot Waveform Graphs

You can show more than one plot on a waveform graph by creating an array of the data types used in the single-plot examples. The next two illustrations detail two methods for wiring multiple-plot waveform graphs. As in previous examples, the graph terminal assumes the data type to which it is wired.

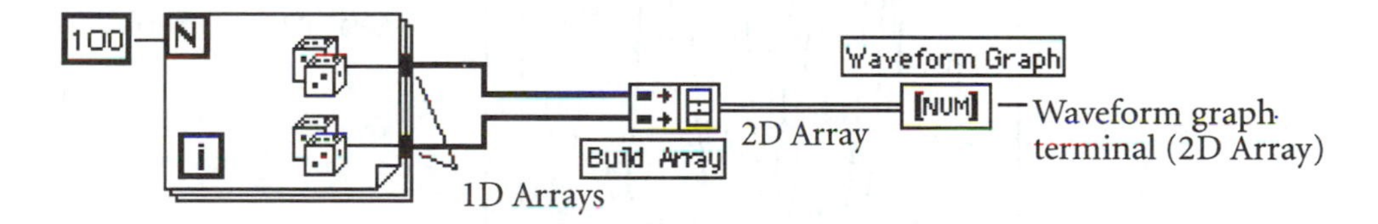

The preceding illustration assumes the initial X value is 0 and the delta X value is 1 for both arrays. Notice that this two-dimensional array has two rows with 100 columns per row—a 2 × 100 array. Graphs always plot the *rows* of a two-dimensional array. If your data is organized by column, you must make sure to transpose your array when you plot it!

In the following illustration, the initial X value and a delta X value for each array is specified. These X parameters do not need to be the same for both sets of data.

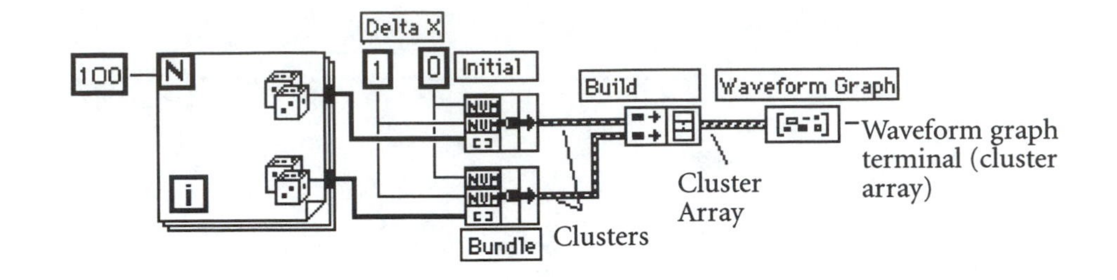

The **Build Array** function (**Array & Cluster** menu) creates a 2D array from the 1D array inputs or creates a cluster array from the cluster inputs.

EXERCISE 8.2
Graphing a Sine on a Waveform Graph

You will build a VI that generates an array containing a sine wave and plots it in a waveform graph. You will also modify the VI to graph multiple plots.

Front Panel

1. Open a new VI, and build the front panel shown. Be sure to modify the controls and indicators as depicted.

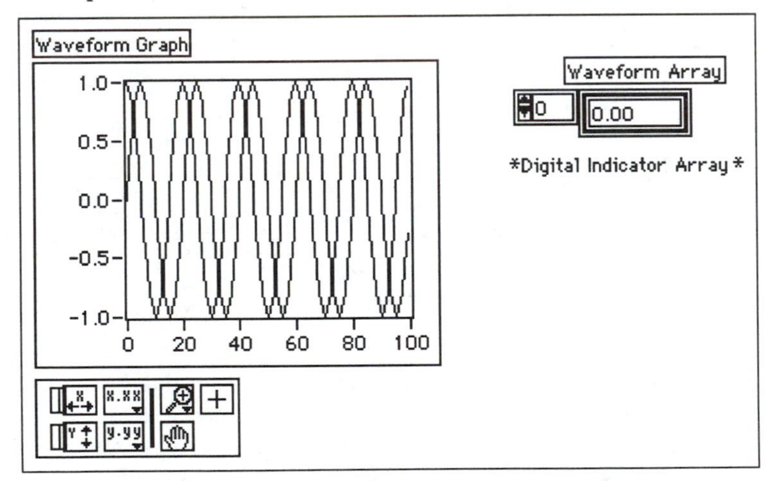

2. Place an array shell (**Array & Graph** menu) in the front panel window. Label the array shell Waveform Array. Place a digital indicator (**Numeric** menu) inside the Data Object window of the array shell using the pop-up menu. This indicator displays the array contents.

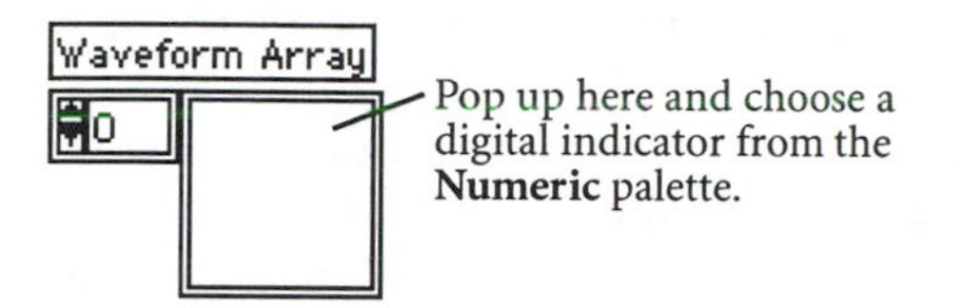

3. Place a waveform graph (**Array & Graph** menu) in the front panel window. Label the graph Waveform Graph. Enlarge the graph by dragging a corner with the Positioning tool.
4. Disable autoscaling by popping up on the graph and choosing **Y Scale ▸ √ AutoScale Y.** Modify the Y axis limits by selecting the scale limits with the Labeling tool and entering the new numbers. Change the Y axis minimum to -1.0 and the maximum to 1.0.

Block Diagram

1. Build the block diagram shown.

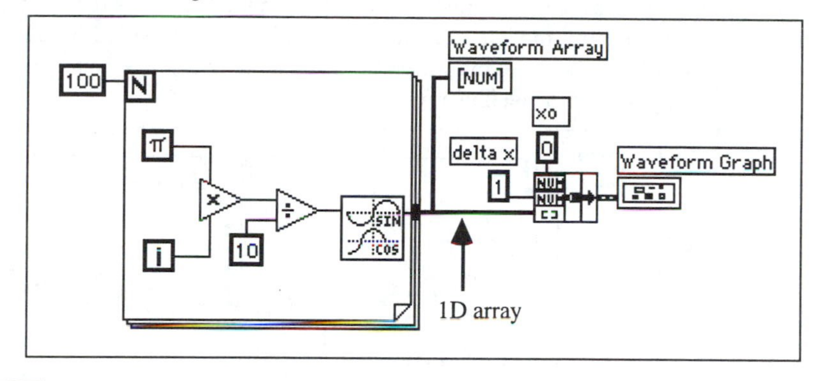

Sine & Cosine function (**Arithmetic** menu) returns one point of a sine wave. The VI requires a scalar index input.

Pi constant (**Structs & Constants** menu).

Bundle function (**Array & Cluster** menu) assembles the plot components into a single cluster. The components include the initial X value (0), the delta X value(1), and the Y array (waveform data). Use the Positioning tool to resize the function by dragging one of the corners.

Each iteration of the For Loop generates one point in a waveform that the VI stores in the waveform array created at the loop border. After the loop finishes execution, the **Bundle** function bundles the initial value of X (Xo), the delta value of X, and the array for plotting on the graph.

2. Return to the front panel and run the VI. The VI plots the auto-indexed waveform array on the waveform graph. The initial X value is 0 and the delta X value is 1.
3. Try changing the delta X value to 0.5 and the initial X value to 20. Notice that the graph now displays the same 100 points of data with a starting value of 20 and a delta X of 0.5 for each point (see the X axis). In a timed test, this graph would correspond to 50 seconds

worth of data starting at 20 seconds. Experiment with several combinations for the initial and delta X values.

4. You can view any element in the array simply by entering the index of that element in the index display. If you enter a number greater than the array size, the display will dim.

   If you want to view more than one element at a time, you can resize the array indicator. Place the Positioning tool on the lower-right corner of the array until the tool becomes a ⊿ and drag. The indicator now displays several elements in an ascending index order, beginning with the element corresponding to the specified index, as illustrated below.

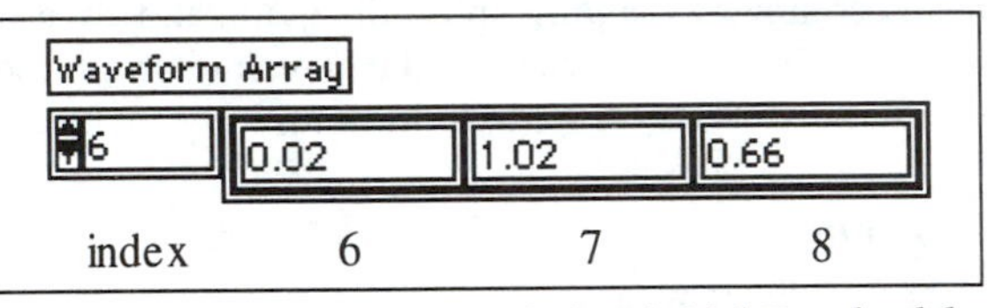

   In the previous block diagram, you specified an initial X and a delta X value for the waveform. Frequently, the initial X value will be zero and the delta X value will be 1. In these instances, you can wire the waveform array directly to the waveform graph terminal, as shown in the following illustration.

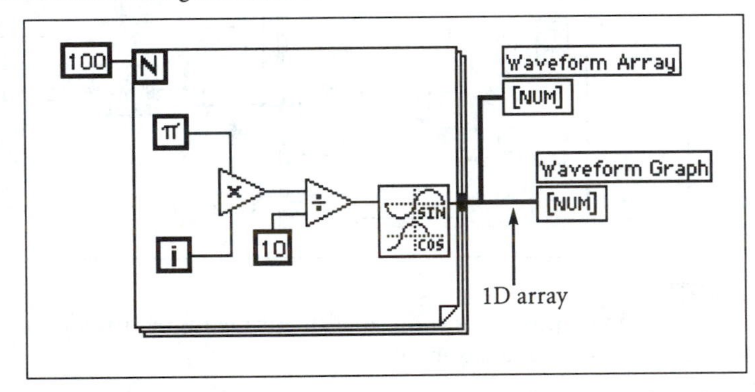

5. Return to the block diagram window. Delete the **Bundle** function and the **Numeric Constants** wired to it. Select **Remove Bad Wires** from the **Edit** menu. Finish wiring the block diagram as shown in the preceding illustration.
6. Run the VI. Notice that the VI plots the waveform with an initial X value of 0 and a delta X value of 1.

Multiple-Plot Graphs

You can create multiple-plot waveform graphs by wiring the data types normally passed to a single-plot graph to a **Build Array** function. While you can't build an array of arrays, you *can* create a 2D array.

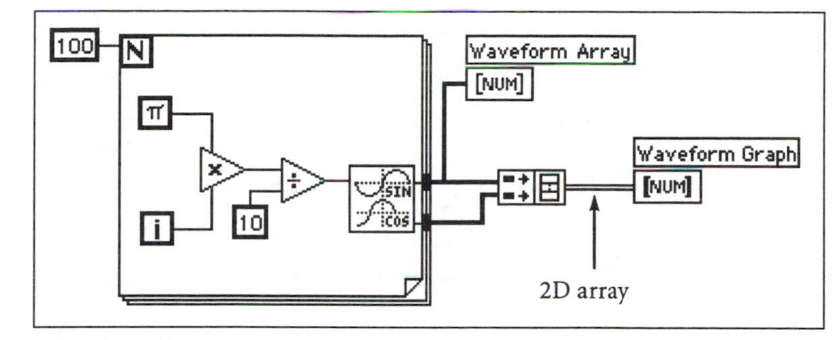

7. Create the block diagram shown in the preceding illustration.

 Build Array function (**Array & Cluster** menu) creates the proper data structure to plot two arrays on a waveform graph. Enlarge the **Build Array** function to include two inputs by dragging a corner with the Positioning tool. Remember that you want to use *element inputs* to the **Build Array** (the default value) rather than *array inputs* in order for output to be a 2D array.

8. Switch to the front panel. Run the VI.

   Notice that the two waveforms both plot on the same waveform graph. The initial X value defaults to 0 and the delta X value defaults to 1 for both data sets.

   Using the palette, change the precision so that the graph shows three decimal places on the Y scale. Now click on the zoom button, select a zooming mode, and zoom in on the graph.

9. Save and close the VI. Name it **Graph Sine Array.vi** in MYWORK.LLB.

Single- and Multiple-Plot XY Graphs

The waveform graphs you have been using are ideal for plotting evenly sampled waveforms. However, if you sample at irregular intervals or are plotting a complicated math function, you will need to specify points using their (X,Y) coordinates. XY graphs plot this different type of data—they require input of a different data type than waveform graphs. A single plot XY graph and its corresponding block diagram are shown in the following illustration.

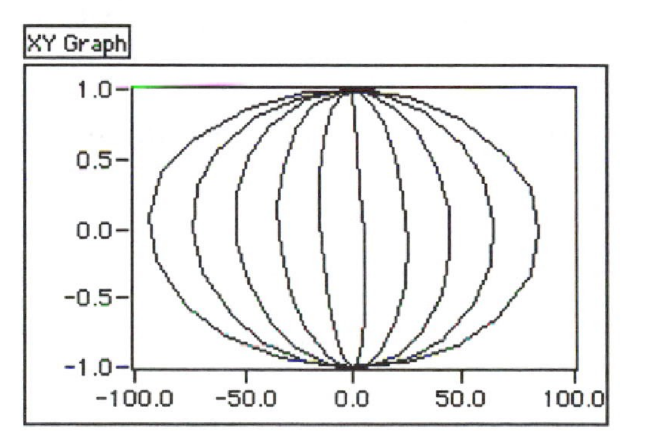

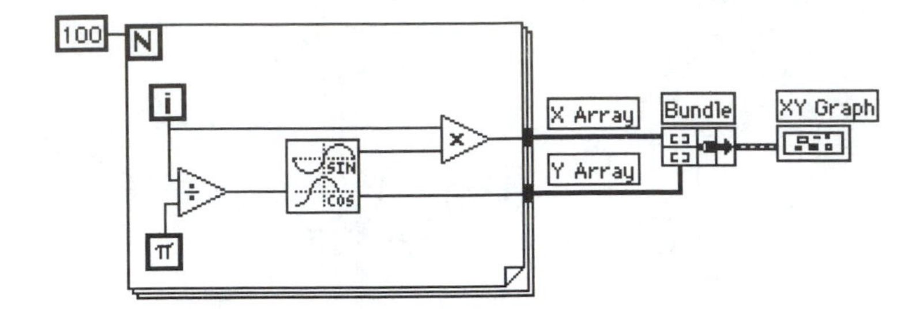

The **Bundle** function (**Array & Cluster** menu) combines the X and Y arrays into a cluster wired to the XY graph. The XY graph expects an input of a bundled X array (the top input) and a Y array (the bottom input). The XY graph terminal now appears as a cluster indicator.

For a multiple-plot XY graph, simply build an array of the clusters of X and Y values used for single plots as shown.

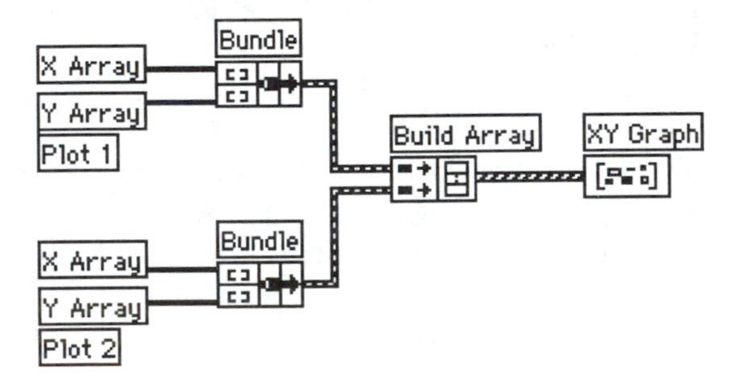

Note
For some online graph examples, look at Waveform Graph.vi and X vs. Y Graph.vi in* GRAPH.LLB *in the* EXAMPLES *folder or directory.

EXERCISE 8.3
Using an XY Graph to Plot a Circle

You will build a VI that plots a circle on an XY graph using independent X and Y arrays.

Front Panel

1. Open a new panel. You will re-create the panel shown.

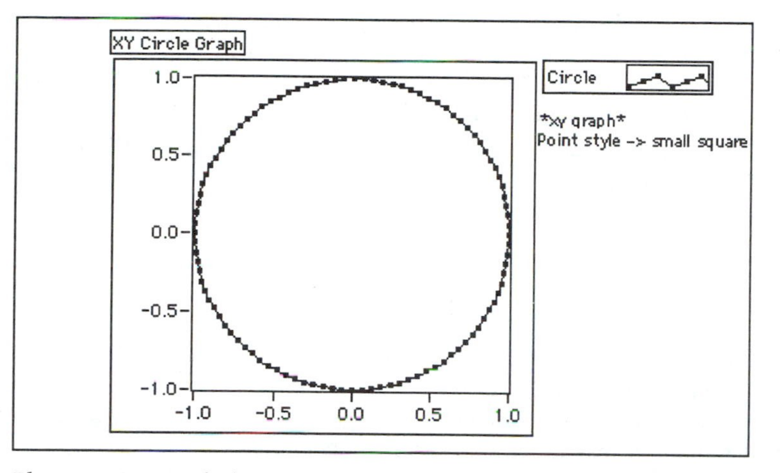

2. Place an XY Graph (**Array & Graph** menu) in the Panel window. Label the graph XY Circle Graph.
3. Enlarge the graph by dragging a corner with the Positioning tool. Try to make the plot region approximately square.
4. Pop up on the graph and select **Show ▸ Legend**. Resize the legend from the left side and enter the Circle label using the Labeling tool. Pop up on the line in the legend and select the small square from the **Point Style** palette. Then select a new plot color from the **Color** palette.

Block Diagram

1. Build the block diagram shown.

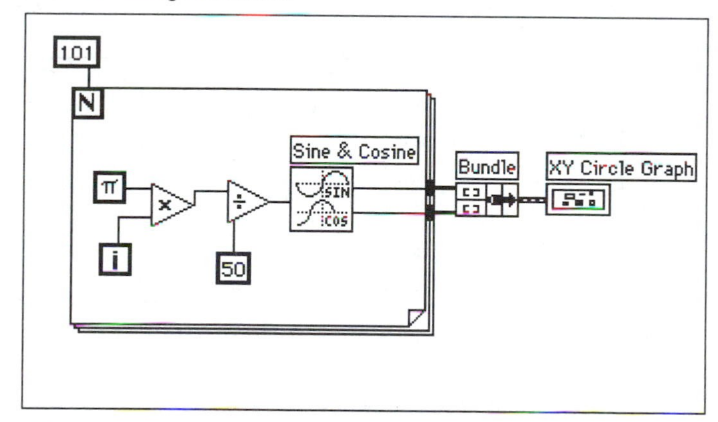

Sine & Cosine function (**Arithmetic** menu). In this exercise, you use the function in a For Loop to build an array of points that represents one cycle of a sine wave and one cycle of a cosine wave.

Bundle function (**Array & Cluster** menu) assembles the sine array and the cosine array to plot the sine array against the cosine array.

Pi constant (**Structs & Constants** menu).

Using a **Bundle** function, you can graph the one-cycle sine array versus the one-cycle cosine array on an XY graph, which produces a circle. The XY graph is useful for cases where the data plotted is a multivalued function, like the circle, or where the data is a waveform with a nonuniform timebase.

2. Return to the front panel. Run the VI. Save it as **Graph Circle.vi** in MYWORK.LLB. Congratualtions!

~ ~ ~

EXERCISE 8.4
Temperature Analysis

You will build a VI that measures temperature approximately every 0.25 seconds for 10 seconds. During the acquisition, the VI displays the measurements in real time on a waveform chart. After the acquisition is complete, the VI plots the data and a best-fit curve on a graph and calculates the minimum, maximum, and average temperatures.

Front Panel

1. Open a new panel and build the panel shown.

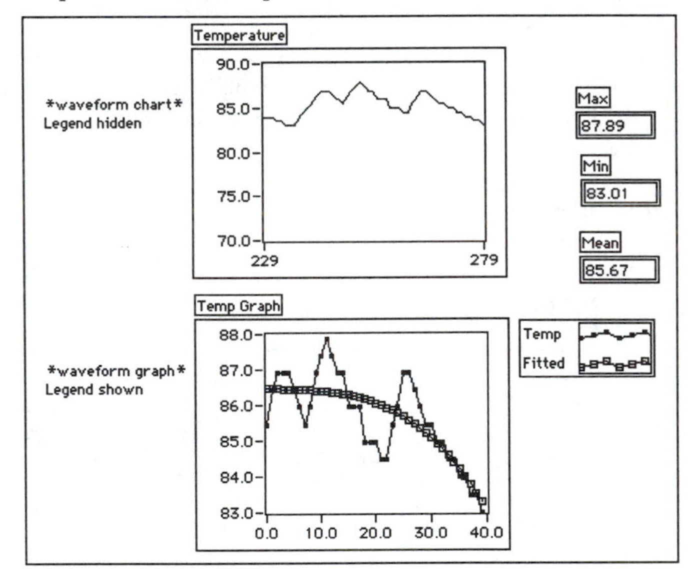

2. Rescale the chart so that it ranges from 70.0 to 90.0. Also, make sure autoscaling is on for both axes of the graph.
3. Show the legend on the graph. Resize it downward to accommodate two plots; then resize it from the left side to make the text area larger. Using the Labeling tool, type in Temp and Fitted as shown. Now pop up on the Temp plot representation in the legend and change the **Point Style** to "small squares."

   The Temperature chart displays the temperature as it is acquired. After acquisition is finished, the VI plots the data and a best-fit curve in Temp Graph. The Mean, Max, and Min digital indicators will display the average, maximum, and minimum temperatures, respectively.

Block Diagram

1. Build the block diagram shown. Make sure to use the Help window to show you the inputs and outputs of these functions, or you will almost certainly wire to the wrong terminal!

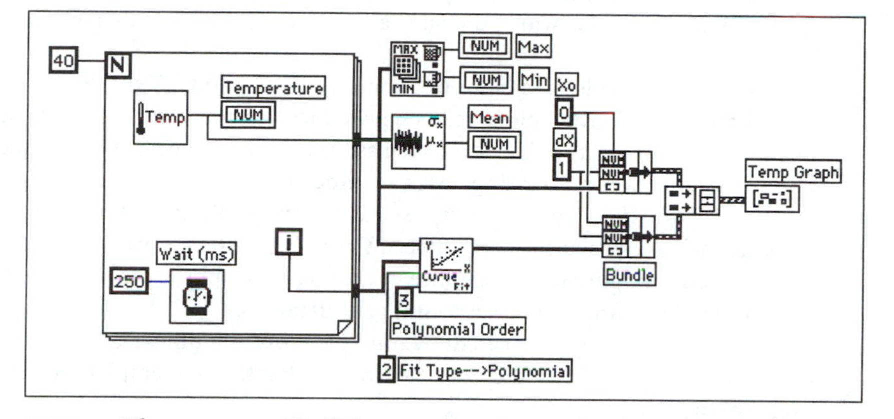

Thermometer VI (**VI...** menu, or **Tutorial** palette if you didn't build it). Returns one temperature measurement.

Wait (ms) function (**Time & Dialog** menu) causes the For Loop to execute every 0.25 seconds (250 ms).

Array Max & Min function (**Array & Cluster** menu) returns the maximum and minimum temperature measured during the acquisition.

Mean & Standard Deviation VI (**Analysis ▸ Statistics** menu) returns the average of the temperature measurements.

Curve Fit VI (**Analysis ▸ Statistics** menu) returns an array that is a polynomial fit (specified by fit type = 2) to the temperature array. This exercise uses three as the polynomial order.

Build Array function (**Array & Cluster** menu) creates a cluster from the temperature array and the "best fit" array. You can increase the number of inputs for the function by dragging down a corner with the Positioning tool.

The For Loop executes 40 times. The **Wait (ms)** function causes each iteration to take place approximately every 250 ms. The VI stores the temperature measurements in an array created at the For Loop border (auto-indexing). After the For Loop completes, the array passes to various nodes. The **Array Max & Min** function returns the maximum and minimum temperature. The **Mean** VI returns the average of the temperature measurements. The **Curve Fit** VI determines the best fit for the points in the temperature array. The **Build Array** function assembles data for the multiplot graph into an array.

2. Return to the front panel and run the VI.
3. Close and save the VI. Name it **Temperature Analysis.vi** and place it in MYWORK.LLB.

~ ~ ~

8.4 Wrap It Up!

You can display plots of data using LabVIEW's charts or graphs. *Charts* append new data to old data, interactively plotting one point (or one set of points) at time, so you can see a current value in context with previous values. *Graphs*, on the other hand, display a full block of data.

LabVIEW provides two kinds of graphs: *waveform* and *XY graphs.*

The *waveform graph* plots only single-valued functions with points that are evenly distributed with respect to the x axis, such as time-varying waveforms. In other words, the waveform graph plots a Y array against a set timebase.

The *XY graph* is a general-purpose, Cartesian graph that lets you plot multivalued functions such as circular shapes. It plots a Y array against an X array.

You can configure the appearance of charts and graphs using the *legend* and the *palette.* You can also change the scales to suit your data.

Both charts and graphs can draw multiple plots at a time. Data types can get tricky, so you may want to refer to the examples in this chapter as a template while writing your own graphing VIs.

8.5 Additional Exercises

EXERCISE 8.5
Temperature Limit

Build a VI that continuously measures the temperature once per second and displays the temperature on a chart in scope mode. If the temperature goes above or below the preset limits, the VI turns on a front panel LED. The chart should plot the temperature as well as the upper and lower temperature limits. You should be able to set the limit from the front panel. See front panel at the right for a start. Name the VI **Temperature Limit.vi**.

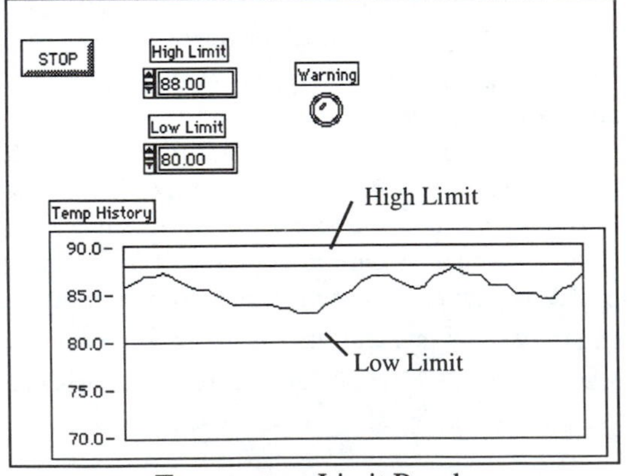

Temperature Limit Panel

EXERCISE 8.6
Max/Min Temperature Limit

Modify the VI you created in Exercise 8.5 to display the maximum and minimum values of the temperature trace. Hint: You must use shift registers and the **Max & Min** function (**Comparison** menu). Name the VI **Temp Limit (max/min).vi.**

EXERCISE 8.7
Plotting Random Arrays

Build a VI that generates a 2D array (3 rows by 10 columns) containing random numbers. After generating the array, index each row and plot each row on its own graph. (Your front panel should contain three graphs.) Name the VI **Extract 2D Array.vi**.

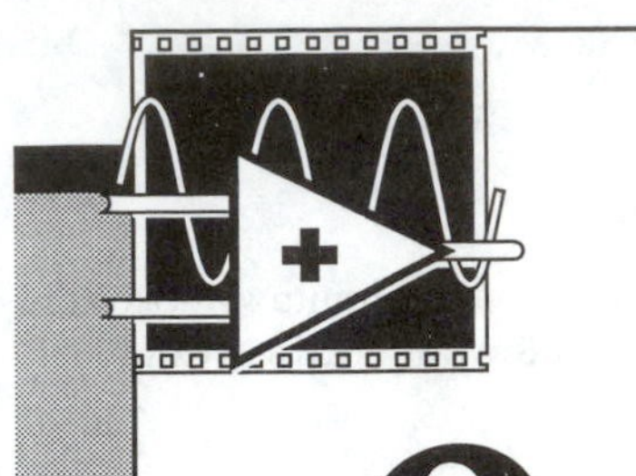

9

Strings and File I/O

Overview

This chapter introduces some of the powerful things you can do with strings. LabVIEW has many built-in string functions, similar to its array functions, that let you manipulate string data. You will also learn how to save data to and retrieve data from a disk file.

Your Goals

- Understand how to use LabVIEW's string functions
- Know how to convert a number to a string, and vice versa
- Be able to use the file I/O VIs to save data to a disk file and then read it back into LabVIEW

Key Term

spreadsheet file

9.1 Using String Functions

We introduced strings in Chapter 4—a string is simply a collection of ASCII characters. Often, you may use strings for more than simple text messages. For example, in instrument control, you pass numeric data as character strings. You then convert these strings to numbers to process the data. Storing numeric data to disk also requires strings—in the file I/O VIs, LabVIEW first converts numeric values to string data before it saves them to a file.

The following functions, available from the **String** palette of the **Functions** menu, facilitate any string activity you need to do.

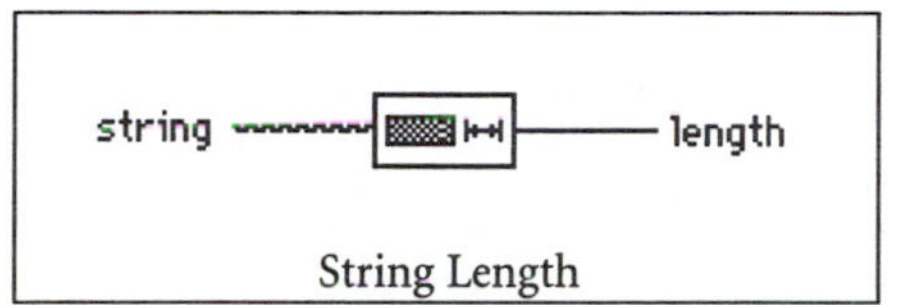

String Length

String Length returns the number of characters in a given string.

Concatenate Strings concatenates all input strings into a single output string.

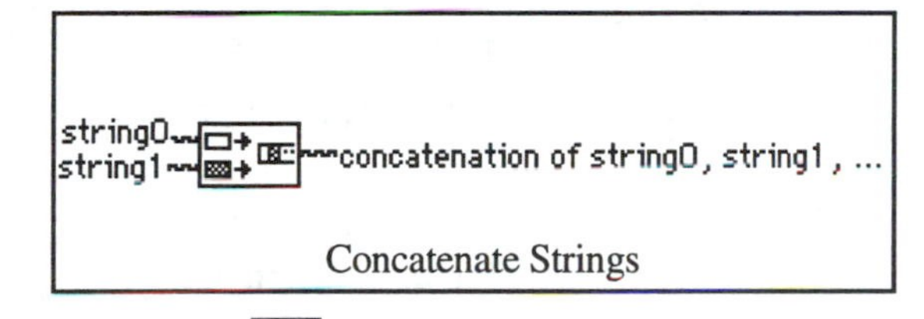

Concatenate Strings

The function appears as [icon] when you place it in the Diagram window. You can resize the function with the Positioning tool to increase the number of inputs.

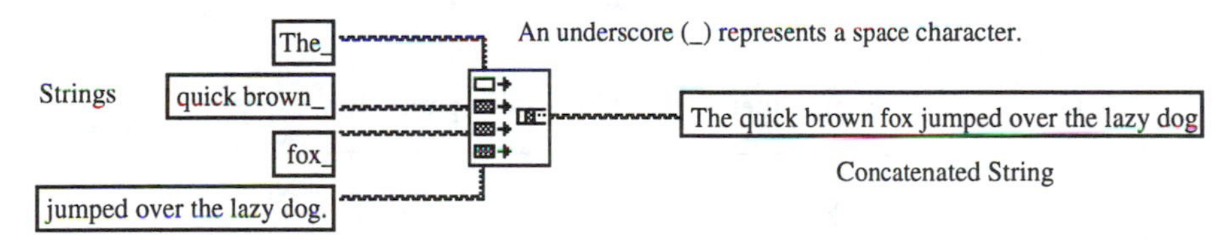

In many instances, you must convert strings to numbers or numbers to strings. The **Format & Append** and **Format & Strip** functions have these capabilities. **Format & Strip** is discussed later.

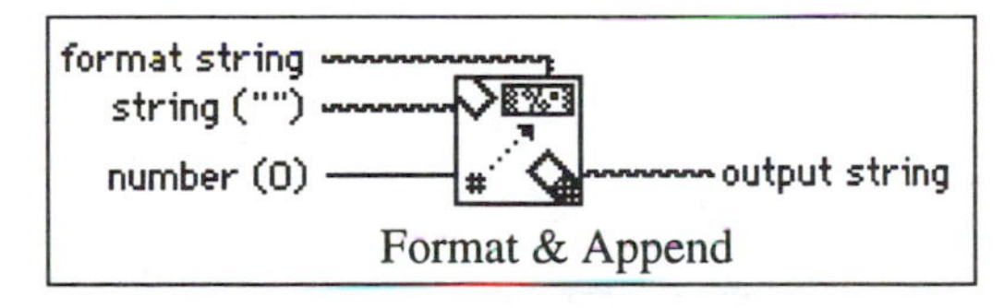

Format & Append

Format & Append formats the input **number** as a string, according to the specifications in **format string**. These specifications are listed in the Function Reference section. It appends the converted string to the input wired to **string**, if there is one. The following table gives some examples of **Format & Append's** behavior. In these examples, the underline character (_) represents a terminating space character. "%f" formats the input number in as a floating-point number with fractional format, "%d" formats it as a decimal integer, and "%e" formats it as a floating-point number with scientific notation.

string	**format string**	**number**	**output string**
(empty)	score=%2d%%	87	score=87%
score= _	%2d%%	87	score=87%
(empty)	level=%7.2e V	0.03642	level=3.64E–2 V
(empty)	%5.3f	5.67 N	5.670 N

Get Date/Time String outputs **date string**, which contains the current date, and **time string**, which contains the current time. This function is useful for time stamping your data. Note that you don't have to wire any of the inputs to **Get Date/Time String** unless you need to use a value other than the default.

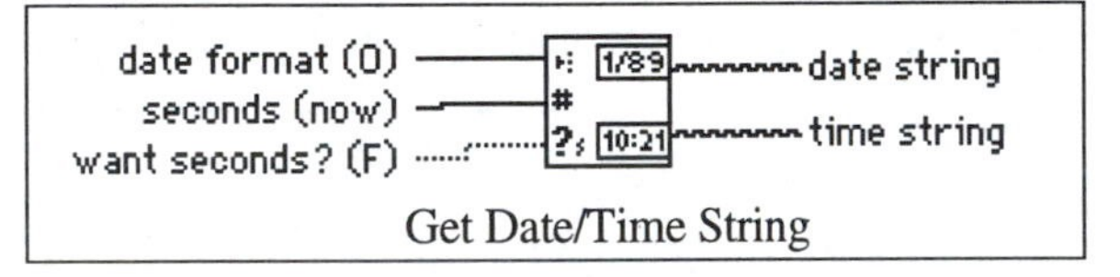

Get Date/Time String

EXERCISE 9.1
String Construction

You will build a VI that converts a number to a string and concatenates that string with other strings to form a single output string. The VI also determines the length of the output string.

Front Panel

1. Open a new front panel and build the front panel shown.

Header
SET
string control

Combined String
SET 5.5000 VOLTS
string indicator

Number
5.50
digital control

Length of String
16
digital indicator

Trailer
VOLTS
string control

The VI will concatenate the input from the two string controls and the digital control into a single output string and display the output in the string indicator. The digital indicator will display the string length.

Block Diagram

1. Build the block diagram shown.

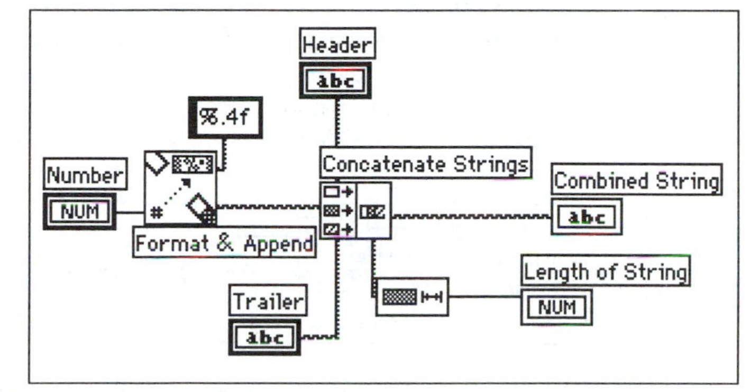

Format & Append function (**String** menu) converts the number you specify in the Number digital control to a string.

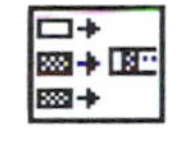

Concatenate Strings function (**String** menu) combines all input strings into a single output string. To increase the number of inputs, stretch the icon using the Positioning tool.

String Length function (**String** menu) returns the number of characters in the concatenated string.

2. Return to the front panel and type text inside the two string controls and a number inside the digital control. Make sure to add spaces at the end of the header and at the beginning of the trailer strings, or your output string will run together. Run the VI.
3. Save and close the VI. Name it **Build String.vi** and place it in `MYWORK.LLB`.

~ ~ ~

Parsing Functions

String Subset returns the substring beginning at **offset** and containing **length** number of characters. Remember, the first character's offset is zero.

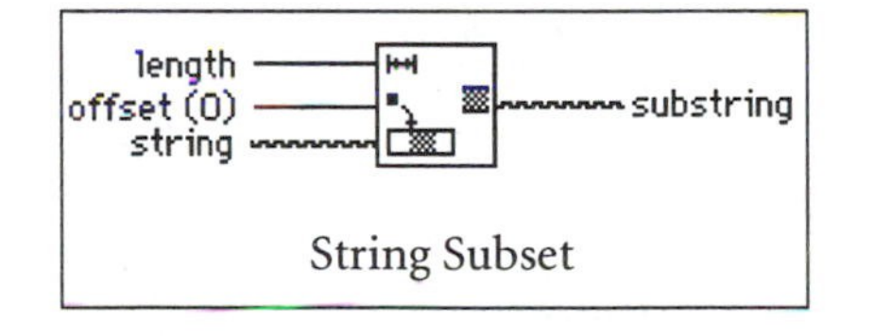

String Subset

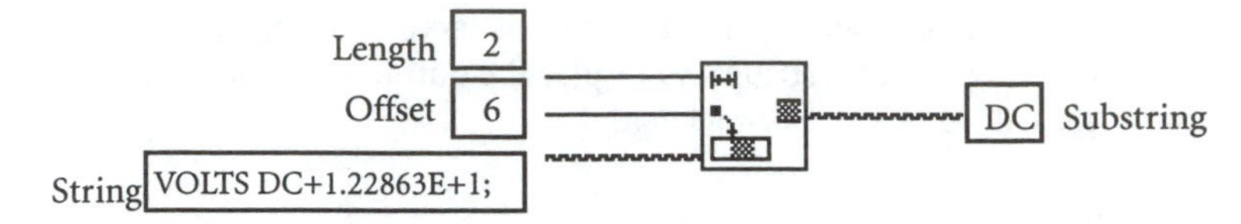

Format & Strip searches at the beginning of **string** for a pattern that matches **format string.** It then converts any number characters in this section of **string** to numeric format, according to the specifications in **format string.** The function outputs the converted numeric data in **number** and any leftover string data (such as a trailer string) in **output string.** If a match between **string** and **format string** is not found, the function will return the value specified in **default.** More information on **Format & Strip** is available in the Function Reference section.

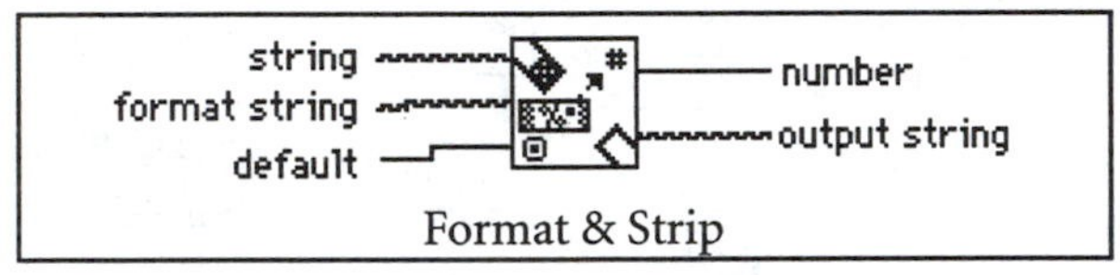

Format & Strip

string	format string	number	output string
Q+1.279E3 trailer	Q%f tr	1279.00	ailer
set49.4.2	set%d	49	.4.2

EXERCISE 9.2
More String Parsing

You will create a VI that parses information out of a longer string—it takes a subset of a string and converts the numeric characters in that subset into a numeric value.

Front Panel

1. Build the front panel shown.

Input String
VOLTS DC +1.345E+02
Subset Offset
6
Subset Length
2
String Subset
DC
Number Offset
9
Number
134.50

Block Diagram

2. Create the block diagram shown.

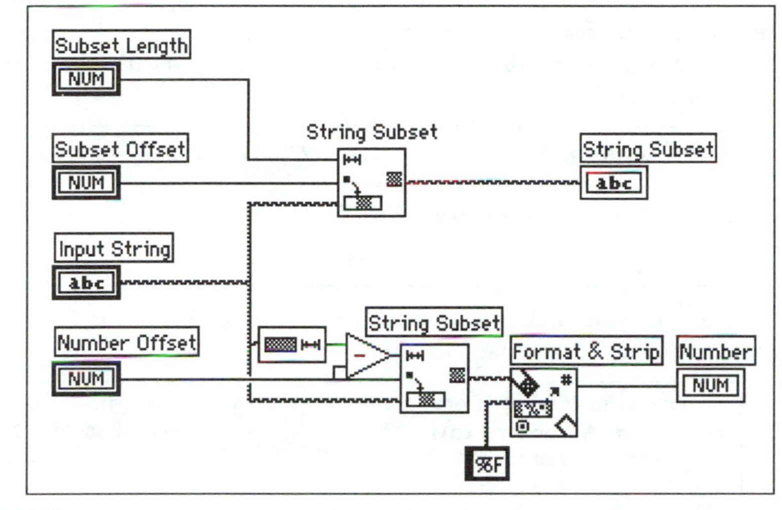

The **Format & Strip** function, located in the **String** palette of the **Functions** menu, converts a string containing valid numeric characters (0 to 9, +, –, e, E, and period) to a number.

3. Run the VI with the default inputs. Notice that the string subset of DC is picked out of the input string. Also notice that the numeric part of the string was also parsed out and converted to a number. You can try different control values if you want (remember that strings, like arrays, are indexed starting at zero).
4. Close the VI by selecting **Close** from the **File** menu. Save the VI in `MYWORK.LLB` as **Parse String.vi**.

9.2 File I/O

File input and output operations store and retrieve information from a disk file. LabVIEW supplies you with simple functions that take care of almost all aspects of file I/O. These functions are located in the **File & Error** palette of the **Functions** menu.

How They Work

The File functions expect a file path input. If you don't wire a file path, the function will pop up a dialog box asking you to select or enter a filename. When called, the File functions open or create a file, read or write the data, and then close the file. The created files are just ordinary text files. Once you have written data to a file, you can open the file using any word processing program to see your data.

One very common application for saving data to file is to format the text file so that you can open it in a spreadsheet program. In most spreadsheets, tabs separate columns and EOL (End of Line) characters separate rows. **Write To Spreadsheet File** and **Read From Spreadsheet File** deal with files in spreadsheet format.

Write Characters To File writes a character string to a new file or appends the string to an existing file.

Write Characters To File

Read Characters From File reads a specified number of characters from a file beginning at a specified character offset.

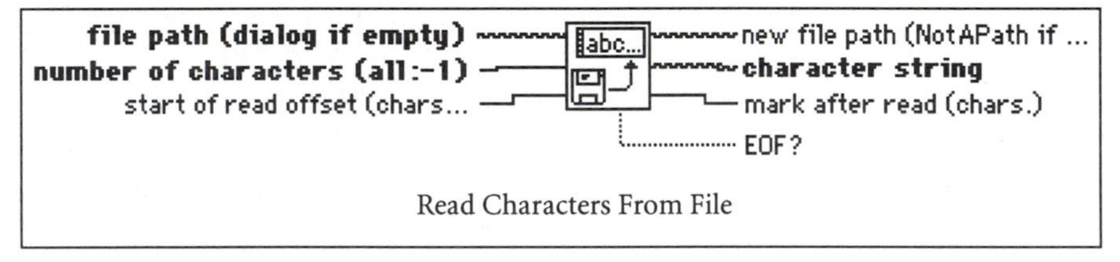

Read Characters From File

Read Lines From File reads a specified number of lines from a file beginning at a specified character offset.

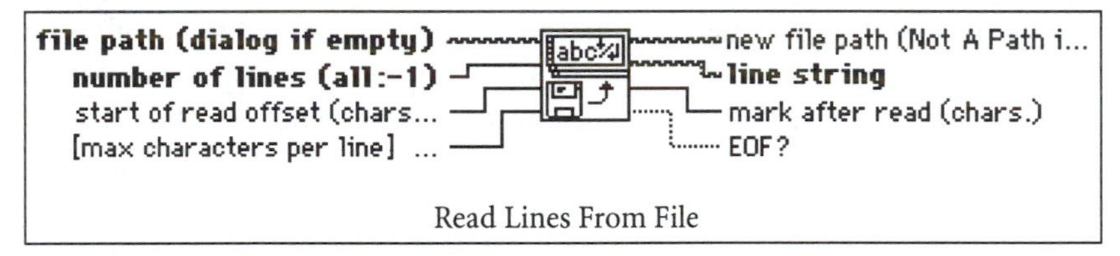

Read Lines From File

Write To Spreadsheet File converts a 2D or 1D array of single-precision numbers to a text string and writes the string to a new file or appends the string to an existing file. You can optionally transpose the data. Do not wire inputs for both 1D and 2D data (or one will be ignored). The text files created by this VI are readable by most spreadsheet applications.

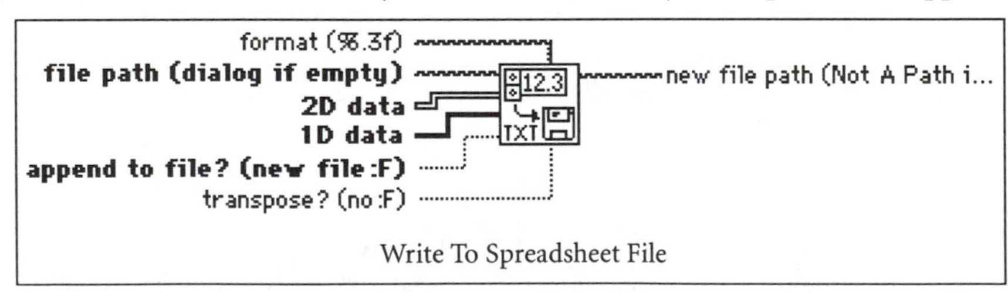

Write To Spreadsheet File

Read From Spreadsheet File reads a specified number of lines or rows from a numeric text file beginning at a specified character offset and converts the data to a 2D single-precision array of numbers. You can optionally transpose the array. This VI will read spreadsheet files saved in text format.

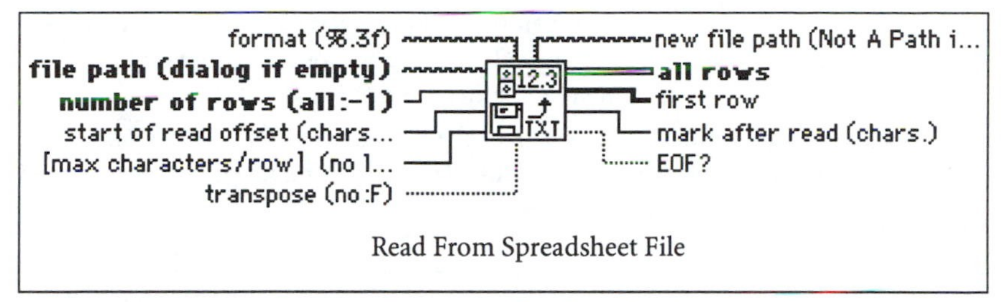

Read From Spreadsheet File

All file functions are described in more detail in the Function Reference section.

Macintosh only. **Write To HiQ Text File** and **Read From HiQ Text File**, also located in the **File** palette, function similarly to the spreadsheet file VIs. They are useful if you need to import data from or export data to HiQ analysis software.

EXERCISE 9.3
Writing to a Spreadsheet File

You will modify an existing VI to save data to a new file in ASCII format. Later you can access this file from a spreadsheet application.

1. Open the **Graph Sine Array.vi** you built in Chapter 8. As you recall, this VI generates two data arrays and plots them on a graph. You will modify this VI to write the two arrays to a file where each column contains a data array.
2. Open the diagram of **Graph Sine Array.vi** and modify the VI by adding the diagram code inside the oval.

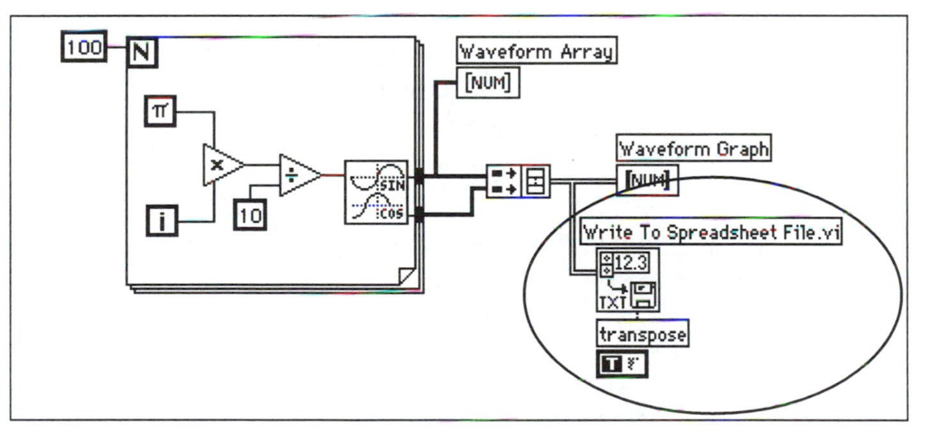

The **Write To Spreadsheet File VI** (**File & Error** palette) converts the two-dimensional array to a spreadsheet string and writes it to a file. If no path name is specified, then a file dialog box will pop up and prompt you for a file name.

The **Boolean Constant** (**Structs & Constants** palette) controls whether or not the two-dimensional array is transposed before it is written to file. To change it to TRUE, click on the constant with the Operating tool. In this case, you do want the data transposed because the data arrays are row specific (each row of the two-dimensional array is a data array). Since you want each column of the spreadsheet file to contain a data array, the two-dimensional array must first be transposed.

3. Return to the front panel and run the VI. After the data arrays have been generated, a file dialog box will prompt you for the file name of the new file you are creating. Type in a file name and click the OK button.

Watch Out!
Do not attempt to write data in VI libraries such as `MYWORK.LLB`. Doing so may result in overwriting your library and losing your previous work.

---

4. Save the VI in `MYWORK.LLB`, name it **Graph Sine Array to File.vi**, and close the VI.
5. You now can use spreadsheet software or a simple text editor to open and view the file you just created. You should see two columns of 100 elements.

EXERCISE 9.4
Reading from a Spreadsheet File

You will read in the data written in the last exercise and plot it on a graph.

1. Open a new VI and place a waveform graph on its front panel.
2. In the block diagram, shown below, use the **Read From Spreadsheet File** function to bring in data and display it on the graph.

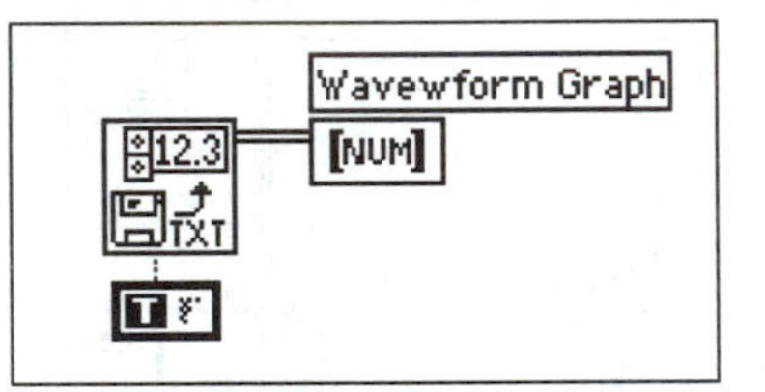

3. Using the TRUE **Boolean Constant**, you must transpose the array when you read it in, because graphs plot data by to row.
4. Since you are not providing a file path, a dialog box will prompt you to enter a filename. Select the file you created in Exercise 9.2.
5. The VI will read the data from the file and plot both waveforms on the graph.
6. Save the VI in `MYWORK.LLB` as **Read File.vi**.

9.3 Wrap it Up!

LabVIEW contains many functions for manipulating strings. These functions can be found in the **String** palette of the **Functions** menu. With them, you can determine string length, combine two strings, peel off a string subset, and convert a string to a number (or vice versa).

Using the functions in the **File & Error** palette of the **Functions** menu, you can write data to or read data from a disk file. **Write Characters To File** will save a text string to a file. **Read Characters From File** and **Read Lines From File** can then read that file back into LabVIEW. If you want to save an array of numbers, you must use the **Write To Spreadsheet File** function. You can read that data back in and convert it to numeric format using the **Read From Spreadsheet File** function.

9.4 Additional Exercises

EXERCISE 9.5
Temperatures and Time Stamps

Build a VI that takes a 50 temperature readings, once every 0.25 second, inside a loop and plots each on a chart. It also converts each reading to a string, then concatenates that string with a Tab character, a time stamp, and an End of Line character. The VI writes all of this data to a file. Save the VI as **Temperature Log.vi.**

Hints:

- Use the **Tab** and **End of Line** constants in the **Structs & Constants** palette.
- Use **Concatenate Strings** to put all of the strings together.
- Use **Write Characters To File** to save the data.

Extra Hint: You can write data to file one line at a time, but it is much faster and more efficient to collect all of the data in one big string using shift registers and **Concatenate Strings**, and then write it all to file at one time.

You can look at your file using any word processing program, but it should look something like this:

```
78.9 ▶    11:34:48
79.0 ▶    11:34:49
79.0 ▶    11:34:50
```

EXERCISE 9.6
Spreadsheet Exercise

Build a VI that generates a 2D array (3 rows x 100 columns) of random numbers and writes the *transposed* data to a spreadsheet file. The file should contain a header for each column as shown below. Use the VIs from the **File** palette for this exercise. Save the VI as **Spreadsheet Exercise.vi.**

Hint: Use the **Write Characters To File** VI to write the header and then the **Write to Spreadsheet File** VI to write the numerical data to the same file.

	A	B	C
1	Waveform 1	Waveform 2	Waveform 3
2	0.281	0.078	0.874
3	0.402	0.647	0.597
4	0.011	0.62	0.731
5	0.605	0.435	0.889
6	0.049	0.259	0.78
.			
.			
.			
99	0.89	0.933	0.54
100	0.864	0.312	0.343
101	0.541	0.134	0.487

— Header

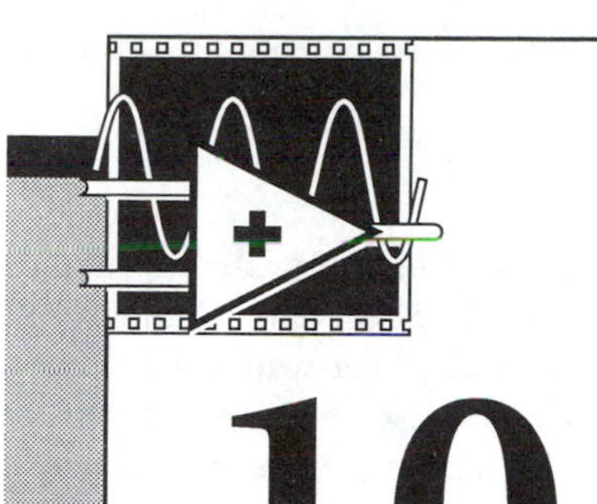

10

Miscellaneous Concepts

Overview

This chapter covers a few important LabVIEW topics that don't quite fit in any other section. You will learn how to configure the appearance and behavior of VIs and subVIs, how to print from LabVIEW, and how to specify the amount of memory you want LabVIEW to use.

Your Goals

- Understand how to use VI Setup options and SubVI Node Setup options and the difference between them
- Be able to configure a subVI to pop up its front panel while it executes and then close the panel when it completes
- Know how to safeguard your VI by hiding Execution palette buttons
- Learn how to print from LabVIEW, both manually and programmatically, and set printing preferences

Key Terms

VI Setup
SubVI Node Setup
reentrant

10.1 Configuring Your VI

Sometimes you may want to set options that affect the execution of a VI or the appearance of its windows. You can configure a VI's setup in two places: *VI Setup* options and *SubVI Node Setup* options. While some of the these option do overlap, you will notice a very important behavioral difference between VI Setup and SubVI Node Setup. *Options set under VI Setup affect every instance of the VI, while SubVI Node Setup options affect only a single subVI call.*

SubVI Node Setup Options

There are several setup options on a subVI node that you can modify. These options are available by popping up on the subVI icon (in the block diagram of the calling VI), and choosing **SubVI Node Setup** from the pop-up menu. They affect only that subVI node—other calls to the subVI will not be affected. A dialog box showing the setup options appears, as shown.

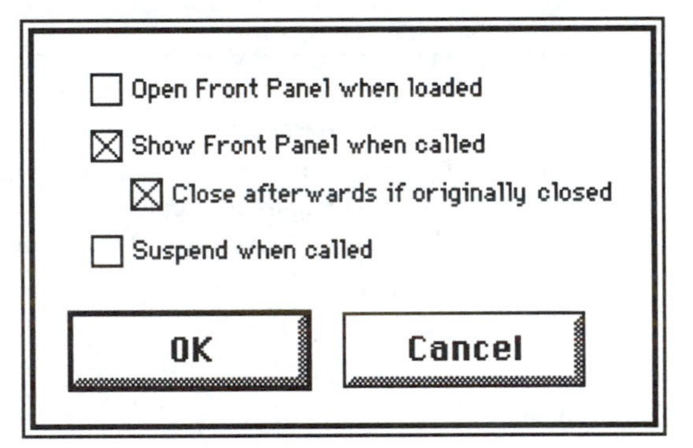

The options in this dialog box include:

- **Open Front Panel when loaded**—If selected, the VI front panel pops open when the VI is *loaded* into memory as a subVI or a main VI.
- **Show Front Panel when called**—If selected, the VI front panel pops open when the VI is *executed* as a subVI.
- **Close afterwards if originally closed**—If **Show Front Panel when called** is selected, the VI front panel pops open when the VI is executed as a subVI. Selecting **Close Afterwards if originally closed** causes the VI front panel to close when VI execution is complete, giving a "pop-up window" effect.
- **Suspend when called**—If selected, VI execution suspends when the subVI is called. This option has the same effect as setting a breakpoint.

EXERCISE 10.1
Pop-Up SubVIs

You will build a VI that acquires a temperature once every 0.5 second for 10 seconds. After the acquisition is complete, the VI pops open another front panel and plots the acquired data in a graph. This pop-up panel remains open until you click on a Boolean button.

First, you must build a VI that pops open its front panel, displays the graph, and waits until you click on a Boolean button. You then will use this VI as a subVI in the block diagram of the VI that acquires the temperature.

Front Panel

1. Open a new panel and build the front panel shown. Be sure to modify the controls and indicators as depicted.

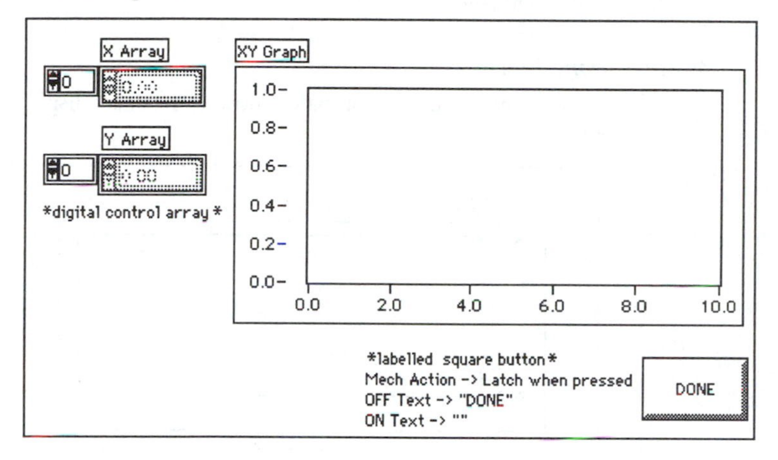

Block Diagram

2. Build the block diagram shown.

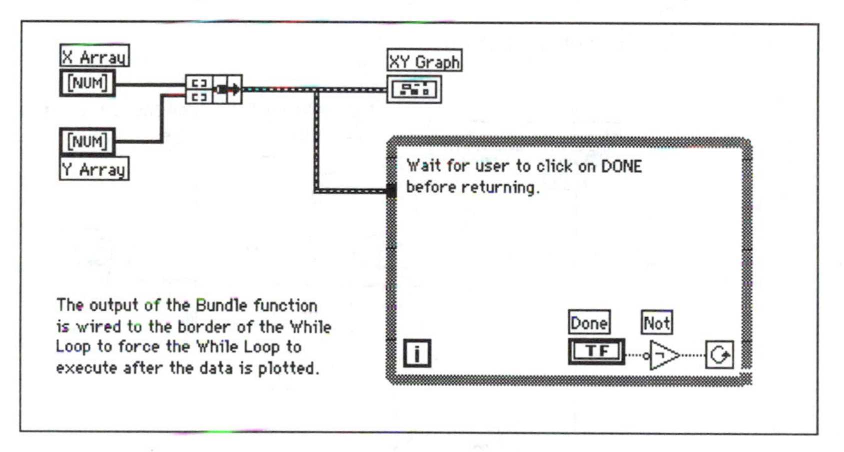

Bundle function (**Array & Cluster** menu). In this exercise, bundles the data passed by the calling VI so the VI can plot the data on an XY graph.

Not function (**Arithmetic** menu). In this exercise, inverts the value of the DONE button; thus, the While Loop executes repeatedly until you click on the button. (The button's default state is FALSE.)

3. Create the icon and connector for the VI. If you don't remember how, go back and read Section 5.5, *Creating SubVIs.*

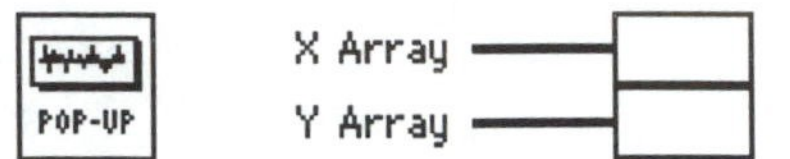

4. Save and close the VI. Name it **Pop-up Graph.vi** and place it in MYWORK.LLB. You will use the VI as a subVI shortly.

Another Front Panel

5. Open a new panel and build the front panel shown. The chart displays the temperature as it is acquired.

Temperature
90.0
85.0
80.0
75.0
70.0
waveform chart

Another Block Diagram

6. Build the block diagram shown.

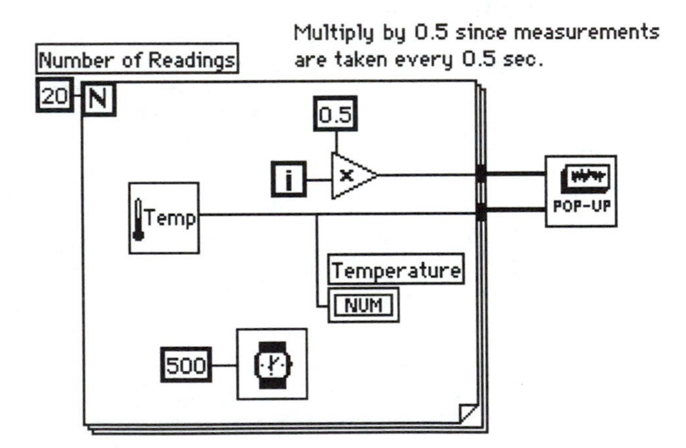

POP-UP

Pop-up Graph VI (**VI...** menu). Pops open its front panel and plots the temperature array against the time array.

7. Configure the **Pop-up Graph** subVI to pop open its front panel when called by the higher-level VI. Pop up on its icon and choose **SubVI Node Setup** from the pop-up

menu. Configure the dialog box to **Show Front Panel when called** and **Close afterwards if originally closed**.

8. Save the VI in MYWORK.LLB. Name it **Use Pop-up Graph.vi**.
9. Run the VI. After the VI acquires the temperature data, the front panel of **Pop-up Graph** pops open and plots the temperature data. Click on DONE to return to the calling VI.
10. Close all windows.

VI Setup Options

You access VI Setup options, another way to configure your VIs, by popping up on the Icon Pane in the upper-right corner of the Panel window and choosing **VI Setup** from the pop-up menu. As illustrated in the following figure, a dialog box appears showing all setup options. To set any option, click in the square next to that option. An X appears in the box by the selected option. To clear an option, click on the X next to that option and the X will disappear. Options set under VI Setup affect all calls to the VI.

- **Show Front Panel When Loaded** – If selected, the front panel of the VI opens *when you load the VI into memory*, even if the VI is a subVI.
- **Show Front Panel When Called** – If selected, the VI front panel opens when the VI *executes* as a subVI.

Note
If you select "Show Front Panel When Called" from the VI Setup menu of VI "xyz," then xyz's front panel pops open any time xyz is called as a subVI. This option affects the execution of any VI that uses xyz as a subVI. If you select "Show Front Panel When Called" from the SubVI Node Setup menu, then the front panel of xyz opens only if that node on that diagram is executed. This option does not affect the execution of other VIs that use xyz as a subVI.

- **Close Afterwards if Originally Closed**—If you select **Show Front Panel When Called**, the VI front panel pops open when the VI executes as a subVI. Selecting **Close Afterwards if Originally Closed** causes the VI front panel to close when the VI execution is complete. You can use these two options to configure a "pop-up" window during subVI execution, just like in SubVI Node Setup.
- **Run When Opened**—If selected, the VI runs automatically whenever you open it. You do not need to click on the run button.
- **Suspend When Called**—Selecting this option is the equivalent of clicking on the breakpoint button on the Execution palette. A subVI will suspend execution when it is called.
- **Reentrant Execution**—If you plan to use the VI as a subVI and make multiple calls to it, you must be careful that the calls don't share the same data and thus corrupt each other. If a VI is reentrant, it will create its own separate data space each time it runs. This way you can run the same VI several times at once.
- **Print Panel When VI Completes Execution**—When this option is checked, a VI's front panel will print when it finishes running. You also have options to **Print Header, Scale to Fit** on a page, and **Surround Panel with Border.**

10.2 Printing

You can print from LabVIEW both manually and programmatically. Manual printing can print either all or specified parts of a VI, while programmatic printing prints only the appropriate front panel.

If you just want to print the active window, you can select **Print Window** from the **File** menu. To manually print all parts of a VI, select **Print Documentation** from the **File** menu. You will see the following dialog box.

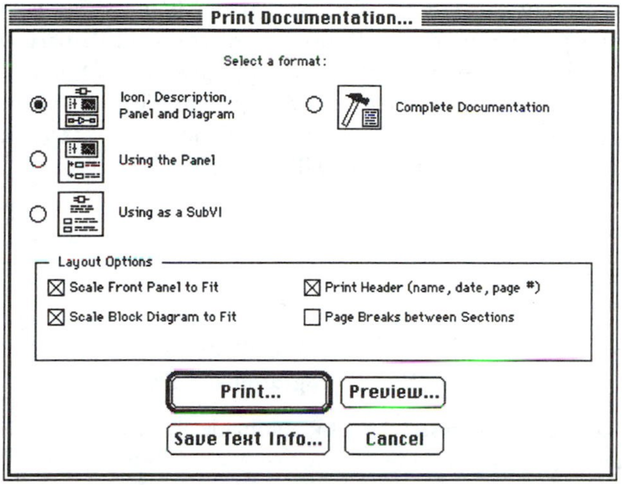

Print Documentation...

Select a format:

Icon, Description, Panel and Diagram

Complete Documentation

Using the Panel

Using as a SubVI

Layout Options

Scale Front Panel to Fit

Print Header (name, date, page #)

Scale Block Diagram to Fit

Page Breaks between Sections

Print... Preview...

Save Text Info... Cancel

From this dialog box, you can specify which parts of the VI you want to print and determine page layout options. You can also **Preview** what your printout will look like.

If you select **Preferences** from the **Edit** menu, you can choose between **Standard printing** and **Postscript printing**. Under Windows, you have a third option of **Non-bitmap printing**. If you select **PostScript printing**, you have the option to select **Level 2 PostScript**, which can support color printing.

Programmatic Printing

You can enable programmatic printing by clicking on the **Print mode** button in the Run mode execution palette or by selecting **Print Panel When VI Completes Execution** from the VI Setup dialog box. When enabled, the button changes to . If programmatic printing is on, a VI will print its front panel, including all of the visible data, after the VI has finished executing. If you only want to print a certain part of a front panel, you must send that data to a subVI's front panel and enable programmatic printing on the subVI. This method also lets printing occur while the main VI is still executing.

10.3 Wrap It Up!

You can modify VI execution characteristics with LabVIEW's setup options. These modifications include running the VI automatically when it loads, opening the front panel when a VI is called, specifying reentrant execution, and so on. You can modify some of these options in two ways: using the VI Setup pop-up dialog box or using the SubVI Node Setup pop-up dialog box.

Any execution characteristic of a VI modified using the VI Setup pop-up dialog box will affect *every* execution of that VI as a main VI or as a subVI. Any execution characteristic of a VI modified using the SubVI Node Setup pop-up dialog box will affect only that node. Other nodes of the same VI will not be affected.

You can print all or specified parts of a VI manually, or you can send a front panel to the printer programmatically. To print all or specified VI parts, select **Print Documentation** from the **File** menu. If you just want to print a front panel, you can enable the **Print mode** button in the run mode execution palette. The VI will then send its front panel to the printer each time it finishes execution. If you just want to print the active window, you can select **Print Window** from the **File** menu.

10.4 Additional Exercises

EXERCISE 10.2
Programmatic Printing

Modify **Graph Sine Array.vi** so that it automatically prints its front panel when it has finished executing.

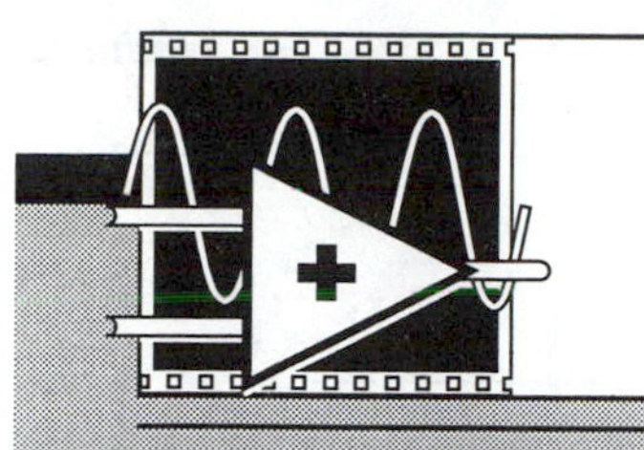

Examples

This chapter describes the software examples that ship with the *LabVIEW Student Edition.* You will find them valuable sources of information on wiring, structure, and programming techniques. Open them up, glance through them, and run them—you will learn a lot about how LabVIEW works. You can modify and run all of these examples as applications, although you must have the proper hardware configured and installed to use the DAQ and GPIB VIs. The examples are located in the EXAMPLES folder or directory, which is located inside the LabVIEW Student Edition folder or directory.

Application Examples

Frequency Response.vi

path: EXAMPLES\APPS\FREQRESP.LLB

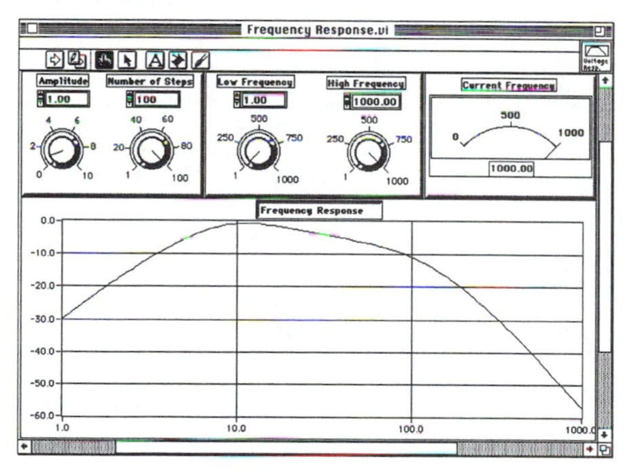

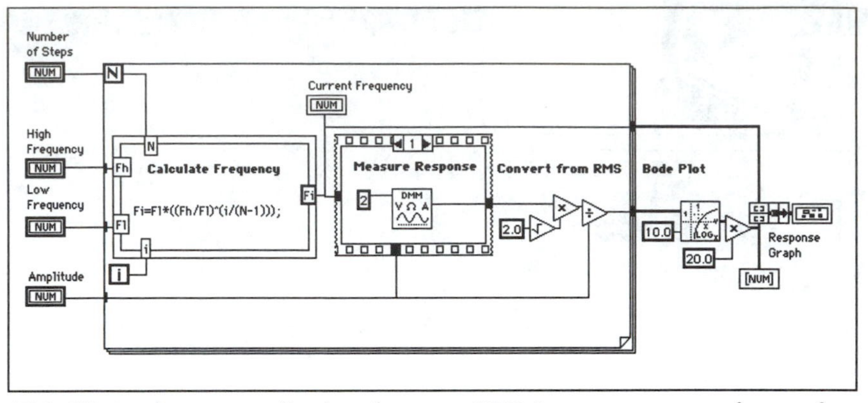

This VI simulates an application that uses GPIB instruments to perform a frequency response test on a "black box" of unknown content. A function generator supplies a sinusoidal input to the "black box" (a bandpass filter in this example), and a digital multimeter measures the output voltage of the box. You can determine the amplitude of the input sine wave and the number of steps the VI uses to find the frequency response by changing the Amplitude control and the Number of Steps control. You can also determine the frequency sweep by specifying the upper and lower limits using the Low Frequency and High Frequency knobs.

The LabVIEW concepts of the For Loop, the Formula Node, the graph, and arrays are shown in this example.

Temperature System Demo.vi

path: EXAMPLES\APPS\TEMPSYS.LLB

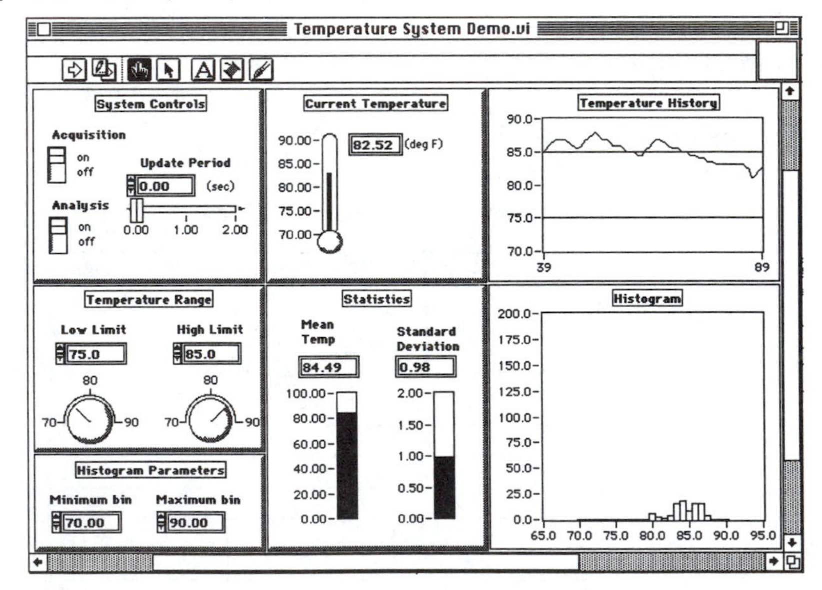

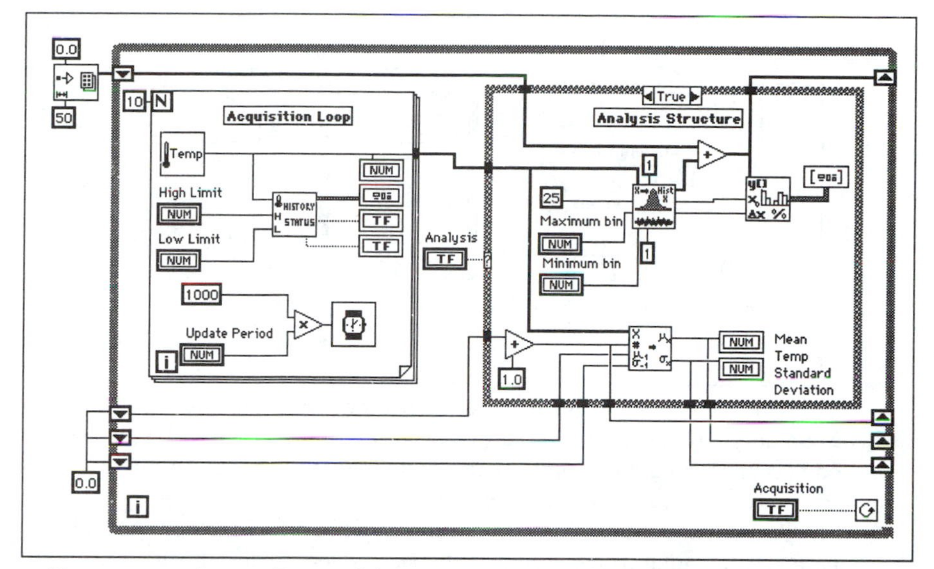

Temperature System Demo.vi demonstrates a temperature analysis application. This VI reads a simulated temperature, alarms if it is outside a given range, and performs a statistical mean, standard deviation, and histogram of the temperature history. The temperature, along with the input High and Low Limits, are plotted on a chart. The Update Period slide control determines how fast the VI acquires new temperature readings.

Array Examples

Building Arrays.vi

path: EXAMPLES\GENERAL\ARRAYS.LLB

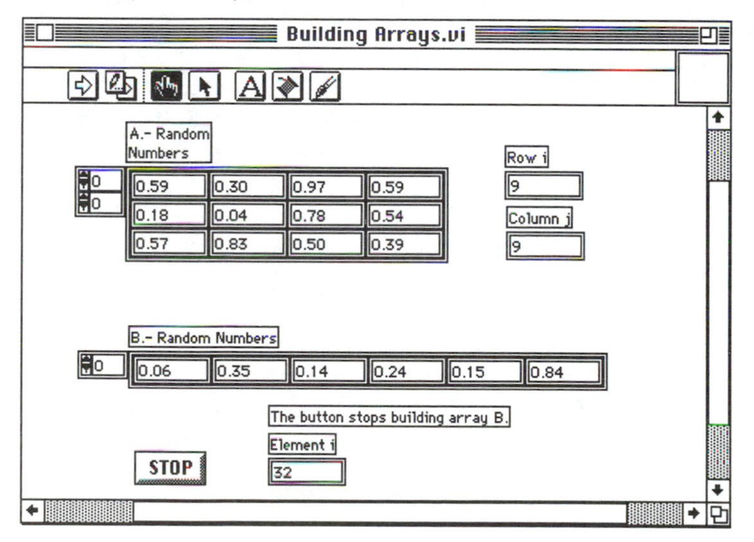

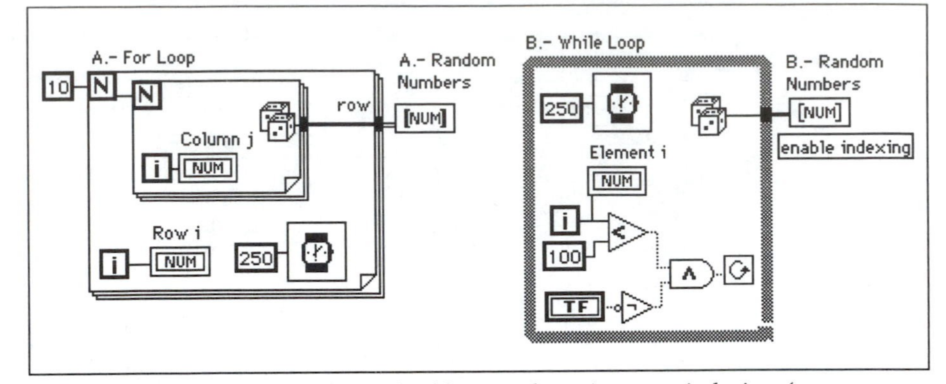

This VI shows two methods to build arrays by using auto-indexing (a pop-up menu option on the tunnel). The For Loop generates a 10 rows by 10 columns 2D array. The While Loop generates a random number either 100 times or until the user pushes STOP; it creates an array of these numbers when it completes.

A. The For Loop has indexing enabled by default. The method shown is the best method to use for generating arrays when the number of values in the array is known. Wire the variable inside the loop subdiagram directly to an array terminal. To create a 2D array, include one loop inside another one. The inner loop creates the column elements, while the outer loop creates the row elements.

B. The While Loop has indexing disabled by default. It is useful for array generation when the number of array values is unknown, such as when the user or a program variable determines the size of the array. Wire the variable inside the loop subdiagram directly to an array terminal, then pop up on the tunnel (the black box on the loop border) and select **Enable Indexing** from the pop up menu. As implemented here, the While Loop may overrun the original array buffer created in memory (since it doesn't know how long it's going to execute or how large the array will be) and have to make one or more new memory buffers. The While Loop shown here would be more efficient if a set of shift registers were used to hold an initialized array of 100 elements, then the **Replace Array Element** function put the new values into the array.

Separate Array Values.vi

path: EXAMPLES\GENERAL\ARRAYS.LLB

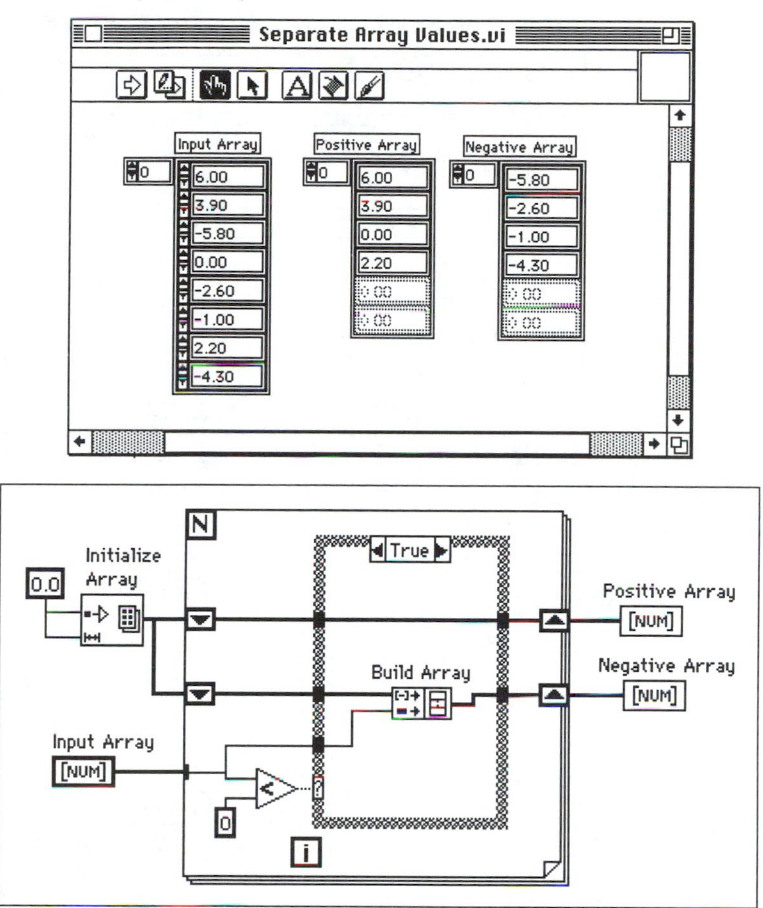

This VI takes an input array that contains a mixture of positive and negative values and separates that array into two smaller arrays—one containing only the negative values and one containing only the positive values.

The **Initialize Array** and **Build Array** functions are used and the For Loop sets its count by auto-indexing on the input array.

This VI could be implemented more efficiently if the **Input Array** initialized the two shift registers, **Replace Array Element** took the place of **Build Array**, and the excess array space was truncated at the end of the VI. This improvement primarily affects large arrays.

Temperature Analysis.vi

path: EXAMPLES\GENERAL\ARRAYS.LLB

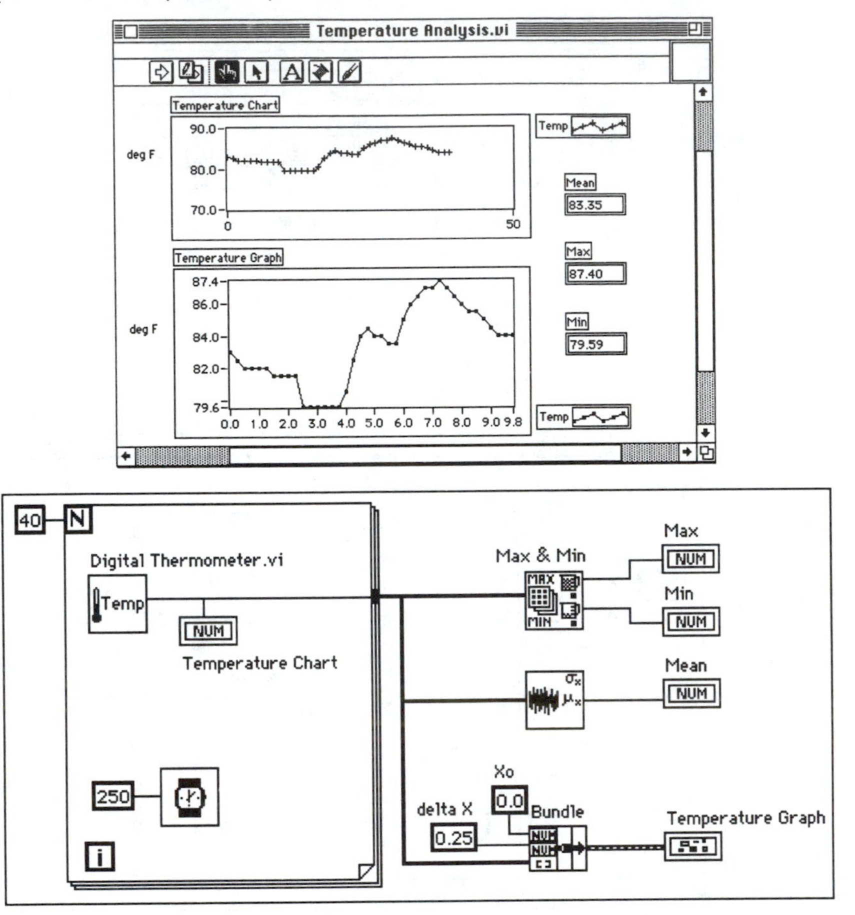

This VI generates an array of temperature data and then calculates the maximum, minimum, and average temperatures. A waveform chart shows the real-time acquisition of the data points, and a waveform graph shows the temperature data after all the points are collected. The temperature array is generated using auto-indexing on the For Loop.

Global Example

One of the characteristics of data flow programming is that structures and subVIs do not output data until they finish executing. Occasionally you may want to pass data between VIs that are not connected by a wire or to pass data into or out of a structure while it executes. You can use a global variable to accomplish this.

A global variable in LabVIEW can be simulated using a While Loop containing an uninitialized shift register, as shown in **Global Switch.vi**. Notice that nothing is wired to the conditional terminal of the While Loop, so it will execute only once.

If the mode is set to "write," the value of the Input control is written to the shift register. If the mode is "read", the content of the shift register is recirculated. In either case, the content of the shift register is written to the Output indicator. Since the shift register is left uninitialized, each time the VI executes, the shift register contains the value left over from the last time the VI executed. If this VI is used as a subVI, it will behave as a global variable.

You should understand three key points in order to use global variables.

1. You must write a value to a global variable before you read from it. If you attempt to read a value from a global variable before you write to it, the output is the default for that data type, which may not be what you and your VI expect or want.
2. You must write to the global variable at a separate location from where you read it. In other words, the same global storage VI cannot be written to and read from the same VI call.
3. For each global value you wish to store, you must create a distinct VI to store the global (similar to **Global Switch.vi**). The easiest way to do this is to make copies of the global variable VI and give them different names.

Global Switch.vi

path: EXAMPLES\GENERAL\GLOBAL.LLB

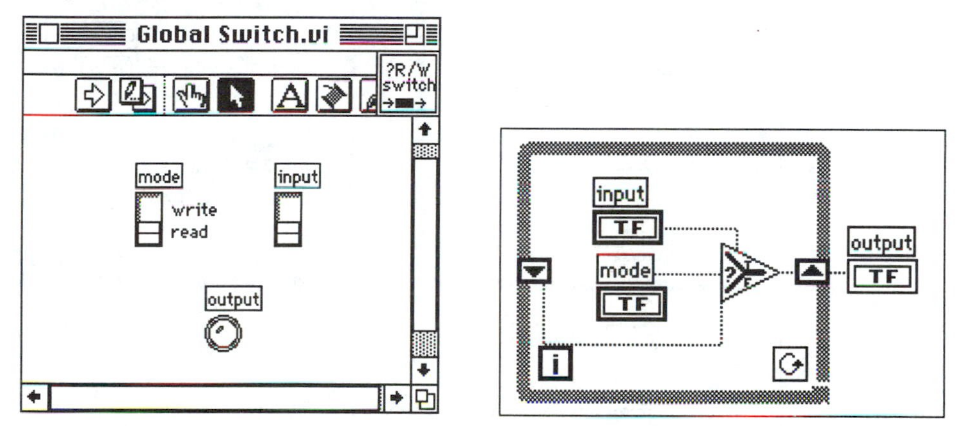

This VI functions as a global Boolean variable. If the mode input is TRUE (write), the VI stores the new input value in an uninitialized shift register. If the mode is FALSE (read), the VI reads and outputs the value stored in the shift register.

Using Globals.vi

path: EXAMPLES\GENERAL\GLOBAL.LLB

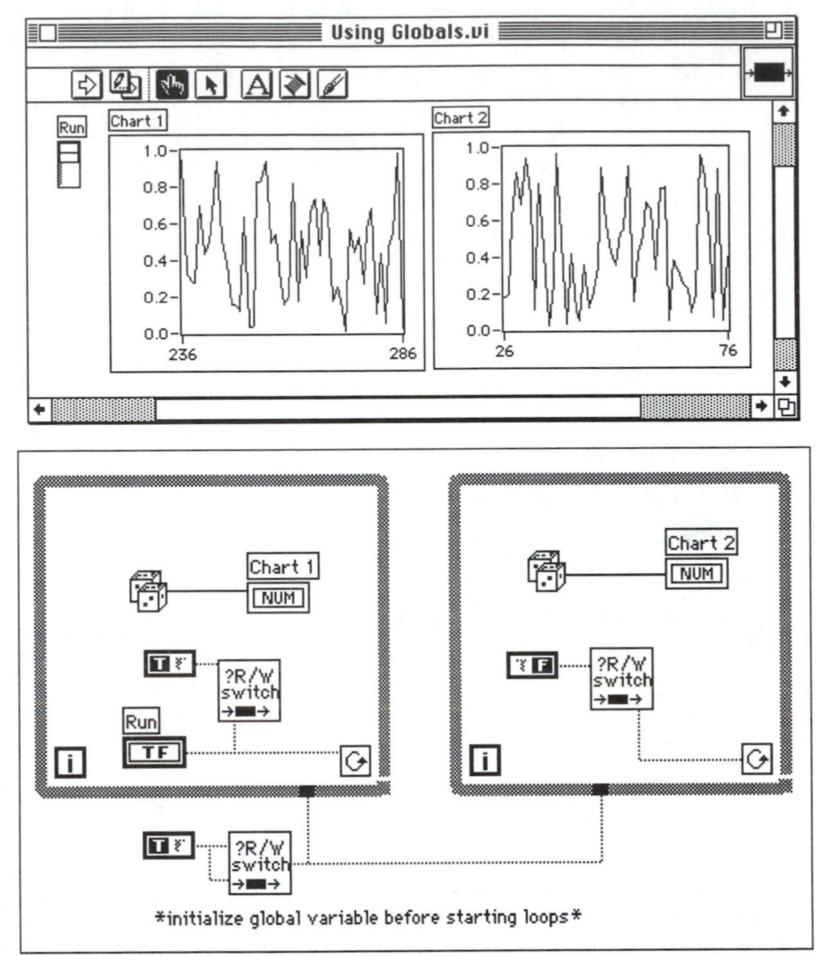

This VI demonstrates the use of a Boolean global variable to control the execution of two charts. The **Global Switch.vi** subVI monitors the Run switch and is used to stop the execution of both While Loops. It is placed outside the two While Loops to initialize the global variable by writing a TRUE to it. The output of the **Global Switch.vi** is wired to the border of the loops to force **Global Switch.vi** to execute before the loops. This technique demonstrates a classic example of artificial data dependency. While the output of the subVI is not really needed by the loops, that output is wired to the loop borders to make LabVIEW think that it must execute the subVI first. The initialization of the global variable prevents the right While Loop from reading an uninitialized global containing invalid data if it happens to execute before the left While Loop.

Graph Examples

Array to Bar Graph Demo.vi

path: EXAMPLES\GENERAL\GRAPH.LLB

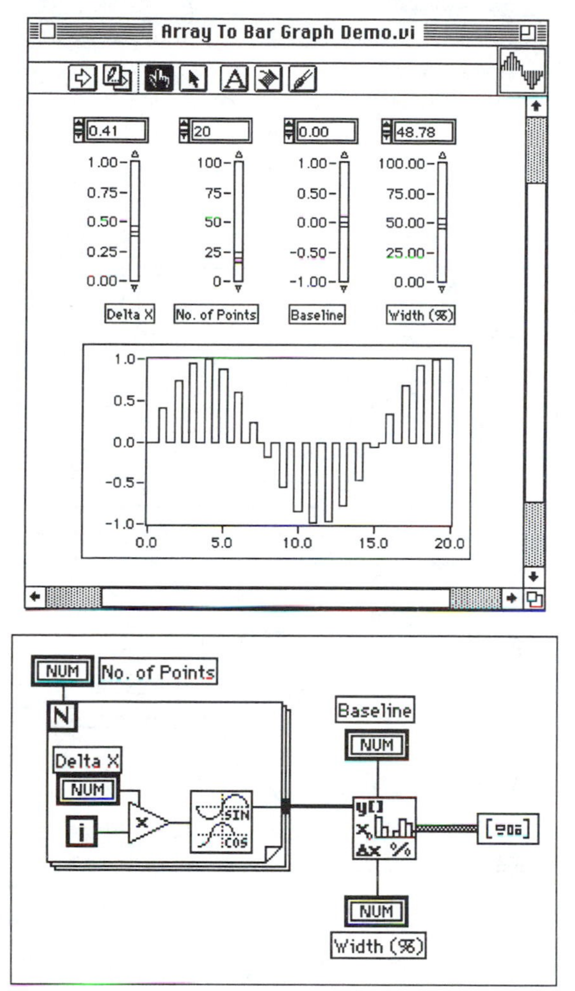

This VI demonstrates how to display an array of values in bar graph form using the **Array to Bar Graph** subVI. This subVI creates a bar graph output for an XY graph given the array of values, the starting point, the distance between points and the percent width. You can copy this VI or subVI to create bar graphs in your own applications.

Charts.vi

path: EXAMPLES\GENERAL\GRAPH.LLB

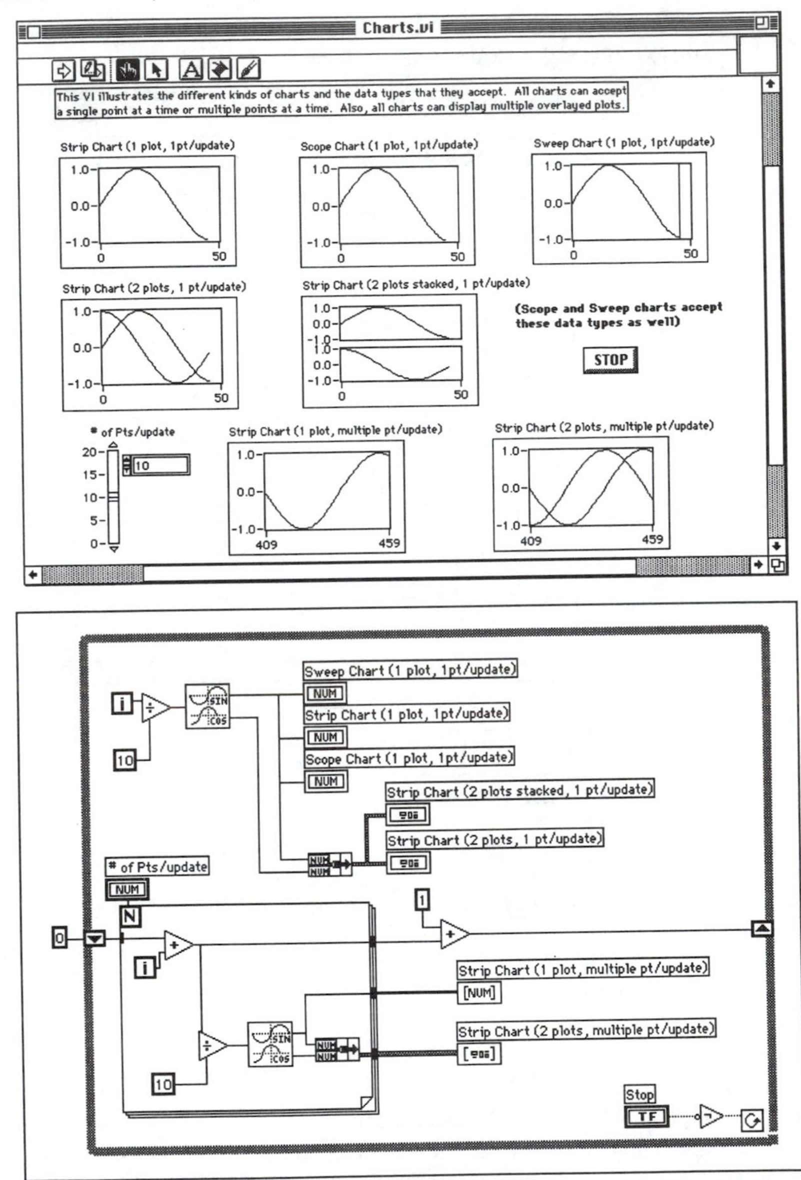

This VI illustrates the different uses of charts and the data types that they can accept. It plots a sine wave (and sometimes a cosine wave) using several different chart configurations. Charts can accept a single point at a time or multiple points at a time. They can also display multiple overlaid plots. Remember that a strip chart is simply a waveform chart in strip mode.

Waveform Graph.vi

path: EXAMPLES\GENERAL\GRAPH.LLB

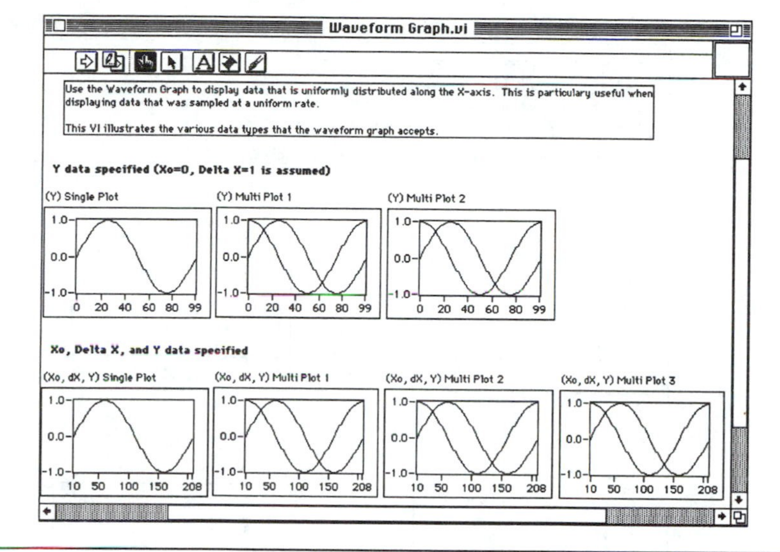

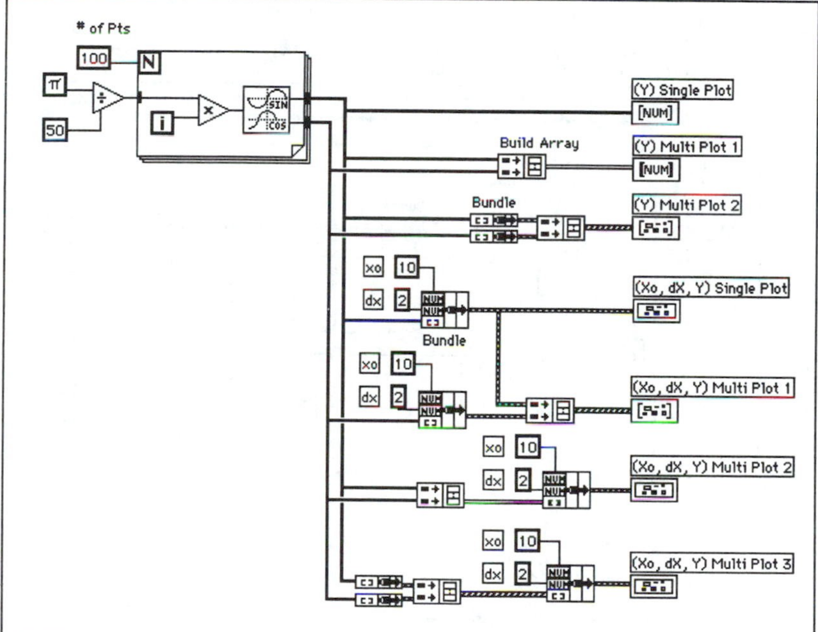

This VI illustrates the various single-plot and multiple-plot data types that the waveform graph accepts. It uses a For Loop with auto-indexing enabled to generate a sine wave and a cosine wave, then plots those waves on graphs using several different methods.

You should use the waveform graph to display data that is uniformly distributed along the X-axis, such as when displaying data that was sampled at a uniform rate. See **X vs. Y Graph.vi** for examples of how to use the XY graph.

X vs. Y graph.vi

path: EXAMPLES\GENERAL\GRAPH.LLB

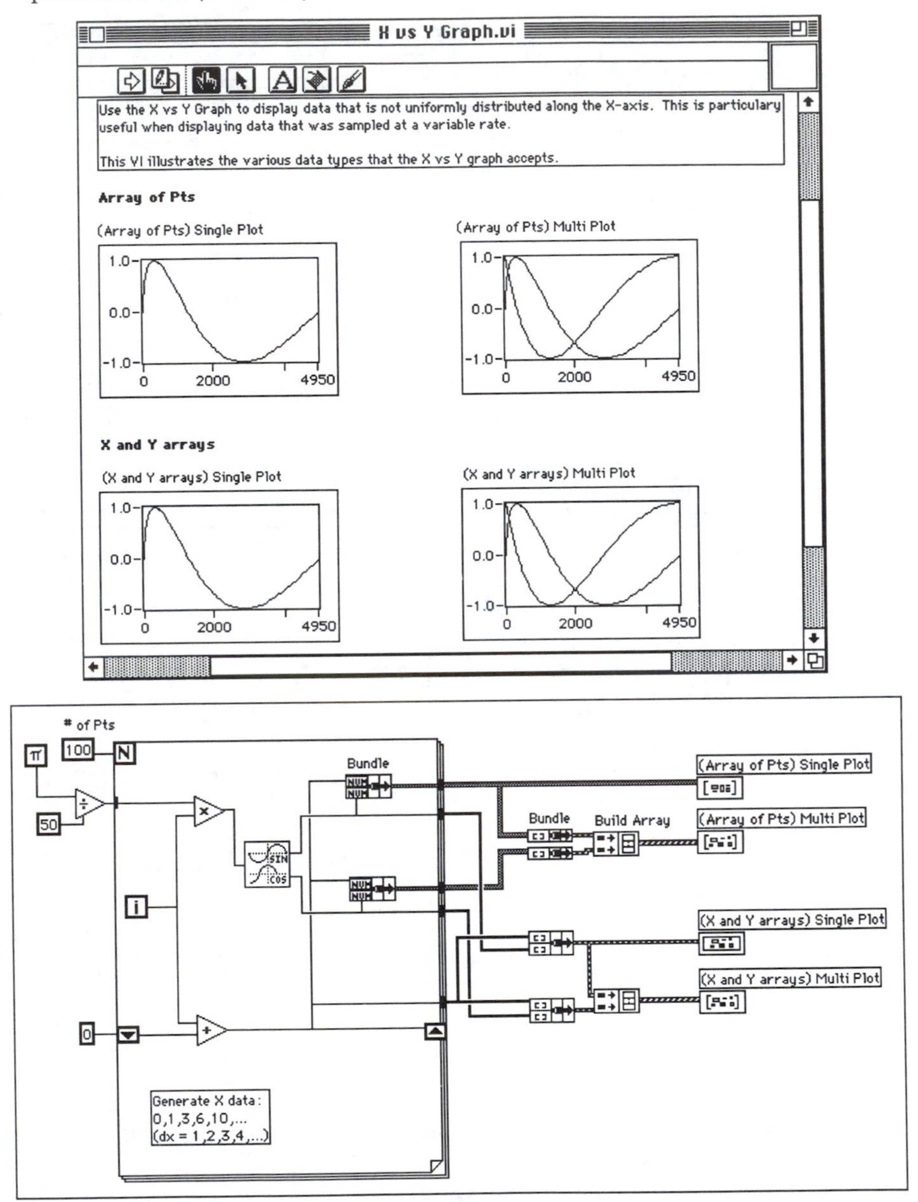

This VI illustrates the various data types that the XY Graph accepts. Use the XY Graph to display data that is not uniformly distributed along the X-axis or to plot complicated math functions. This graph plots points using their (X,Y) coordinates and is particularly useful when you need to display data that was sampled at a variable rate. See the **Waveform Graph.vi** for an example of how to use the waveform graph.

String Examples

Build String.vi

path: EXAMPLES\GENERAL\STRINGS.LLB

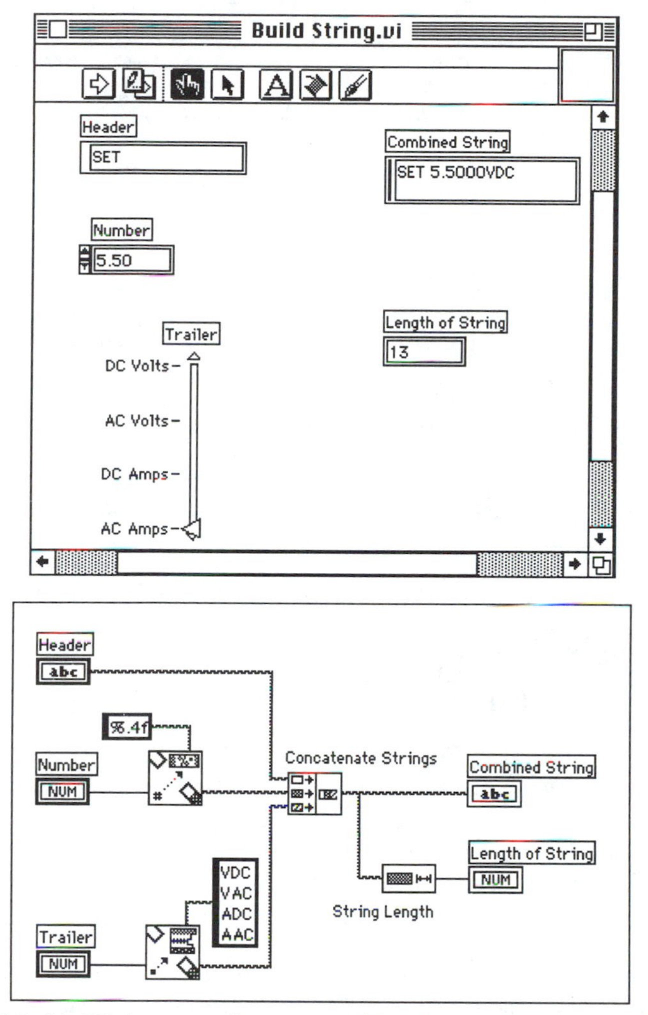

The **Build String** VI shows you how to use the various string functions in LabVIEW to assemble a string combining header information, a numeric value, and trailer information. You must use the **Format & Append** function to convert the number to string format. The conversion codes needed for **Format & Append** are detailed in the Function Reference section. The **Pick Line & Append** function accepts a number input and then chooses the corresponding line from its string input. Then the **Concatenate Strings** function puts all of the component strings together. The total length of this combined string, given by the **String Length** function, is also displayed.

Extract Numbers.vi

path: EXAMPLES\GENERAL\STRINGS.LLB

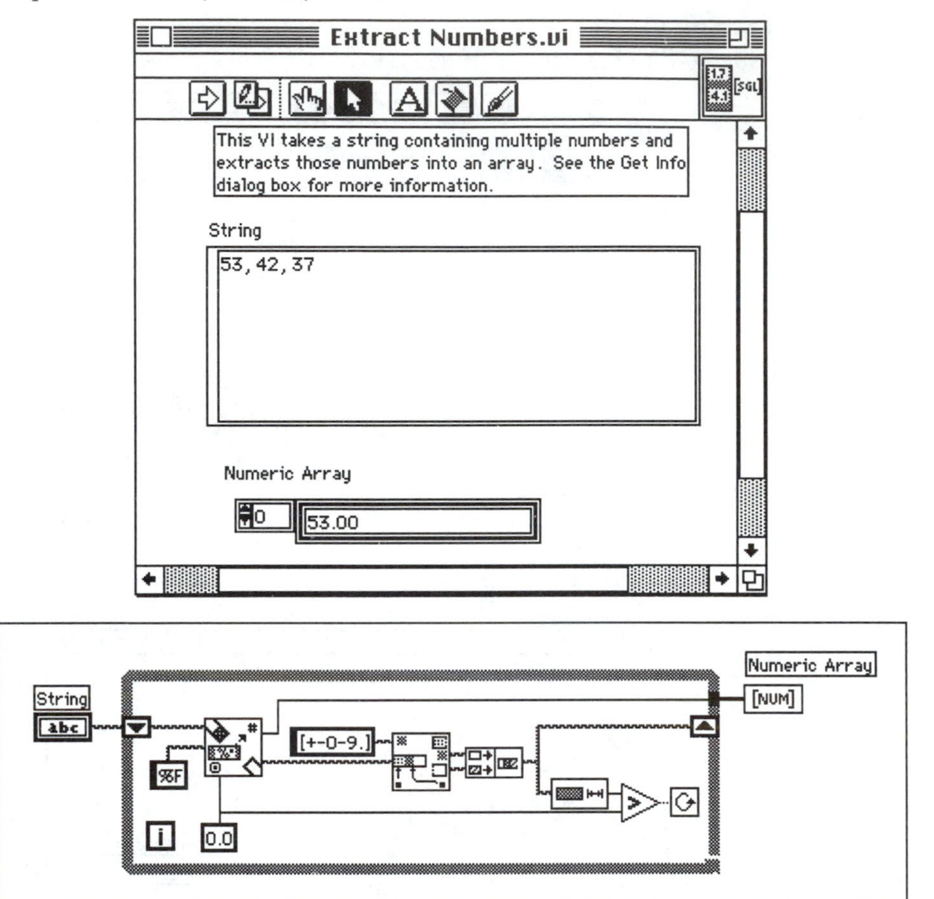

This VI finds all numbers in the given string and puts them into an array of numbers. All of the following formats are recognized.

123 1.23 1.23e2 1.23E2

No spaces or other characters may appear within the number, but any characters may appear before or after the number. This VI is very useful as a subVI when you need to convert a string to an array of numbers.

Parse String.vi

path: EXAMPLES\GENERAL\STRINGS.LLB

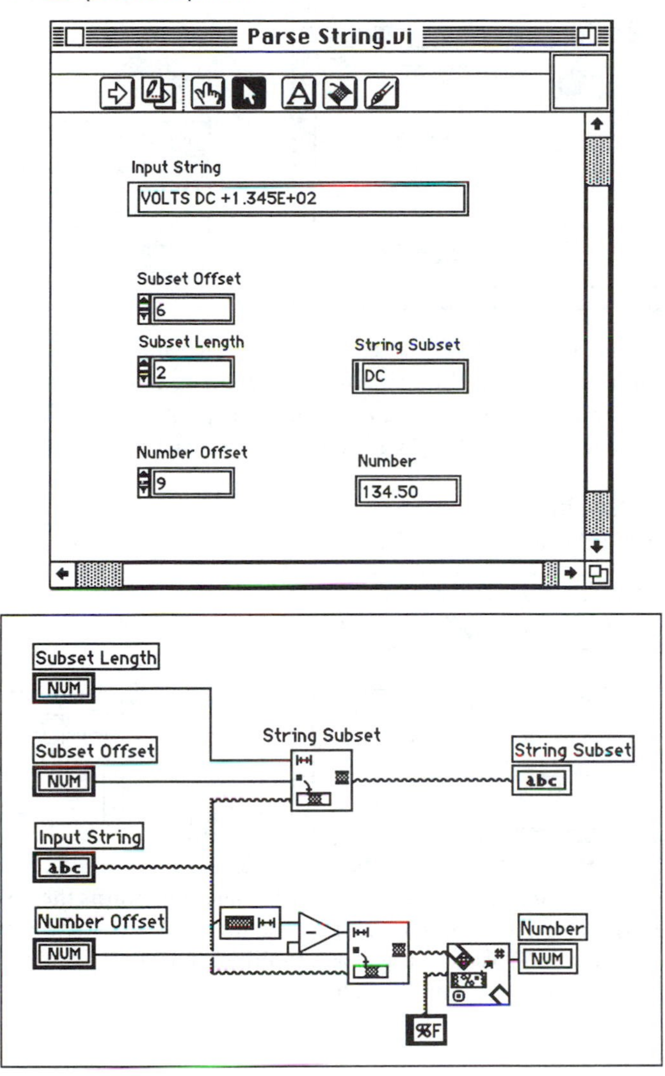

The **Parse String** VI shows you how to use the string functions in LabVIEW to divide an input string into string subsets and to convert numeric ASCII strings to the actual numeric values. The **String Subset**, **String Length**, and **Format & Append** functions are used in this VI to manipulate the input string.

Character to ASCII.vi

path: EXAMPLES\GENERAL\STRINGS.LLB

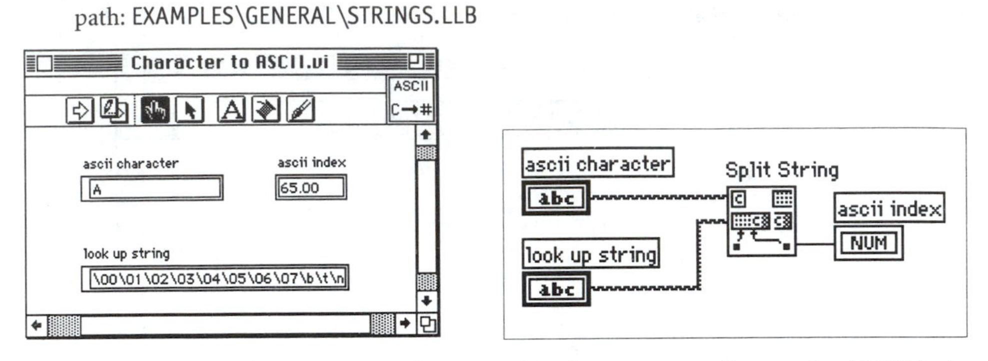

Given a string, this VI returns the numeric value corresponding to the ASCII index of the first character in the string. It uses a look up table and matches the input character with a character in the table, then returns the proper index. This VI is useful as a subVI when you need to find the ASCII index of a character in a string.

ASCII to Character.vi

path: EXAMPLES\GENERAL\STRINGS.LLB

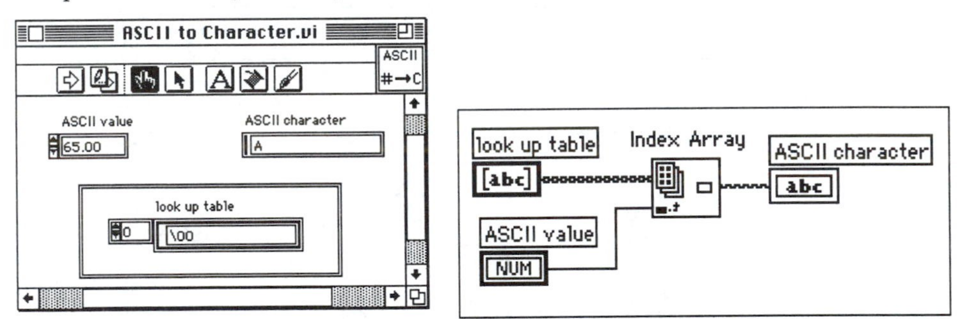

This VI takes an input numeric ASCII value and returns the ASCII character that corresponds to that index. It uses a look up table to find ASCII character that corresponds to the input ASCII index. This VI is useful as a subVI when you need to find the ASCII character that corresponds to a specific numeric value.

Serial Example

Serial Communication.vi

path: EXAMPLES\SERIAL\SMPLSERL.LLB

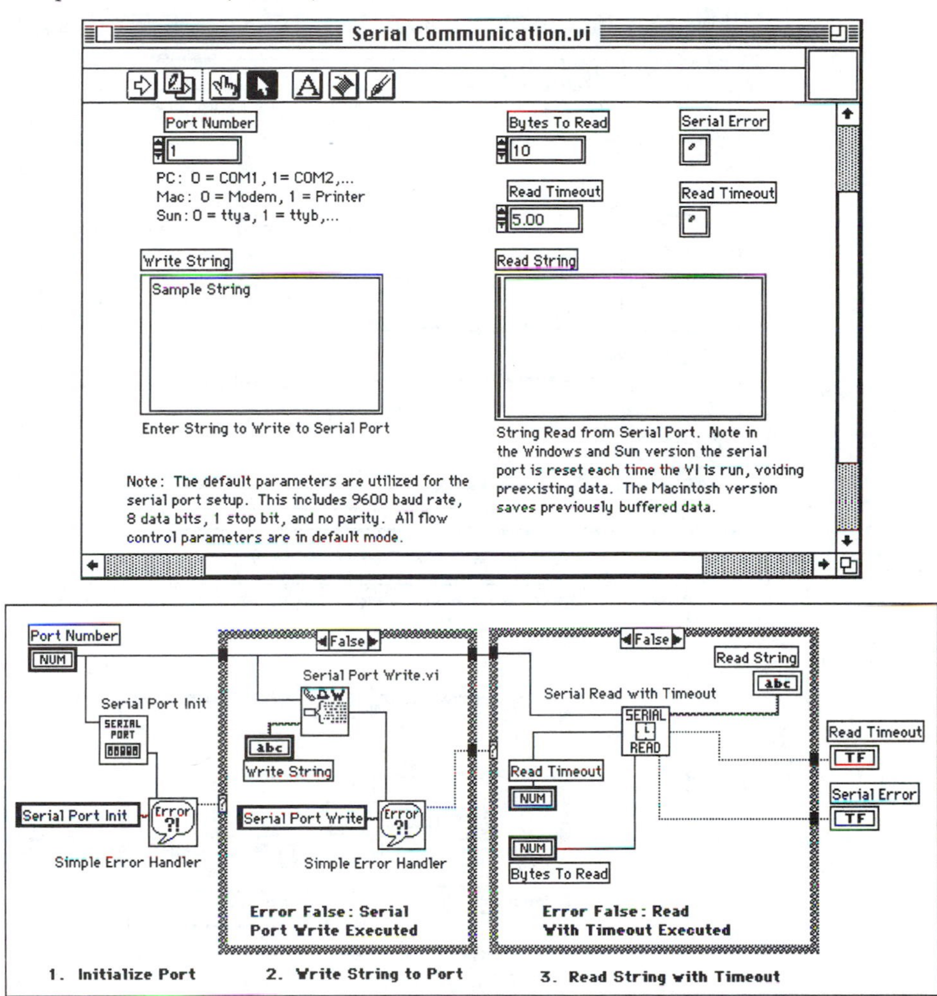

The **Serial Communication** VI performs bidirectional serial communication using a built-in port on the computer. It initializes the port, writes a string to the port, and performs a read with timeout. The **Serial Read with Timeout** VI will wait until the requested bytes are available or the time limit is up, whichever comes first.

This VI can be used as an application in itself to transfer serial data, or as a template to create your own custom serial application. You simply tell it which port number to use, along with the string you wish to send out, the number of bytes to read in, and the number of seconds to wait before timing out on the read.

GPIB Examples

LabVIEW <--> GPIB.vi

path: EXAMPLES\GPIB\SMPLGPIB.LLB

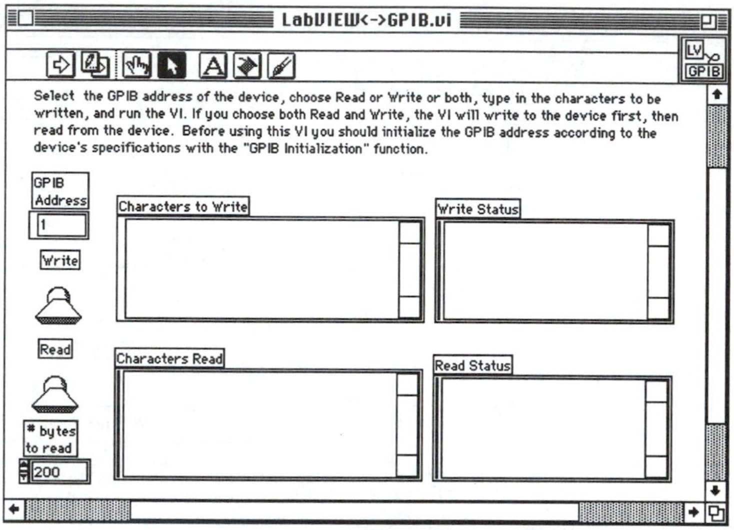

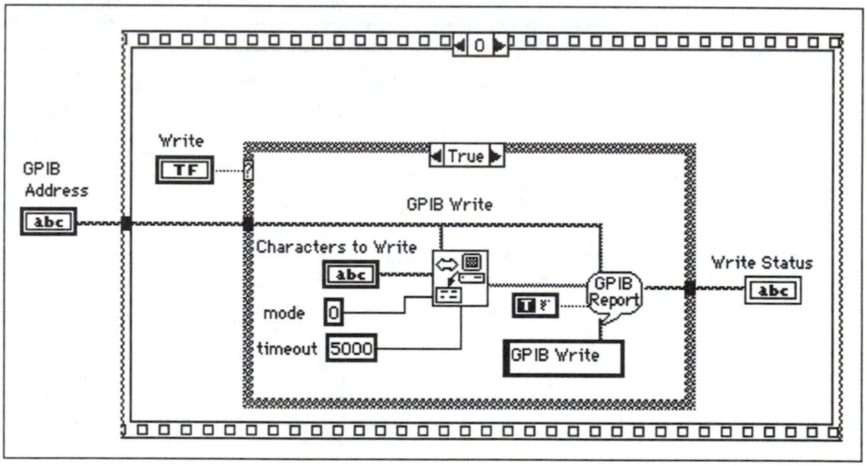

You can write to and read from a device connected to the GPIB bus using this VI. You must have the proper drivers and hardware installed for the VI to work correctly. Select the GPIB address of the device, choose Read or Write or both, type in the characters to be written, and run the VI. If you choose both Read and Write, the VI will write to the device first, then read from the device. Before using this VI you might need to initialize the GPIB address according to the device's specifications with the **GPIB Initialization** function, although this is not always necessary. **LabVIEW<-->GPIB.vi** handles error reporting for you and also displays the status byte returned after each GPIB operation.

You can use this example as a stand-alone GPIB application, or you can modify it to suit your requirements.

Fluke 8840A.vi

path: EXAMPLES\GPIB\FL8840.LLB

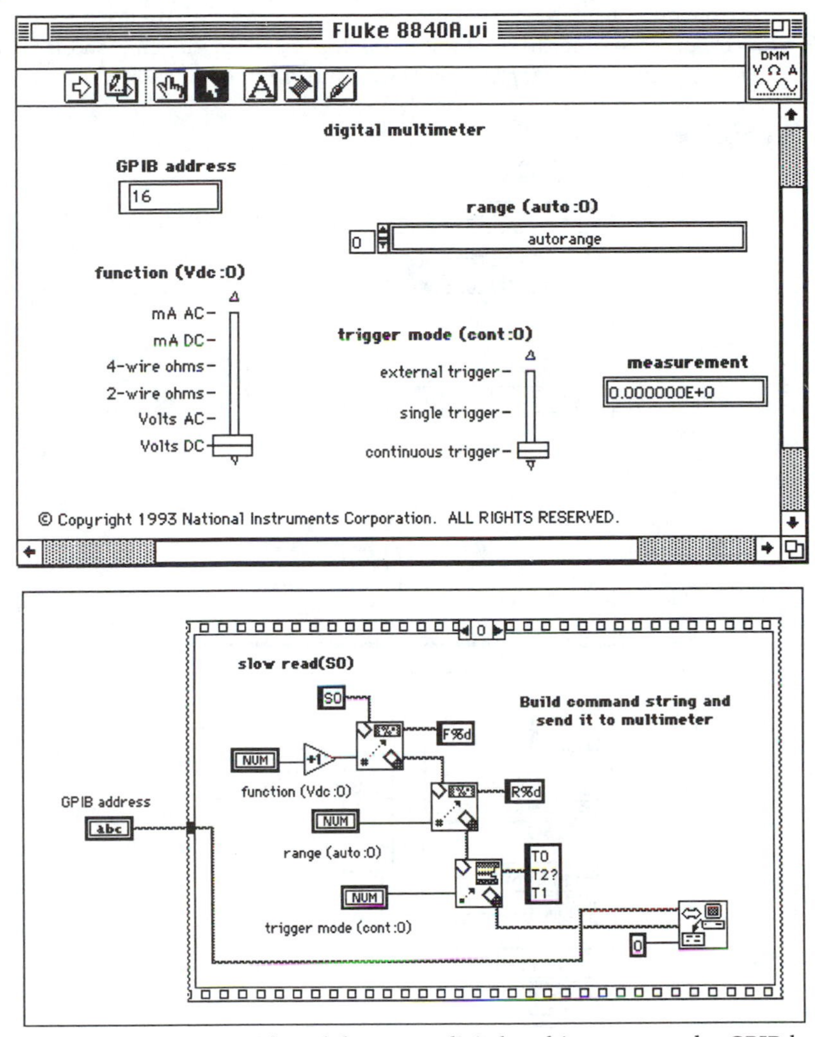

This VI communicates with a Fluke 8840A digital multimeter over the GPIB bus. It can measure volts, amps, or ohms, depending on the user's selection from six possible functions. You must also select the range or use auto range, and select the trigger mode you want to use.

This VI is an example of an instrument driver—a LabVIEW VI that controls a remote instrument. **Fluke 8840A.vi** is written specifically to communicate with the Fluke 8840A using the digital multimeter's expected command set.

DAQ Examples

You must have National Instruments data acquisition hardware correctly installed in your computer in order to run these examples. For more information on data acquisition and how to set it up, refer to *Appendix A, DAQ and GPIB Configuration.*

AI Single Point.vi

path: EXAMPLES\DAQ.LLB

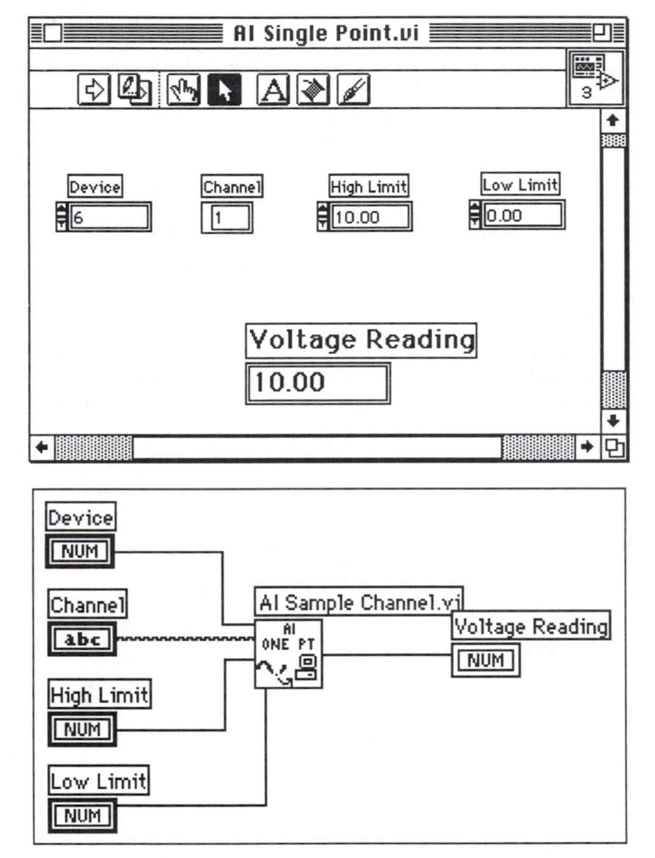

This VI uses the **AI Sample Channel** VI to acquire a single reading from the specified analog input channel. Make sure to input the correct device number of your board. The measured voltage should fall between the High Limit and Low Limit inputs, which are used to calculate the gain automatically applied to the voltage.

AI Multi Point.vi

path: EXAMPLES\DAQ.LLB

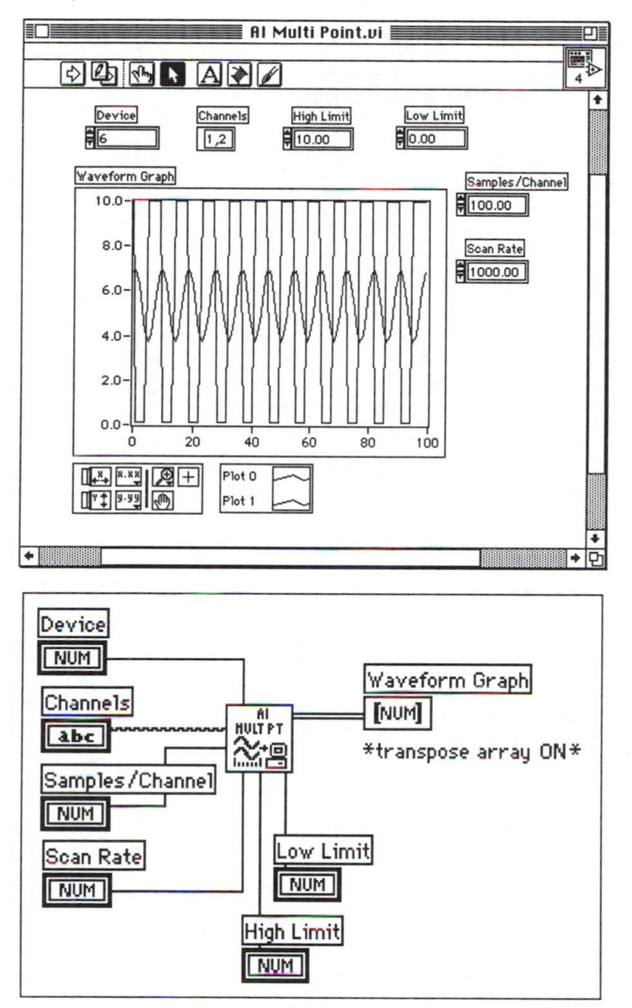

This VI uses the **AI Acquire Waveforms** VI to acquire multiple-point readings from the specified analog input channels and then plots the waveforms on a graph. Make sure to input the correct device number of your board, the number of samples to acquire per channel, and the scan rate per channel you wish to use. The measured voltages should fall between the High Limit and Low Limit inputs, which are used to calculate the gain automatically applied to the voltages.

AO Single Point.vi

path: EXAMPLES\DAQ.LLB

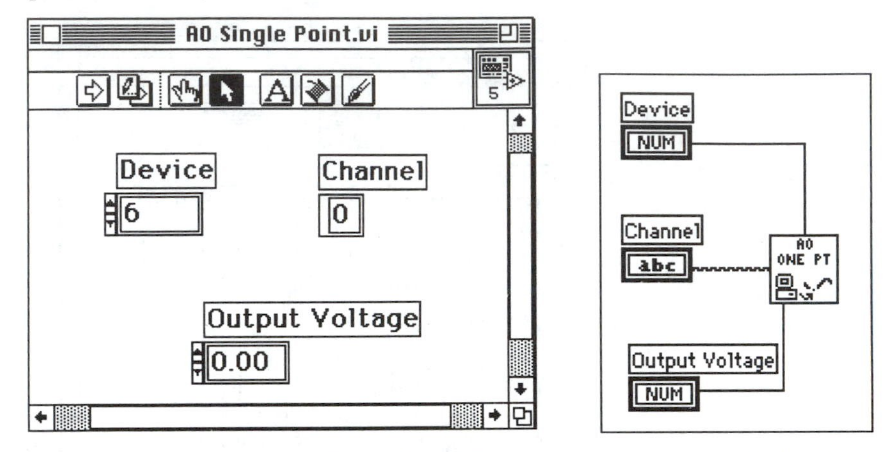

This VI uses the **AO Update Channel** VI to send a voltage out the specified analog output channel. Make sure to input the correct device number of your board.

AO Multi Point.vi

path: EXAMPLES\DAQ.LLB

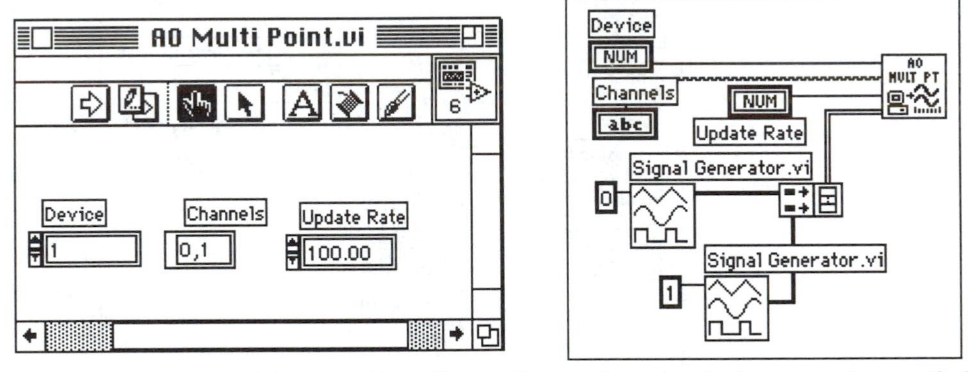

This VI uses the **AO Generate Waveforms** VI to send arrays of voltages out the specified analog output channels. Make sure to input the correct device number of your board.

Note to Mac Users

You must have a National Instruments DMA2800 board connected to your DAQ board in order to do any form of waveform generation. Only the NB-MIO-16X has the capability to generate multiple waveforms at a time, so if you have more than one channel in the channel list, this VI will only work on the 16X board.

Digital Example.vi

path: EXAMPLES\DAQ.LLB

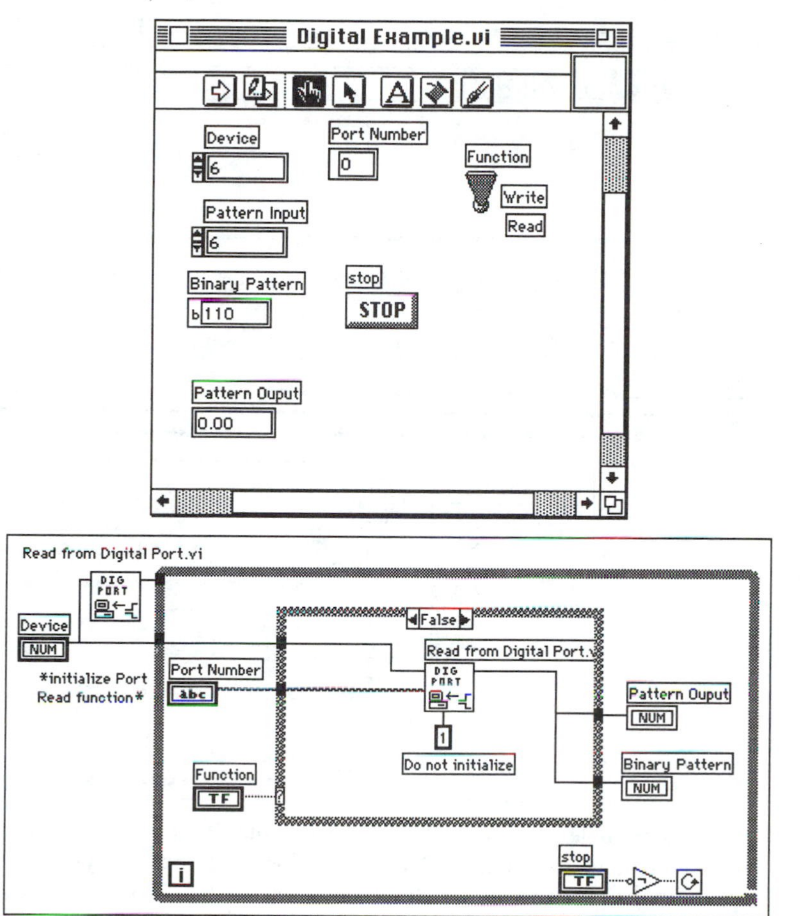

This VI writes to and reads from the digital lines on a DAQ board. To write a value, specify the Pattern Input that will generate the proper 4-bit binary pattern you wish to send out and set Function to "Write." The pattern is displayed in the Binary Pattern control. To read a value from the port, simply set Function to "Read;" you will see the port's pattern in Pattern Output. Make sure to input the correct device number of your board.

Initialization must occur before you can read from a port, but it also clears all values on the port. This example calls the **Read from Digital Port** VI with its default setting of "initialize" before the loop executes. Inside the loop, **Read from Digital Port** does not initialize and therefore the port retains its values, which are read back in.

Note

Digital Example.vi doesn't work on the DAQcard 700 because its digital ports are not bidirectional.

CTR Count Time or Events.vi

path: EXAMPLES\DAQ.LLB

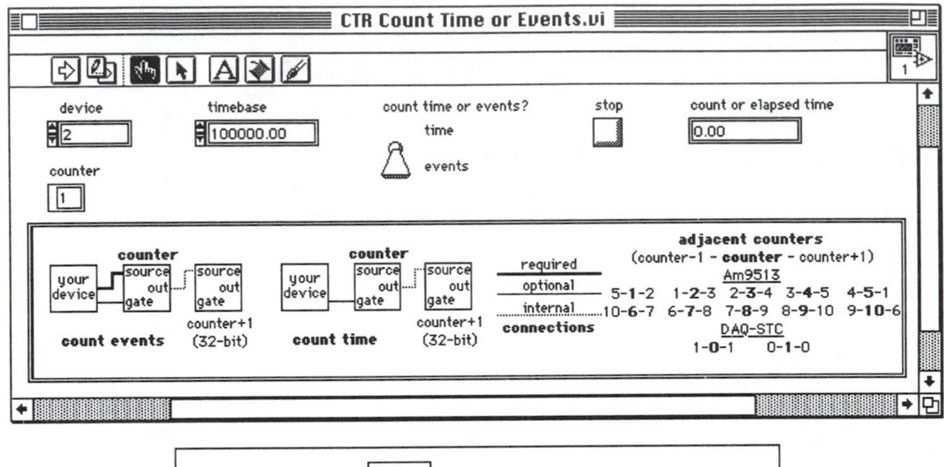

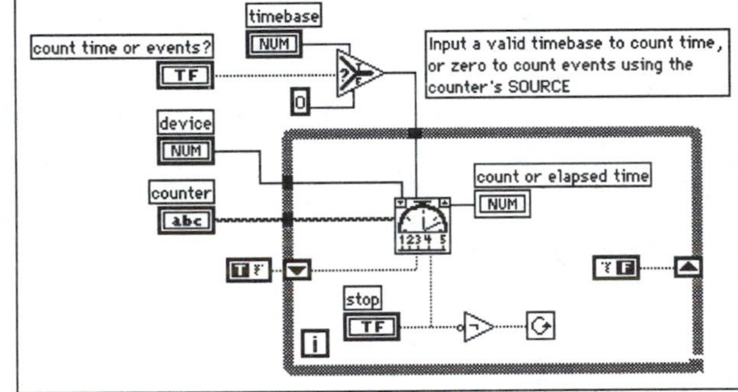

This example uses the counter/timers on your DAQ board to count continuously until you stop the program with the STOP button. The results are displayed in the "count or elapsed time" indicator. If you set the switch to "count events," the VI counts the number of rising edges on counters's SOURCE pin. If you set the switch to "count time," the VI measures elapsed time using the selected timebase. Use the wiring connections shown on the front panel of this VI to hook up your counter/timers properly.

This VI works only with boards containing Am9513 or DAQ-STC counter chips. For other boards ICTR Control.vi must be used.

CTR Generate Pulses.vi

path: EXAMPLES\DAQ.LLB

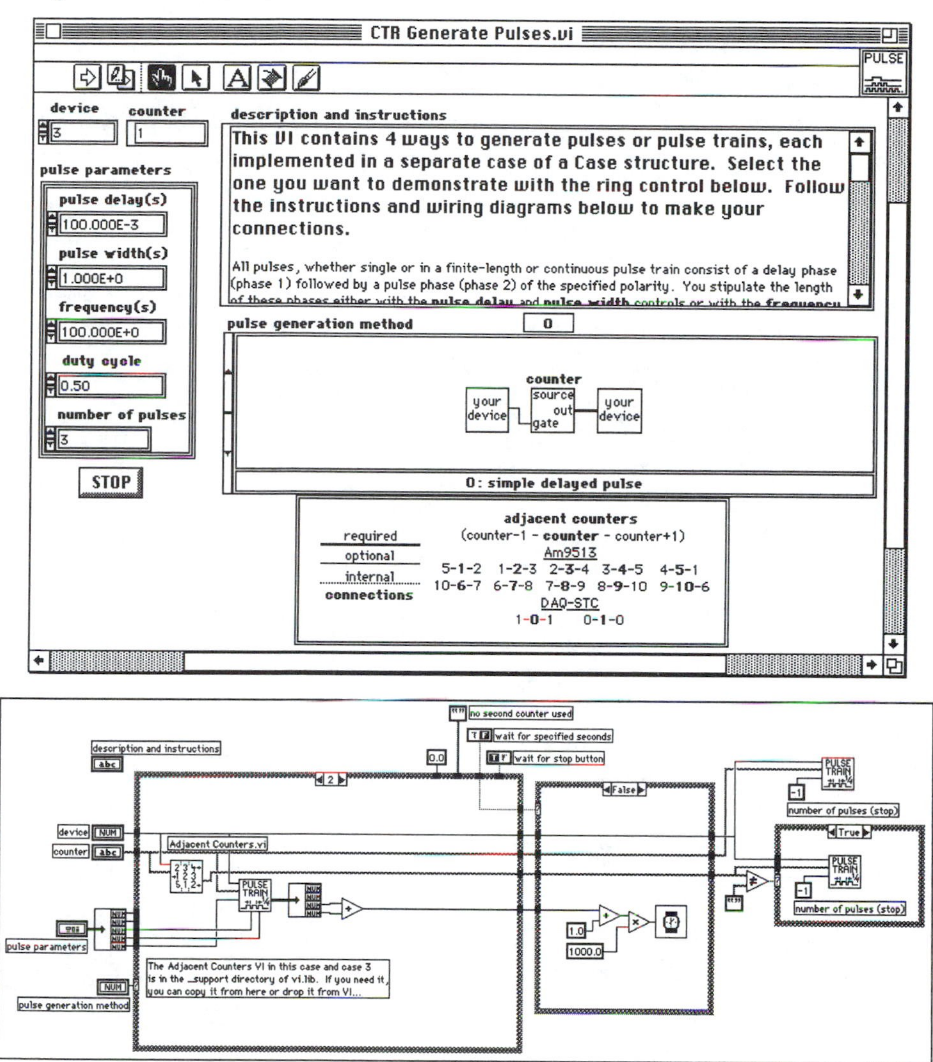

This VI contains four ways to generate pulses or pulse trains using the counter/timers on your DAQ board, each implemented in a separate case of a Case structure. Select the one you want to demonstrate using the ring control; then follow the instructions on the front panel. Use the wiring connections shown on the front panel of this VI to hook up your counter/timers properly.

All pulses, whether single or in a finite length or continuous pulse train consist of a delay phase followed by a pulse phase of the specified polarity. You stipulate the length of these phases either with the pulse delay and pulse width controls or with the frequency and duty cycle controls. A high duty cycle denotes a long pulse phase relative to the delay. You need to modify the program to use any of the optional gate connections shown in the wiring connection diagrams.

This VI works only with boards containing Am9513 or DAQ-STC counter chips. For other boards ***ICTR Control.vi*** *must be used.*

CTR Measure Frequency and Period.vi

path: EXAMPLES\DAQ.LLB

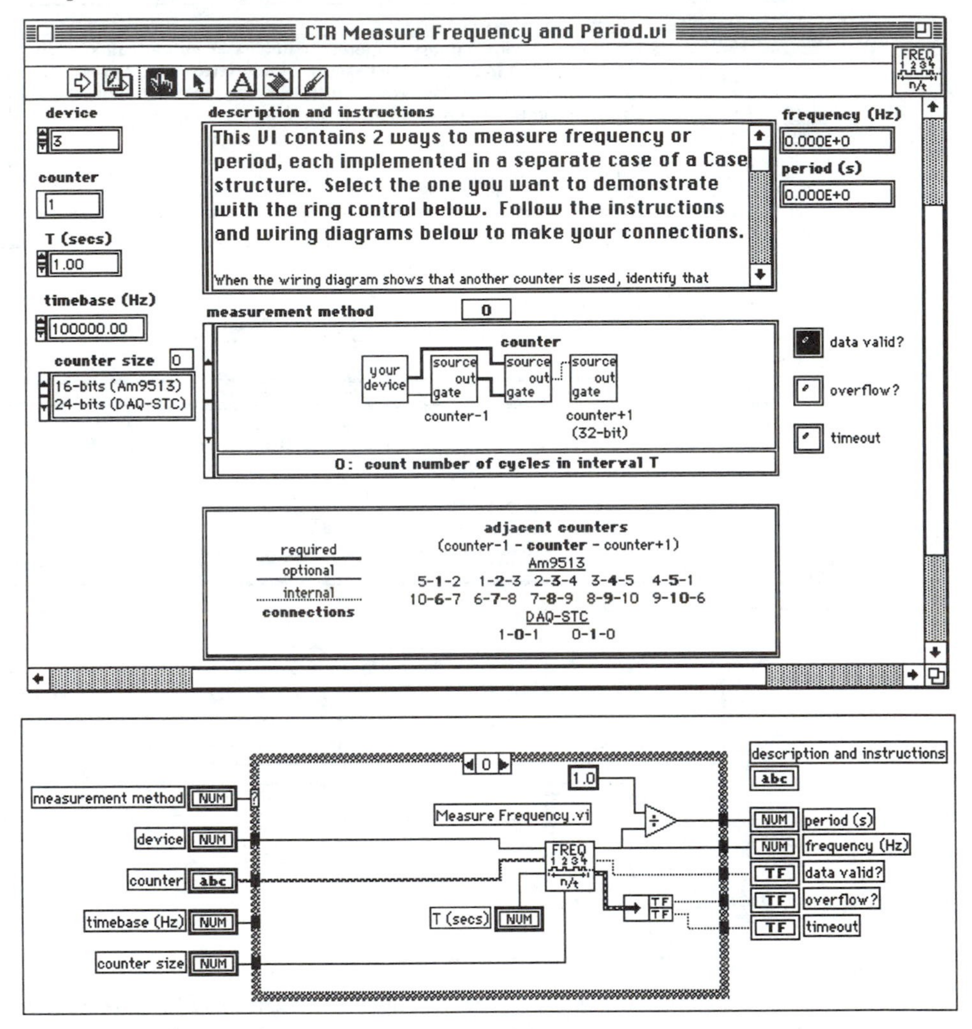

This VI contains two ways to measure frequency or period using the counter/timers on your DAQ board, each implemented in a separate case of a Case structure. Select the one you want to demonstrate using the ring control, then follow the appropriate instructions on the front panel. Use the wiring connections shown on the front panel of this VI to hook up your counter/timers properly.

This VI works only with boards containing Am9513 or DAQ-STC counter chips. For other boards ***ICTR Control.vi*** *must be used.*

Analysis Examples

Pulse Demo.vi

path: EXAMPLES\ANALYSIS.LLB

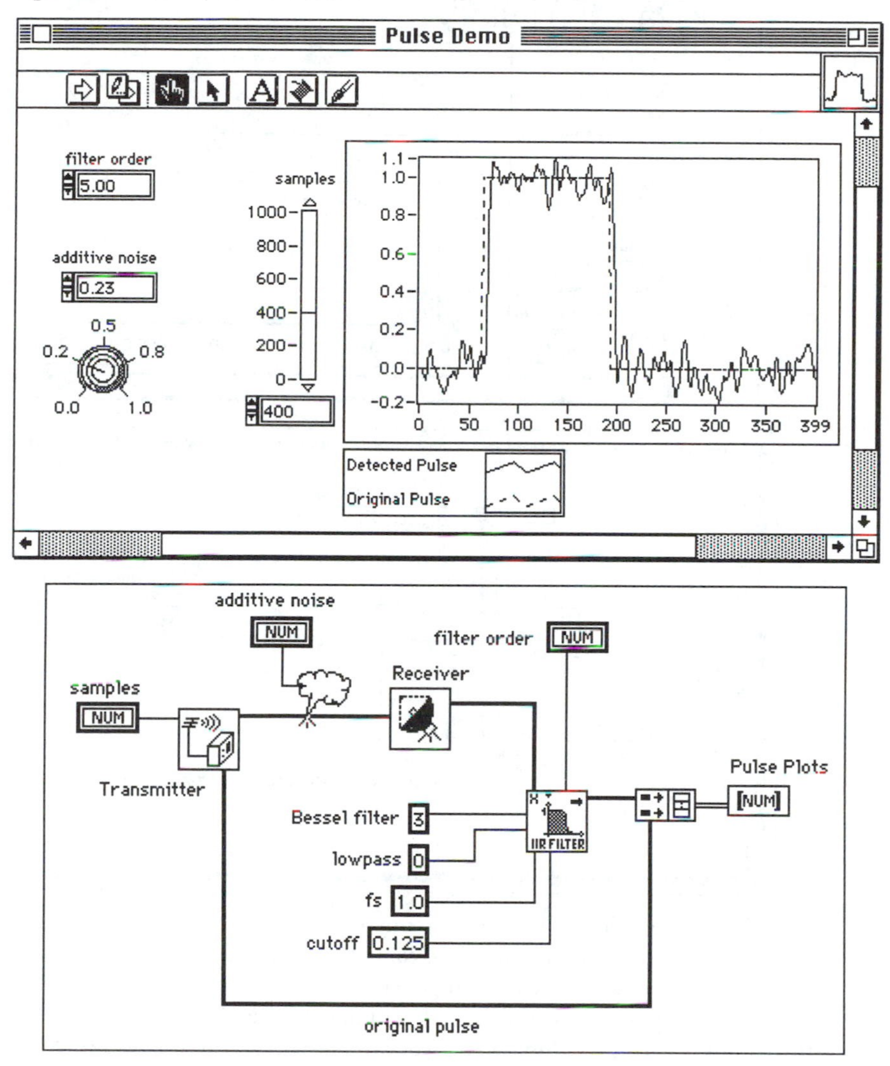

This example simulates a pulse transmission and receiver system. You specify the number of samples in the pulse, the amount of noise to add, and the type of IIR filter you wish to apply to the pulse. Both the original pulse and the detected pulse are displayed on a graph.

Amplitude Spectrum Example.vi

path: EXAMPLES\ANALYSIS.LLB

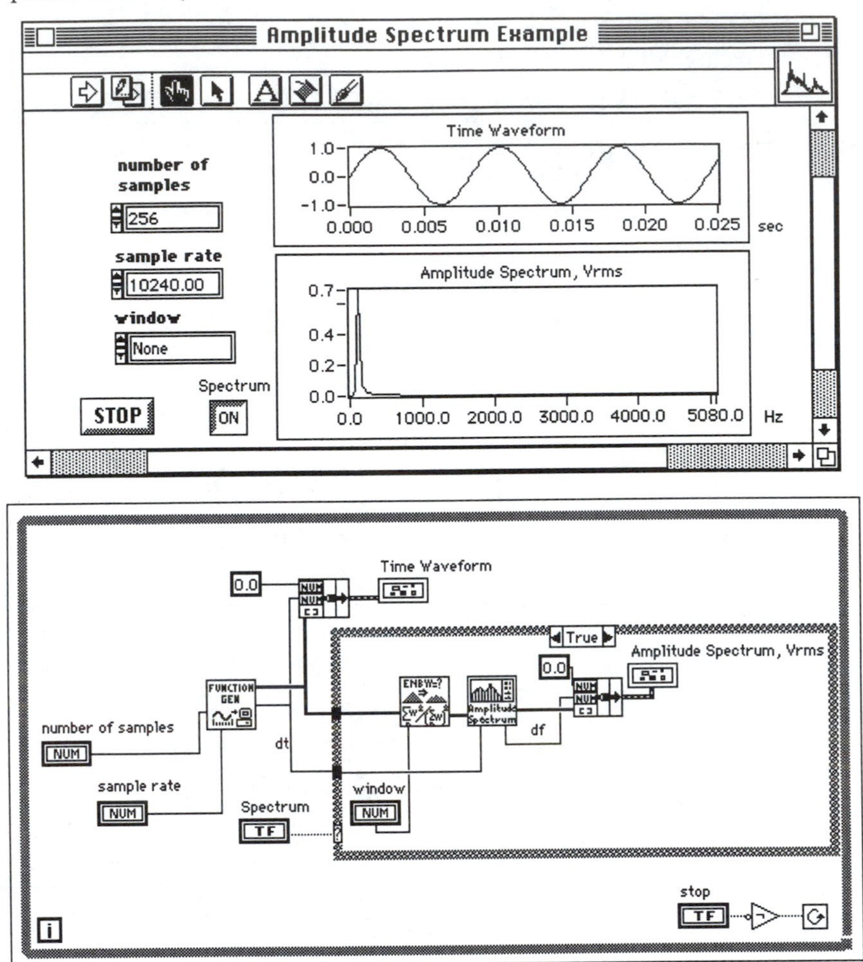

This VI shows how to construct a simple Spectrum Analyzer example. The example simulates a waveform, windows it, and then computes the autopower spectrum and the power and frequency of the peak frequency component. The power spectrum is graphed in the desired units. You specify the number of samples, sample rate, and the type of window to use. Make sure that the Spectrum button is ON if you wish to see the Amplitude Spectrum graph.

Curve Fit Example.vi

path: EXAMPLES\ANALYSIS.LLB

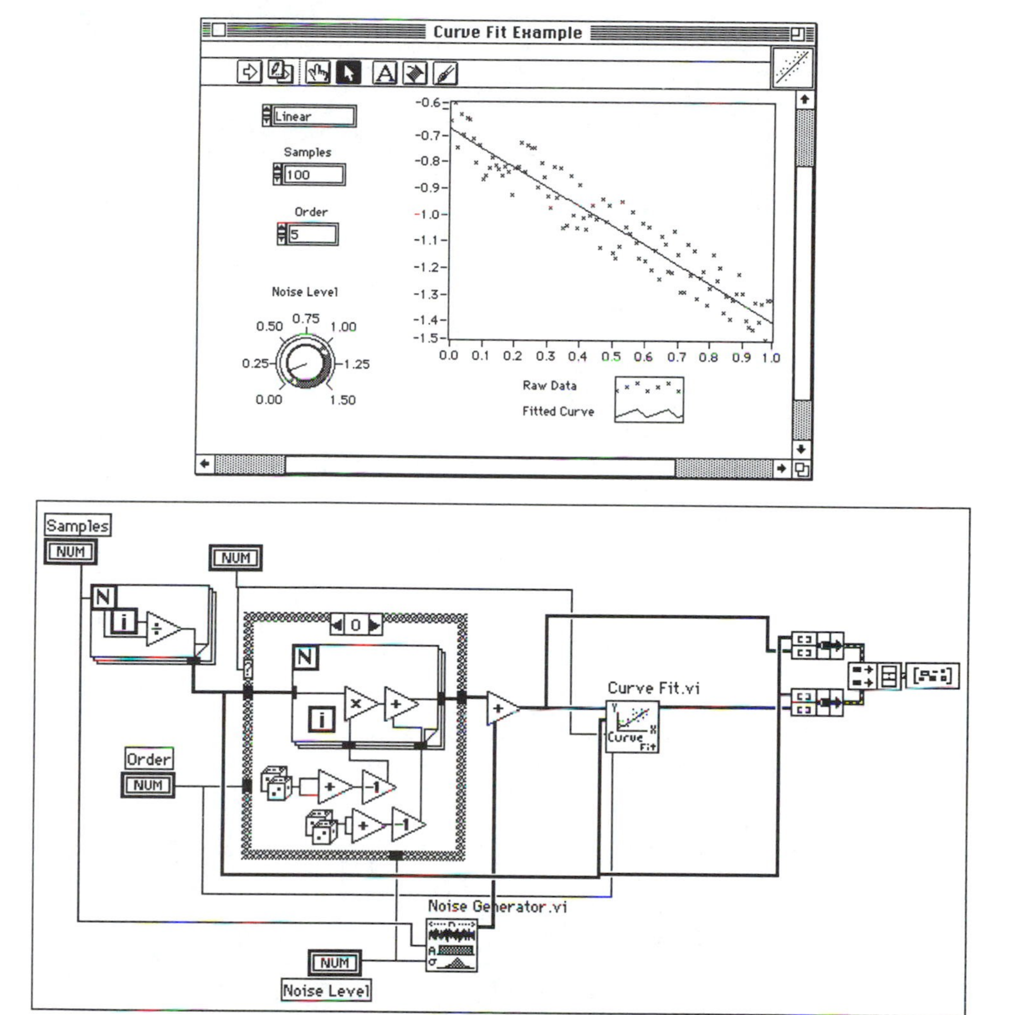

This example demonstrates the use of the **Curve Fit** VI for performing a linear, exponential, or polynomial fit on data sets with random coefficients. You can specify the order used, and you can also choose to add noise to the random data. The **Noise Generator** VI produces the array of noise values, which are added to the random data before the curve fit is performed.

Integral & Derivative.vi

path: EXAMPLES\ANALYSIS.LLB

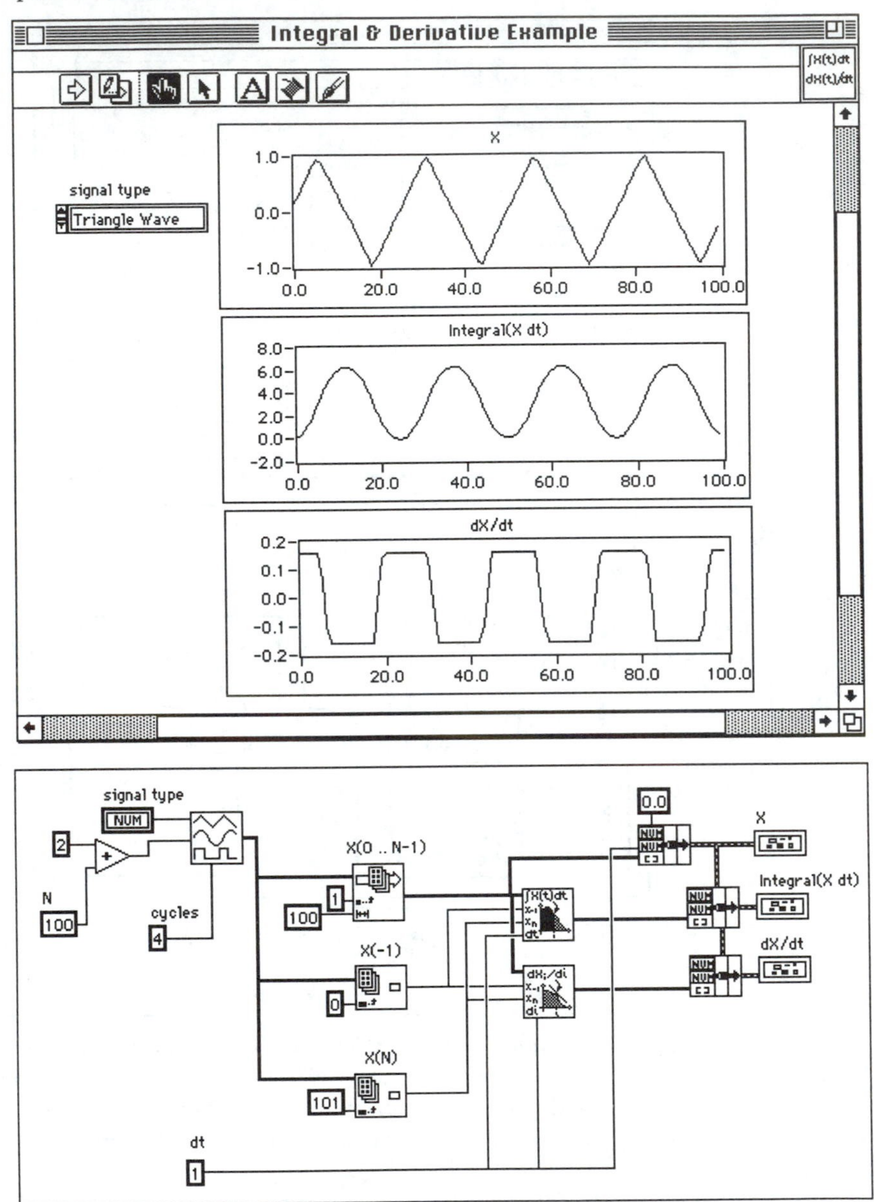

This example demonstrates how to use the **Integral x(t)** and **Derivative x(t)** Analysis VIs. An array is passed from the **Signal Generator** VI to both the **Integral x(t)** and **Derivative x(t)** VIs. Both of these VIs require the array of data points, $X(0)$ through $X(n–1)$. They also require an additional point on each end of the array, $X(–1)$ and $X(n)$. $X(t)$, Integral ($X\ dt$), and dX/dt are each plotted on a graph.

IIR Filter Design.vi

path: EXAMPLES\ANALYSIS.LLB

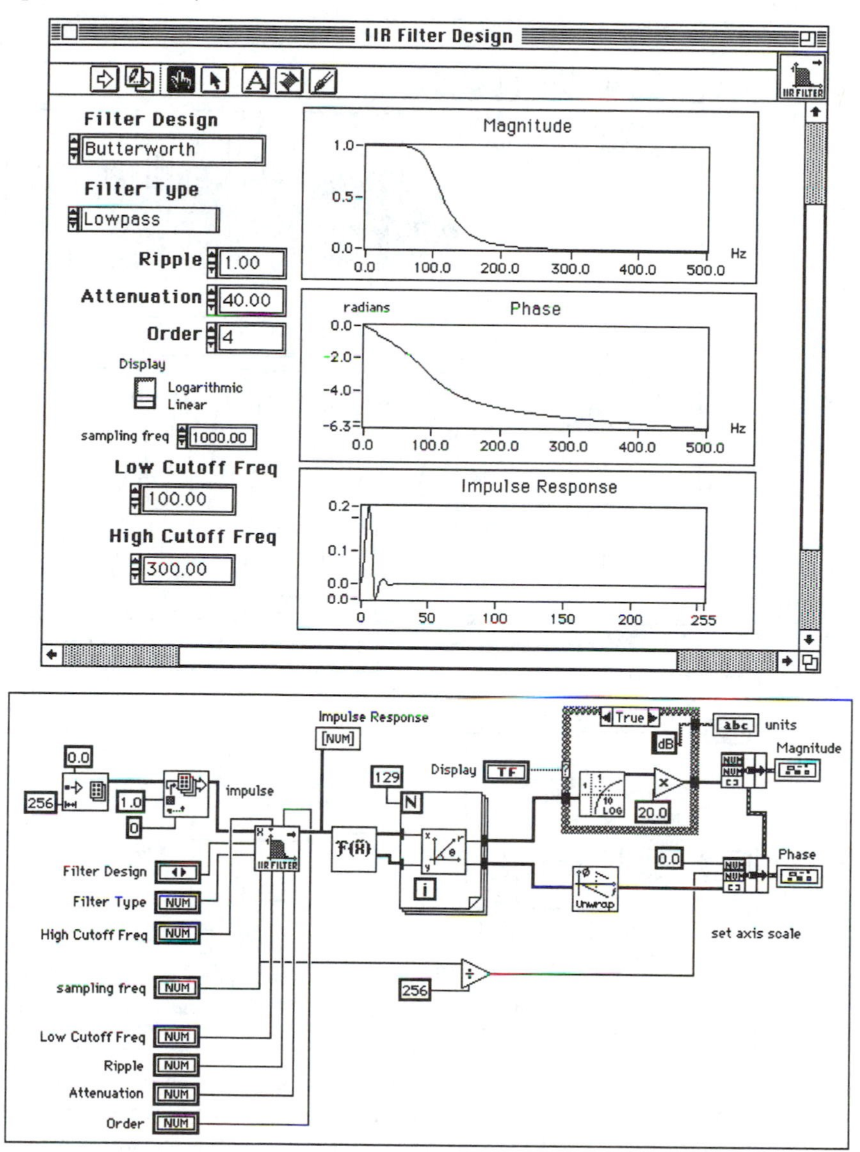

This example displays the frequency response of the filter you choose from the front panel controls. The user may select any of the four IIR filter design methods (Bessel, Butterworth, Chebyshev I and II) and the passband type (lowpass, highpass, bandpass, and bandstop), as well as several other parameters. The Magnitude, Phase, and Impulse Response are displayed on graphs.

Extract the Sine Wave.vi

path: EXAMPLES\ANALYSIS.LLB

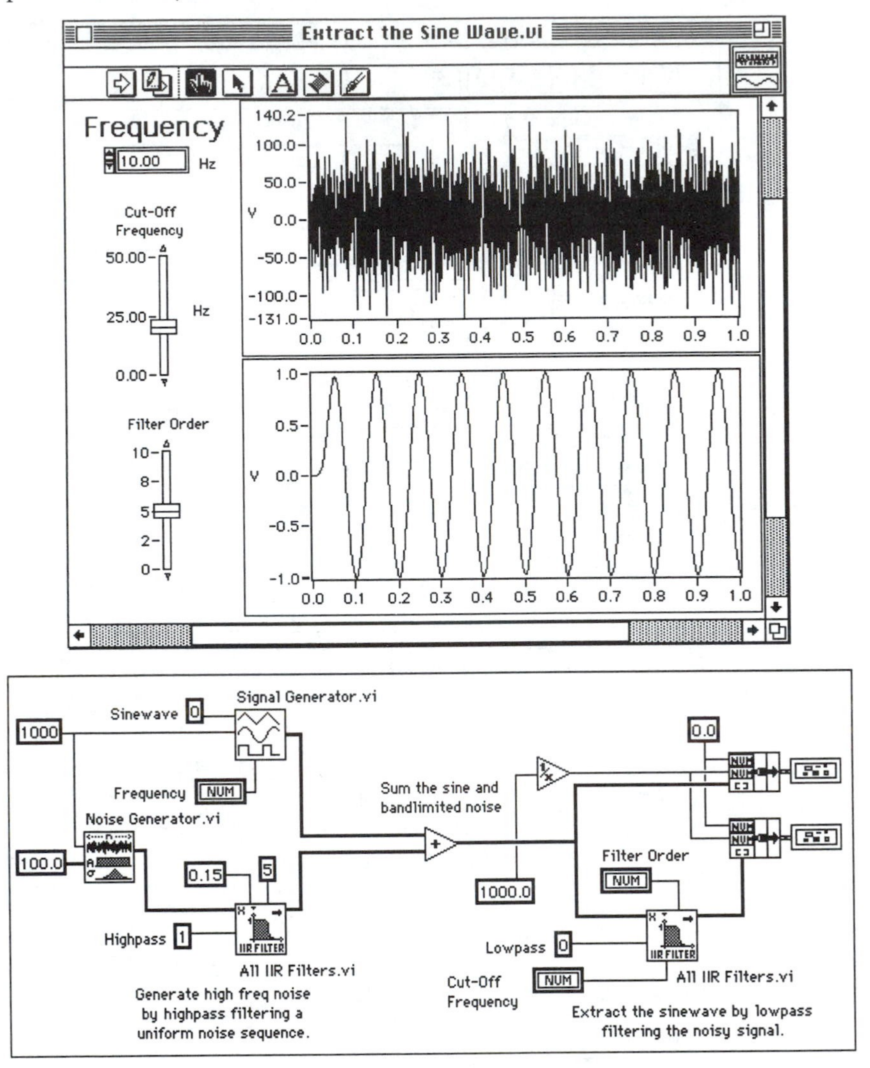

This example demonstrates how a signal can be extracted from bandlimited noise using the filter VIs. It adds high-frequency noise to a sine wave, then uses a lowpass filter to extract the noise from the sine wave. Both the noisy signal and the extracted sine wave are displayed on a graph.

Exercises and Solutions

EXERCISE.LLB and SOLUTION.LLB, found in the EXAMPLES\GENERAL folder or directory, contain the exercises and solutions described throughout this book. Most of the VIs in

EXERCISE.LLB are partially built; you step through how to finish them as part of an exercise. SOLUTION.LLB contains the solutions to most of the VIs presented in the exercises. A few of the other exercises can be found in other parts of the EXAMPLES folder or directory, so if the one you want is not in SOLUTION.LLB, you might browse a little to find it. You can use these solutions and examples to learn LabVIEW programming style, compare your answers, and help you when you get stuck. The VIs shown are ONE way to write the given program. You can often find may other ways, some of them even more efficient, to accomplish the given task.

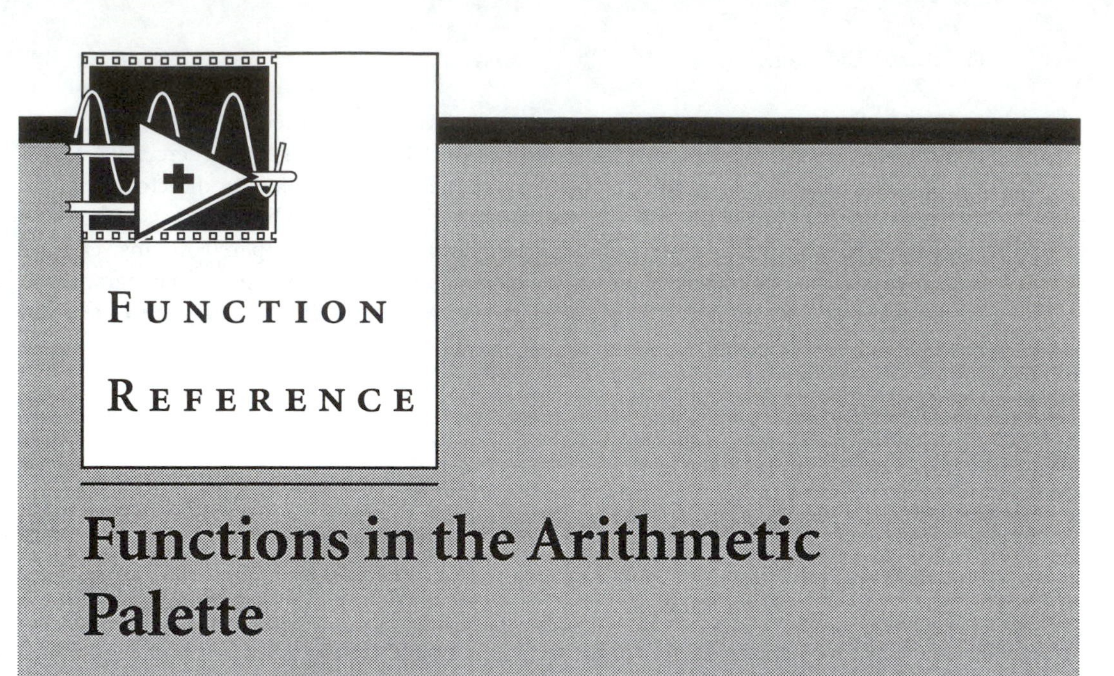

Functions in the Arithmetic Palette

This chapter describes the functions that perform arithmetic and logical operations. To access the **Arithmetic** function palette, select the **Arithmetic** item from the **Functions** menu. This palette contains an assortment of one- and two-input arithmetic and logical functions and a random number generator.

Polymorphism

Polymorphism is the ability of a function to adjust to input data of different types or representations. Most functions are polymorphic. VIs are not polymorphic.

Functions are polymorphic to varying degrees; none, some, or all of their inputs may be polymorphic. Some function inputs accept numbers or Boolean values. Some accept numbers or strings. Some accept not only scalar numbers but also arrays of numbers, clusters of numbers, arrays of clusters of numbers, and so on. Some accept only one-dimensional arrays although the array elements may be of any type. Some functions accept all types of data.

The icon shown to the left indicates that a parameter is polymorphic.

Arithmetic Functions

The arithmetic functions work on numbers, arrays of numbers, clusters of numbers, arrays of clusters of numbers, and so on. A formal and recursive definition of the allowable input type is as follows.

Numeric type = numeric scalar || array [numeric type] || cluster [numeric types]

where numeric scalars can be floating-point, integers, or arrays of clusters. Arrays of arrays are not allowed.

Arrays can have any number of dimensions and any size. Clusters can have any number of elements. The output type is of the same composition as the input type, and the functions operate on each element of the structure, for functions with one input.

For functions with two inputs, LabVIEW allows the following input combinations.

- *Similar*—both inputs have the same structure, and the output has the same structure as the inputs.
- *One scalar*—one input is a numeric scalar, the other is an array or cluster, and the output is an array or cluster.
- *Array of*—one input is an array of some numeric type, the other is the numeric type itself, and the output is an array.

For similar inputs, LabVIEW performs the function on the respective elements of the structures. For example, LabVIEW can add two arrays element by element. Both arrays must have the same dimension and in order to have predictable results, must have the same dimension size. You can add arrays with differing numbers of elements; the output of such an addition has the same number of elements as the smallest input. Both clusters must have the same number of elements, and the respective elements must have the same structure.

For operations involving a scalar and an array or cluster, LabVIEW performs the function on the scalar and the respective elements of the structure. For example, LabVIEW can subtract a number from all elements of an array, regardless of the dimension of the array.

For operations that involve a numeric type and an array of that type, LabVIEW performs the function on each array element. For example, a graph is an array of points, and a point is a cluster of two numeric types, x and y. To offset a graph by 5 units in the x direction and 8 units in the y direction, you can add a point (5, 8), to the graph.

The following figure illustrates some of the possible polymorphic combinations of the **Add** function.

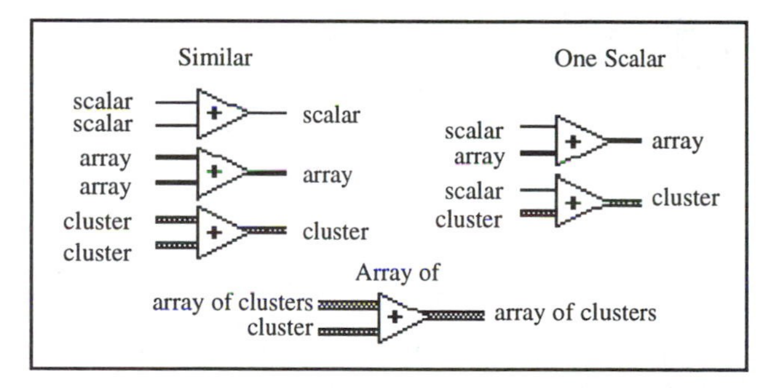

Logical Functions

The logical functions work on Booleans, arrays of Boolean values, clusters of Boolean values, arrays of clusters of Boolean values, and so on. A formal and recursive definition of the allowable input type is as follows.

Logical type = Boolean scalar || array [*logical type*] || cluster [*logical types*]

except that arrays of arrays are not allowed.

Some combinations of Boolean values are shown in the following illustration for the **And** function.

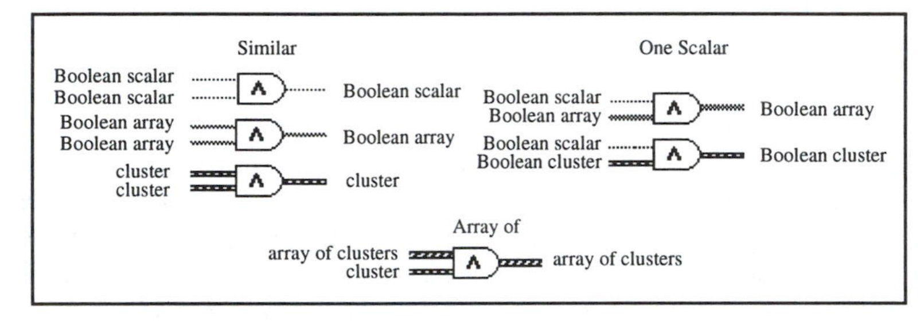

Arithmetic Function Descriptions

Add

Computes the sum of the inputs.

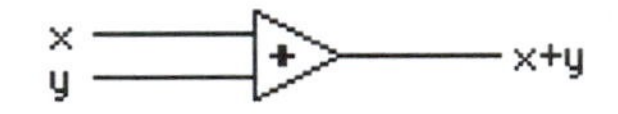

 **x** and **y** can be scalar numbers, arrays or clusters of numbers, arrays of clusters of numbers, and so on.

 **x+y.**

Subtract

Computes the difference of the inputs.

x
y
x-y

 **x** and **y** can be scalar numbers, arrays or clusters of numbers, arrays of clusters of numbers, and so on.

POLY **x-y.**

Multiply

Returns the product of the inputs.

x and **y** can be scalar numbers, arrays or clusters of numbers, arrays of clusters of numbers, and so on.

x*y.

Divide

Computes the quotient of the inputs.

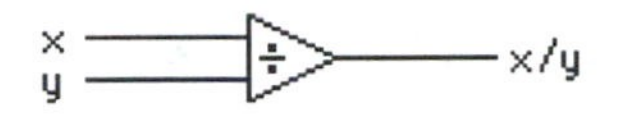

x and **y** can be scalar numbers, arrays or clusters of numbers, arrays of clusters of numbers, and so on.

x/y.

Absolute Value

Returns the absolute value of the input.

x can be a scalar number, array or cluster of numbers, array of clusters of numbers, and so on.

abs(x).

Round To Nearest

Rounds the input to the nearest integer. If the value of the input is midway between two integers (for example, 1.5 or 2.5), the function returns the nearest even integer (2).

number can be a scalar number, array or cluster of numbers, array of clusters of numbers, and so on.

nearest integer value.

Square Root

Computes the square root of the input value.

x sqrt(x)

x can be a scalar number, array or cluster of numbers, array of clusters of numbers, and so on.

sqrt(x) if x ≥ 0. **sqrt(x)** is not a number (NaN) if **x** is less than 0.

Negate

Negates the input value.

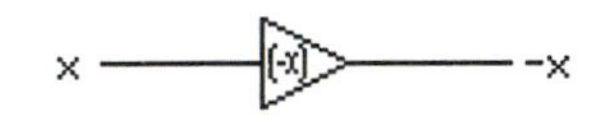

x can be a scalar number, array or cluster of numbers, array of clusters of numbers, and so on.

-x is the negative value of **x**.

Random Number (0-1)

Produces a number between 0 and 1, exclusively. The distribution is uniform.

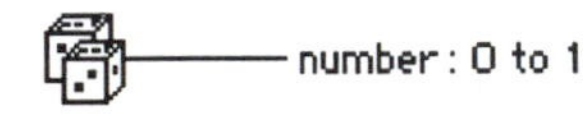

number : 0 to 1 is a number between 0 and 1.

Logical Function Descriptions

And

Computes the logical AND of the inputs.

x
y
x .and. y ?

x and **y** can be Boolean scalars, arrays or clusters of Boolean values, arrays of clusters of Booleans, and so on.

x .and. y?.

Or

Computes the logical OR of the inputs.

x and **y** can be Boolean scalars, arrays or clusters of Boolean values, arrays of clusters of Booleans, and so on.

x .or. y?.

Exclusive Or

Computes the logical Exclusive OR of the inputs.

x .xor. y ?

x and **y** can be Boolean scalars, arrays or clusters of Boolean values, arrays of clusters of Booleans, and so on.

x .xor. y?.

Not

Computes the logical NEGATION of the input.

x .not. x?

x can be a Boolean scalar, an array or cluster of Boolean values, an array of clusters of Booleans, and so on.

.not. x?.

Sine & Cosine

Computes both the sine and cosine of **x**, where **x** is in radians.

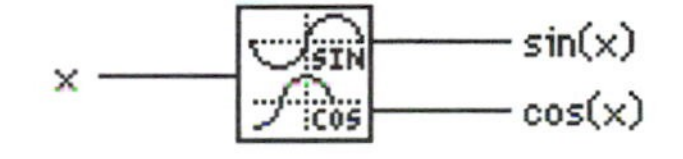

x can be a scalar number, an array or cluster of numbers, an array of clusters of numbers, and so on.

sin(x) and **cos(x)**.

Inverse Tangent

Computes the arctangent of **y/x** (which can be between $-\pi$ and π) in radians. Thus, this function can compute the arctangent for angles in any of the four quadrants of the xy plane.

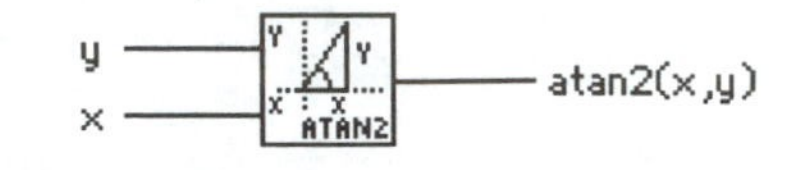

x and y can be scalar numbers, arrays or clusters of numbers, arrays of clusters of numbers, and so on.

atan2(x,y).

Natural Logarithm

Computes the natural logarithm of **x**; that is, the logarithm of **x** to the base e. If **x** is 0, **ln(x)** is -∞. If **x** is not complex, **ln(x)** is NaN.

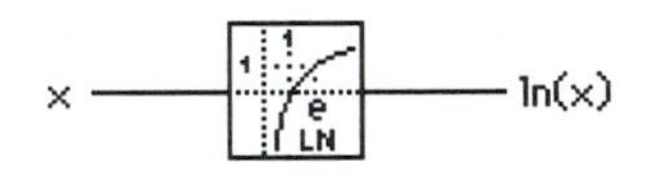

x can be a scalar number, an array or cluster of numbers, an array of clusters of numbers, and so on.

ln(x).

Exponential

Computes the value of e raised to the **x** power.

x can be a scalar number, an array or cluster of numbers, an array of clusters of numbers, and so on.

exp(x).

Power Of X

Computes **x** raised to the **y** power. **x** must be greater than zero unless **y** is an integer value. Otherwise, the result is NaN. If **y** is zero, **x^y** is 1 for all values of **x**, including zero.

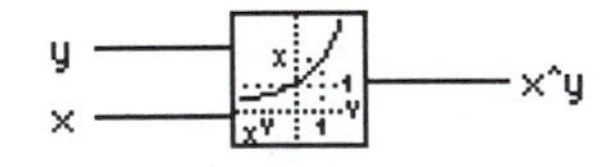

y and **x** can be scalar numbers, arrays or clusters of numbers, arrays of clusters of numbers, and so on.

x^y.

Logarithm Base X

Computes the logarithm of **y** to the base **x** (**x**>0, **y**>0). If **y** is 0, the output is -∞. When **x** is less than or equal to 0, or **y** is less than 0, the output is NaN.

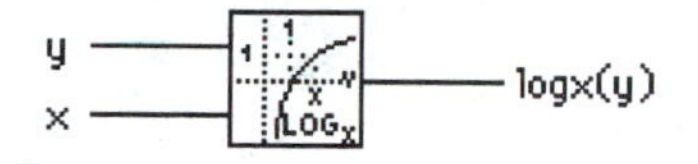

y can be a scalar number, an array or cluster of numbers, an array of clusters of numbers, and so on. **y** must be greater than 0.

x must be greater than 0. **x** can be a scalar number, an array or cluster of numbers, an array of clusters of numbers, and so on.

logx(y).

Comparison Functions

This chapter describes the functions that perform comparisons or conditional testing. You access the **Comparison** palette by selecting **Comparison** from the **Functions** menu.

Rules for Comparison

Most of the comparison functions test one input or compare two inputs and return a Boolean value. You get unpredictable results if you try to compare a value of NaN (not a number) with a number or another NaN.

The Boolean value TRUE is greater than the Boolean value FALSE.

These functions compare strings according to the numerical equivalent of the ASCII characters. Thus, a (with a decimal value of 97) is greater than A (65), which is greater than the numeral 0 (48), which is greater than the space character (32). These functions compare characters one by one from the beginning of the string until an inequality occurs, at which time the comparison ends. For example, LabVIEW evaluates the strings abcd and abef until it finds c, which is greater than e. The presence of a character is greater than the absence of one. Thus, the string abcd is greater than abc because the first string is longer.

The comparison functions compare clusters the same way they compare strings, one element at a time starting with the 0th element until an inequality occurs. Clusters must have the same number of elements in order to be compared.

Some of the comparison functions have two modes for comparing arrays or clusters. In the **Compare Aggregates** mode, if you compare two arrays, the function returns a single value. In the **Compare Elements** mode, you can compare the elements individually, and the function returns an array or cluster of Boolean values. The following illustration shows the two modes.

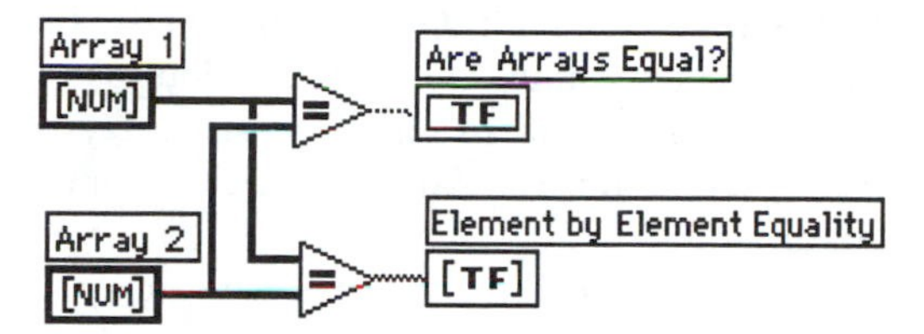

You change the comparison mode by selecting **Compare Aggregates** or **Compare Elements** in the pop-up menu for the node, as shown in the following illustrations.

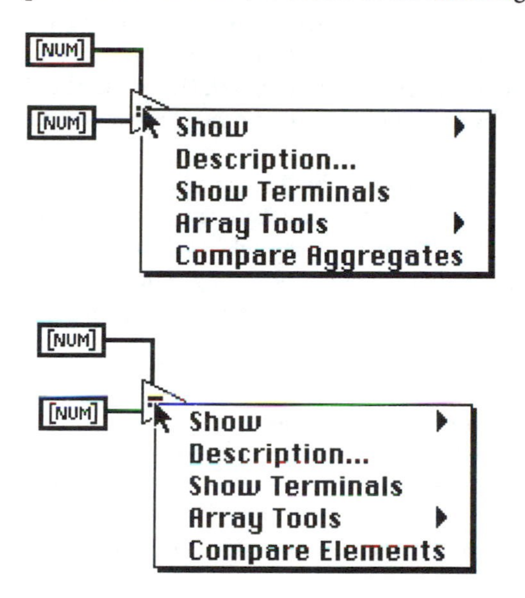

When you compare two arrays of unequal lengths in **Compare Elements** mode, LabVIEW ignores each element in the larger array whose index is greater than the index of the last element in the smaller array. When you use **Compare Aggregates** mode to compare two arrays that are identical except that one has more elements, LabVIEW considers the longer array to be greater than the shorter array. Thus, LabVIEW considers the array [1,2,3] to be greater than the array [1,2] and returns a single Boolean value in **Compare Aggregates** mode.

In **Compare Elements** mode, LabVIEW returns a Boolean for each of the first two elements and ignores the last element of the larger array, as in the above example. Arrays must have the same dimension size (for example, both two-dimensional), and for the comparison between multidimensional arrays to make sense, each dimension should have the same size.

The comparison functions that do not have the **Compare Aggregates** or **Compare Elements** modes compare arrays in the same manner as strings—one element at a time starting with the 0th element until an inequality occurs.

Polymorphism

The functions **Equal?, Not Equal?, Greater or Equal?, Less or Equal?, Less?, Greater?, Max & Min,** and **Select** take inputs of any type, as long as the inputs have the same type.

You can compare numbers, strings, Booleans, arrays of strings, clusters of numbers, clusters of strings, and so on. You cannot, however, compare a number to a string or a string to a Boolean, and so on.

The functions that compare values to zero accept numeric scalars, clusters, and arrays of numbers. These functions output Boolean values in the same data structure as the input.

The **Not A Number/Path/Refnum** function accepts the same input types as functions that compare values to zero. This function also accepts paths and refnums. **Not A Number/Path/Refnum** outputs Boolean values in corresponding structures.

You can use the **Equal?, Not Equal?, Not A Number/Path/Refnum?,** and **Select** functions with paths and refnums, but no other comparison functions accept paths or refnums as inputs.

Comparison functions that use arrays and clusters normally produce Boolean arrays and clusters of the same structure. You can pop up and change to compare aggregates, in which case the function outputs a single Boolean result. The function compares aggregates by comparing the first set of elements to produce the output, unless the first elements are equal, in which case the function compares the second set of elements, and so on.

Comparison Function Descriptions

Equal?

Returns TRUE if **x** is equal to **y**. Otherwise, this function returns FALSE.

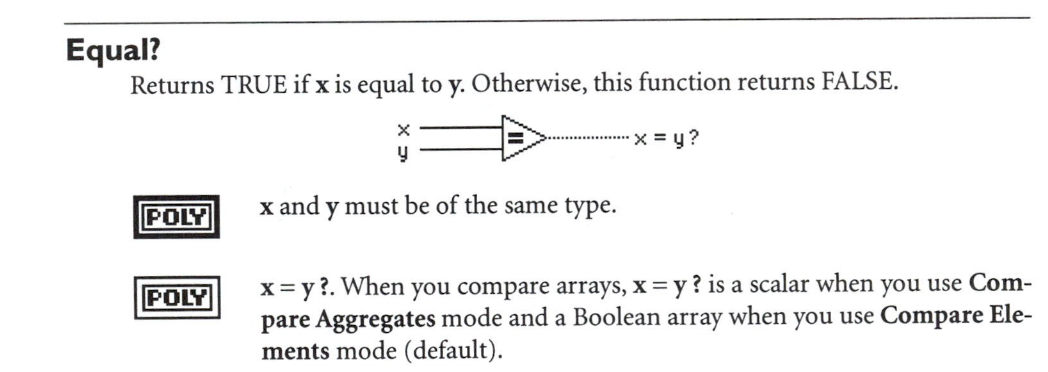

x and **y** must be of the same type.

x = y ?. When you compare arrays, **x = y ?** is a scalar when you use **Compare Aggregates** mode and a Boolean array when you use **Compare Elements** mode (default).

Greater Or Equal?

Returns TRUE if **x** is greater than or equal to **y**. Otherwise, this function returns FALSE.

x and **y** must be of the same type.

POLY

x >= y ?. When you compare arrays, **x >= y ?** is a scalar when you use **Compare Aggregates** mode and a Boolean array when you use **Compare Elements** mode (default).

Less Or Equal?

Returns TRUE if **x** is less than or equal to **y**. Otherwise, this function returns FALSE.

POLY

x and **y** must be of the same type.

POLY

x <= y ?. When you compare arrays, **x <= y ?** is a scalar when you use **Compare Aggregates** mode and a Boolean array when you use **Compare Elements** mode (default).

Not Equal?

Returns TRUE if **x** is not equal to **y**. Otherwise, this function returns FALSE.

x
y
x != y ?

x and **y** must be of the same type.

POLY

x != y ?. When you compare arrays, **x != y ?** is a scalar when you use **Compare Aggregates** mode and a Boolean array when you use **Compare Elements** mode (default).

Greater?

Returns TRUE if **x** is greater than **y**. Otherwise, this function returns FALSE.

x and **y** must be of the same type.

POLY

x > y ?. When you compare arrays, **x > y ?** is a scalar when you use **Compare Aggregates** mode and a Boolean array when you use **Compare Elements** mode (default).

Less?

Returns TRUE if **x** is less than **y**. Otherwise, this function returns FALSE.

x and **y** must be of the same type.

x < y ?. When you compare arrays, **x < y ?** is a scalar when you use **Compare Aggregates** mode and a Boolean array when you use **Compare Elements** mode (default).

Not A Number/Path/Refnum?

Returns TRUE if **number/path/refnum** is not a numeric value, path, or refnum. Otherwise, this function returns FALSE. **NaN** can be the result of dividing by 0, the square root of a negative number, and so on.

number/path/refnum —— NaN/Path/Refnum?

number/path/refnum can be a scalar number, path, or file refnum, or it can be a cluster or array of numbers, paths, or refnums.

NaN/Path/Refnum? is a Boolean value of the same data structure as **number/path/refnum**.

Max & Min

Compares **x** and **y** and returns the larger value at the top output terminal and the smaller value at the bottom output terminal.

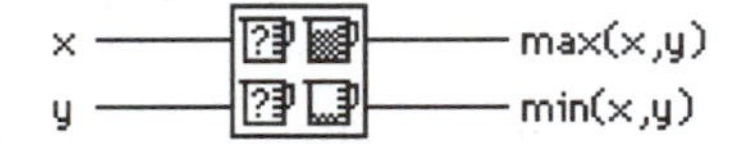

x and **y** must be of the same type.

max(x,y) is the larger value.

min(x,y) is the smaller value.

Select

Returns the value connected to the **t** input or **f** input, depending on the value of **s**. If **s** is TRUE, this function returns the value connected to **t**. If **s** is FALSE, this function returns the value connected to f.

t and **f** must be of the same type.

s determines whether the function returns the value of **t** or **f** in **s? t:f**.

s? t:f is the value wired to **t** if **s** is TRUE, the value wired to **f** if **s** is FALSE.

Boolean To (0,1)

Converts a Boolean value to a number: 0 and 1 for the input values FALSE and TRUE, respectively.

Boolean ?0:1 0, 1

Boolean can be a scalar, an array, or a cluster of Boolean values, an array of clusters of Boolean values, and so on.

0, 1. The output is 0 if **Boolean** is FALSE, 1 if **Boolean** is TRUE. **0, 1** is of the same data structure as **Boolean**.

String Functions

This chapter describes the string functions, including those that convert strings to numbers and numbers to strings. The string function palette can be accessed by selecting **String** from the **Functions** menu.

String Function Descriptions

String Length

Returns in **length** the number of characters (bytes) in **string**.

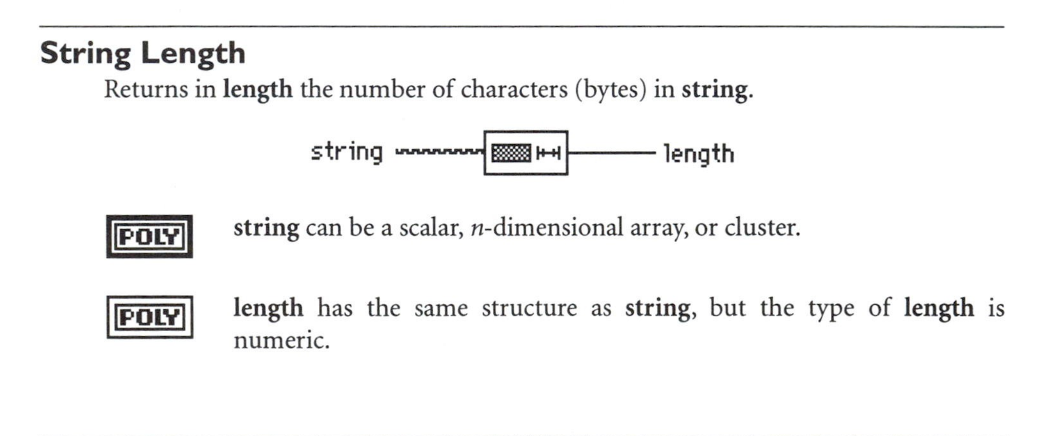

string can be a scalar, *n*-dimensional array, or cluster.

length has the same structure as **string**, but the type of **length** is numeric.

Concatenate Strings

Concatenates all the input strings into a single output string.

string0
string1
concatenation of string0, string1, ...

string0 and **string1** are the default input terminals. You can add as many input terminals as you need by selecting **Add Element** from the node pop-up menu, or by resizing the node with the Positioning tool.

concatenation of string0, string1, ... contains the concatenated input strings in the order you wire them to the node from top to bottom.

String Subset

Returns the **substring** of the original **string** beginning at **offset** and containing **length** number of characters.

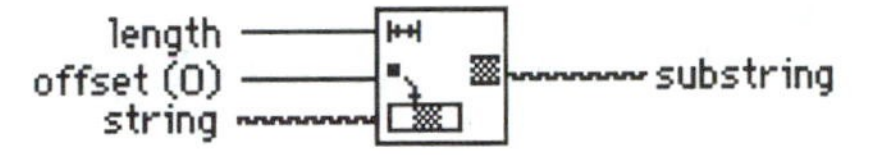

length must be scalar. If **length** is a floating-point number, the function rounds it to an integer value.

offset must be scalar. The **offset** of the first character in the string is 0. If you do not wire **offset**, or if it is less than 0, it defaults to 0. If **offset** is a floating-point number, the function rounds it to an integer value.

string.

substring is empty if **offset** is greater than the length of **string** or if **length** is less than or equal to 0. If **length** is greater than the length of **string** minus **offset**, the length of **string** minus **offset** is the length of **substring**.

Split String

Searchs for the first occurrence of **search char** in **string**, beginning at **offset**, and returns three outputs.

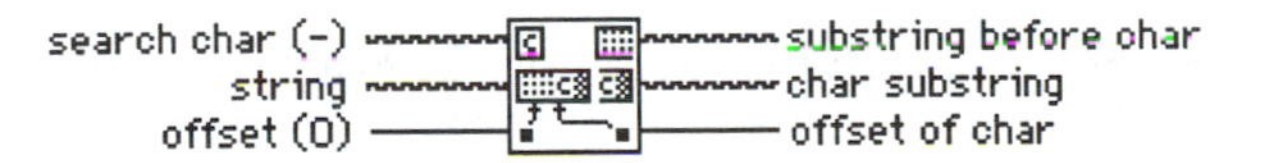

search char. If the function does not find **search char** in **string**, the function sets **offset of char** to –1, **substring before char** to the entire **string**, and **char substring** to an empty string. If you do not wire **search char**, or if it is an empty string, the function splits **string** at **offset**. You must wire either **search char** or **offset**.

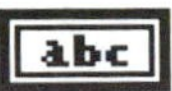

string is the input string that the function searches through or splits.

offset is the starting position for the search. **offset** must be scalar. The **offset** of the first character in **string** is 0. If you do not wire **offset**, or if it is less than 0, it defaults to 0. You must wire either **search char** or **offset.** If **offset** is a floating-point number, the function rounds it to an integer value.

substring before char returns all characters in **string** before **search char.**

char substring returns **search char** and all subsequent characters in **string.**

offset of char returns the position of **search char** in **string.**

Match Pattern

Searches for **regular expression** in **string** beginning at **offset** and returns your outputs.

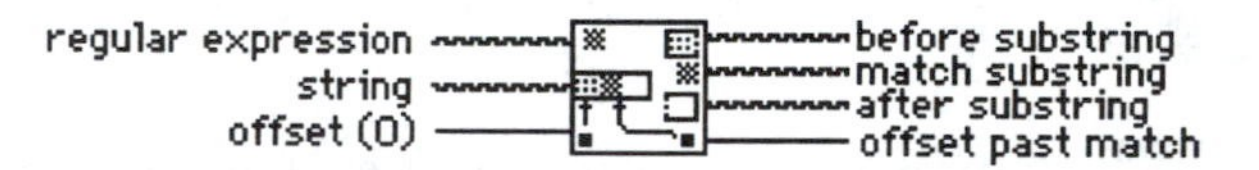

regular expression is the pattern to be matched in **string.** If the function does not find **regular expression**, then **match substring** is empty, **before substring** is the entire **string, after substring** is empty, and **offset past match** is –1. The tables that follow the input and output descriptions of this function describe special characters for **regular expression.**

string is the input string the function searches for.

offset must be scalar. The **offset** of the first character in **string** is 0, and if **offset** is left unwired or is less than 0, it defaults to 0. If **offset** is a floating-point number, the function rounds it to an integer value.

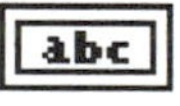

before substring returns a string containing all the characters before the match.

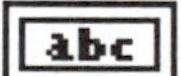

match substring returns the matched string.

after substring returns all characters following the matched pattern.

offset past match returns the index in **string** of the first character following the matched pattern.

The following table describes special characters you can use for regular expression.

Special Character	Interpreted by the Match Pattern Function As...
.	Matches any character.
?	Matches zero or one instance of the expression preceding ?.
\	Denotes the code for special characters (for example, * matches an asterisk). You can also use the following constructions for the space and nondisplayable characters. \b backspace \f form feed \n newline \s space \r carriage return *xx* any character, where *xx* is the hex code using 0 through 9 and upper case A through F \t tab
[ ]	Encloses alternates. For example, [abc] matches a, b, or c. The following character has special significance when used within the brackets in the following manner. - (dash) Indicate a range when used between digits, or lowercase or uppercase letters (for example, 0–5, a–g, or L–Q) The following characters have significance only when they are the first character within the brackets. ~ negates the set with respect to all characters, including the nondisplayable characters. [~0–9] gives any character but 0 through 9 ^ Negates the set with respect to all the displayable characters (and the space character). [^0–9] gives the space characters and all displayable characters except 0 through 9
^	If ^ is the first character of **regular expression**, it anchors the match to the **offset** in **string**. The match will fail unless **regular expression** matches a portion of **string** beginning with the character at **offset**. If ^ is not the first character, it is treated as a regular character.
+	Matches the longest number of instances of the expression preceding +; there must be at least one instance to constitute a match.
*	Matches the longest number of instances of the expression preceding * in **regular expression**, including zero instances.
$	If $ is the last character of **regular expression**, it anchors the match to the last element of **string**. The match will fail unless **regular expression** matches up to and including the last character in the string. If $ is not last, it is treated as a regular character.

The following table gives examples of strings you can wire to the regular expression parameter of the **Match Pattern** function.

Characters to Be Matched	**regular expression**
VOLTS	VOLTS
All uppercase and lowercase versions of volts, that is, VOLTS, Volts, volts, and so on	[Vv][Oo][Ll][Tt][Ss]
A space, a plus sign, or a minus sign	[+-]
A sequence of one or more digits	[0–9]+
Zero or more Spaces	\s* or * (that is, a space followed by an asterisk)
One or more Spaces, Tabs, Newlines, or Carriage Returns	[\t \r \n \s]+
One or more characters other than digits	[~0–9]+
The word Level only if it begins at the offset position in the string	^Level
The word Volts only if it appears at the end of the string	Volts$
The longest string within parentheses	(.*)
The longest string within parentheses but not containing any parentheses within it	([~()]*)

Select & Append

Chooses a string according to a Boolean **selector** and appends that string to **string.**

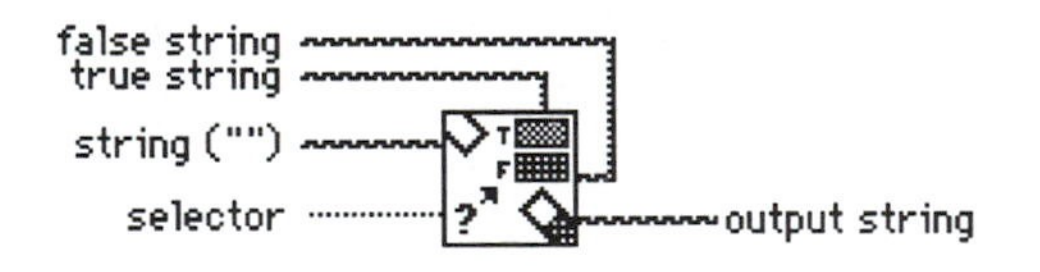

 **false string.**

 **true string.**

string defaults to empty string if you do not wire it.

 **selector**. If **selector** is TRUE, the function appends **true string** to **string.** If **selector** is FALSE, the function appends **false string** to **string.**

output string.

Format & Append

Converts **format string** into a regular LabVIEW string, converts **number** into a numeric field within the **format string**, and then appends the converted string to **string**. This operation is similar to the C programming language function printf and is useful for inserting the ASCII equivalent of a number into a string.

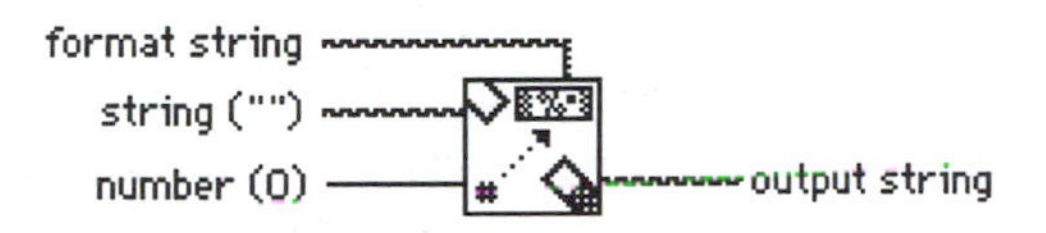

format string. For the values you can use in **format string**, refer to the tables at the end of the **Format & Append** section.

string defaults to an empty string if you do not wire it.

number must be scalar.

output string.

format string has the following syntax:

```
[String]%[-][0][WidthString][.PrecisionString]
ConversionCharacter[String]
```

The following table explains the elements of the preceding syntax.

Syntax Element	Description
String (optional)	Regular string in which you can insert certain characters as described below.
%	Character that begins the formatting specification.
- (dash) (optional)	Character that left justifies rather than right justifies the converted number within its width.
0 (zero) (optional)	Character that pads any excess space to the left of the number with zeros rather than spaces.

Syntax Element	Description
WidthString (optional)	Number specifying the minimum character width of the numeric field that contains the converted number. More characters are used if necessary. LabVIEW pads excess space to the left or right of the number with spaces, depending on justification. If WidthString is missing or if the width is zero, the converted number string is as long as necessary to contain the converted number.
. (period)	Character that separates WidthString from PrecisionString.
PrecisionString (optional)	Number specifying the number of digits to the right of the decimal point in the numeric field when **number** is a floating-point number. If PrecisionString is not followed by a period, a fractional part of six digits is inserted. If WidthString is followed by a period, and PrecisionString is missing or zero, no fractional part is inserted.
ConversionCharacter	Single character that specifies how to convert **number**, as follows. d to decimal integer x to hex integer o to octal integer f to floating-point number with fractional format e,g to floating-point number with scientific notation *Note: **ConversionCharacter** can be upper or lower case.*

To insert nondisplayable characters and the backslash and percent characters within **string**, use the codes described in the following table.

Code	Action
\r	to insert Carriage Return
\t	to insert Tab
\b	to insert Backspace
\n	to insert Newline
\f	to insert Form Feed
\s	to insert a space
\xx	to insert a character with hex code xx using 0 through 9 and uppercase A through F
\\	to insert \
%%	to insert %

The following table gives examples for the **Format & Append** function. In these examples, the underline character (_) represents a terminating space character.

String	Format String	Number	Output String
(empty)	score= %2d%%	87	score= 87%
score=_	%2d %%	87	score= 87%
(empty)	level= %–7.2e V	0.03642	level= 3.64e–2 V
level=_	\r%–7.2e V	0.03642	level= 3.64e–2 V
level= 3.64e–2 V	\r%–6.2e V	1.4	level= 3.64e–2 V 1.40e0 V

Pick Line & Append

Chooses a line from **multi-line string** and appends that line to **string**.

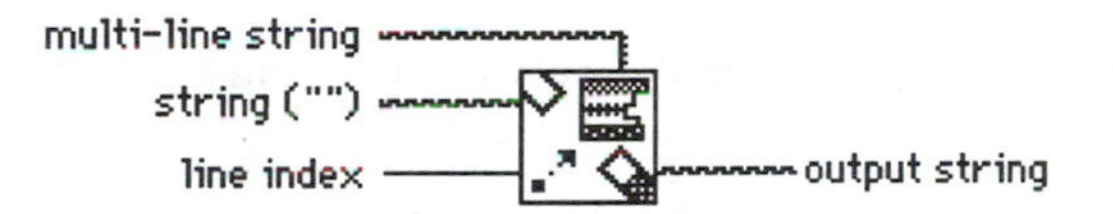

multi-line string contains one or more substrings separated by carriage returns.

string defaults to an empty string if you do not wire it.

line index selects the line the function appends from **multi-line string**. **line index** must be scalar. A **line index** of 0 selects the first line. If **line index** is negative or is greater than or equal to the number of lines in **multi-line string**, the function sets **output string** to **string**. If **line index** is a floating-point number, the function rounds it to an integer value.

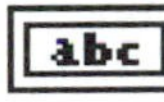

output string.

Index & Strip

Compares each string in **string array** with the beginning of **string** to see if there is a match. If there is a match, this function returns **string** minus the matched characters in **output string** and the **index** of the string in **string array** that matched. If there is no match, **output string** is **string** and **index** is –1.

string.

string array of potential matches to **string.**

index of the string in **string array** where a match was found. **index** is –1 if there is no match.

output string is **string** minus the matched characters if a match was found. **output string** is **string** if there is no match.

Format & Strip

Searches **string** until it finds a match with **format string,** formats any number in this string portion of **string** according to the conversion codes in **format string,** and returns the converted number in **number** and the remainder of **string** after the match in **output string.** This function behaves much like the C function scanf.

The following table shows how **format string** affects **number** and **output string.**

String	Format String	Number	Output String
number 1.2 wins!	number%.2f	1.20	wins!
1.2 wins!	%.2f	1.20	wins!
Q+1.279E3 trailer	Q%f tr	1279.00	ailer
set49.4.2	set%d	49	.4.2

string.

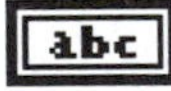

format string has the following syntax:

[String]%[-][0][WidthString][.PrecisionString]
ConversionCharacter[String]

Common conversion characters are

d to decimal integer
x to hex integer
o to octal integer
f to floating-point number with fractional format
e,g to floating-point number with scientific notation

For a more complete list of values you can use in **format string,** see the **Format & Append** function description in this chapter.

For this function only, LabVIEW ignores leading spaces. A space character in the format string such as ... `abc def` ... matches an arbitrary amount of white space including zero or more spaces, tabs, carriage returns, line feeds, or form feeds. To match exactly one space, use the backslash-space construction, (\ ). The conversion characters `e` and `f` both match numbers with exponents. See the previous table for an example.

default is the default value of the output number any object of the appropriate representation to **default**. If you do not wire **default**, the parameter defaults to a value of 0.

number. If there is no match, **number** is the value in **default**.

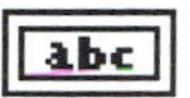

output string. If there is no match, **output string** is **string**.

Spreadsheet String To Array

Converts the **spreadsheet string** to a numeric **array** of the dimension and representation of **array type**. This function works for arrays of strings as well as arrays of numbers.

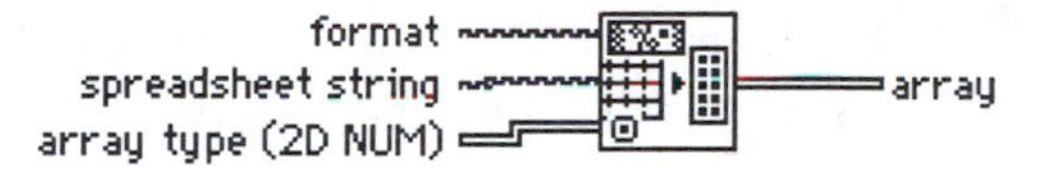

format. For a description of valid **format** conversion characters, see the discussion of the **Format & Append** function earlier in this chapter. For the **Spreadsheet String to Array** function, you can use the percent character followed by an s (%s) to convert **spreadsheet string** to an array of strings.

spreadsheet string. Tabs separate columns in **spreadsheet string**, and an End-of-Line (EOL) character separates rows. The function converts each element in **spreadsheet string** according to **format** and then stores them in **array**.

array type. If **array type** is unwired, it defaults to a two-dimensional array of numbers.

array.

Array To Spreadsheet String

Converts a numeric **array** of any dimension to **spreadsheet string**, which is a table in string form in which tabs separate column elements, an EOL character separates rows, and for arrays of three or more dimensions, pages are separated as described below.

format string
array
spreadsheet string

format string. For a description of valid **format string** conversion characters, see the **Format & Append** function description earlier in this chapter. For the **Array to Spreadsheet String** function, you can use the percent character followed by an s (%s) to convert **array** to the **spreadsheet string**.

array. The function converts each element in **array** according to **format string** and then appends them to **spreadsheet string**.

spreadsheet string.

Build Path

Creates a new path by appending a name to an existing path.

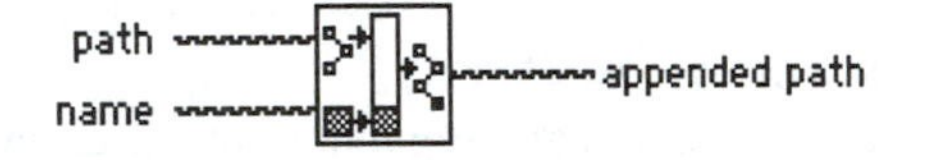

path specifies the base path to which this function appends **name**. It defaults to **empty path** if left unwired. If **path** is invalid, this function sets **appended path** to not-a-path.

name is the new path component appended to the specified path. If **name** is an empty string, this function sets **appended path** to not-a-path.

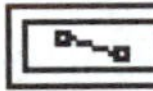

appended path is the resulting path.

Strip Path

Returns the name of the last component of a path and the path that leads to that component.

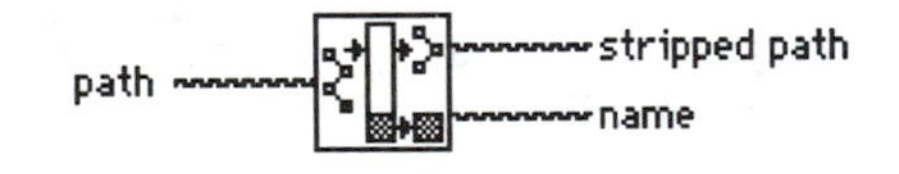

path specifies the path on which you want to operate. If this parameter is an empty path or is invalid, this function returns an empty string in **name** and not-a-path in **stripped path**.

name is the last component of the specified path.

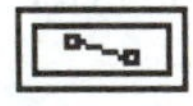

stripped path is the resulting path obtained by removing **name** from the end of **path**.

String To Path

Converts a string describing a path in the standard format for the current platform to a path.

string path

String can be a string, a cluster of strings, an array of strings, an array of clusters of strings, and so on.

path is the platform-independent representation of the path described by **string**. If **string** is not a valid path description on the current platform, the function set **path** to not-a-path. **path** is of the same data structure as **string**.

Path To String

Converts a path into a string describing a path in the standard format for the current platform.

path string

Path can be a path, a cluster of paths, an array of paths, an array of a cluster of paths, and so on.

string is the path description in the standard format for the current platform of **path**. If **path** is not-a-path, the function sets **string** to an empty string. **string** is of the same data structure as **path**.

Array and Cluster Functions

This chapter describes the functions for array and cluster operations. You access the array and cluster function palette by selecting **Array & Cluster** from the **Functions** menu.

Some of the array and cluster functions are also available from the **Array Tools** and **Cluster Tools** palettes of most terminal or wire pop-up menus, if the data type of the terminal or wire is array or cluster. If you select the functions from these palettes, they appear with the correct number of terminals to wire to the object on which you popped up.

General Behavior of Array and Cluster Functions

Some of the array and cluster functions have a changeable number of terminals. When you drop a new function of this kind, it appears on the block diagram with only one or two terminals. You can add and remove terminals by using the pop-up menu **Add** and **Remove** commands (the actual names depend on the function) or by resizing the node vertically from any corner. You cannot shrink the node if doing so would delete wired terminals. The **Add** command inserts a terminal directly after the one on which you popped up. The **Remove** command removes the terminal on which you popped up, even if it is wired. The following illustration shows the two ways to add more terminals to the **Build Array** function.

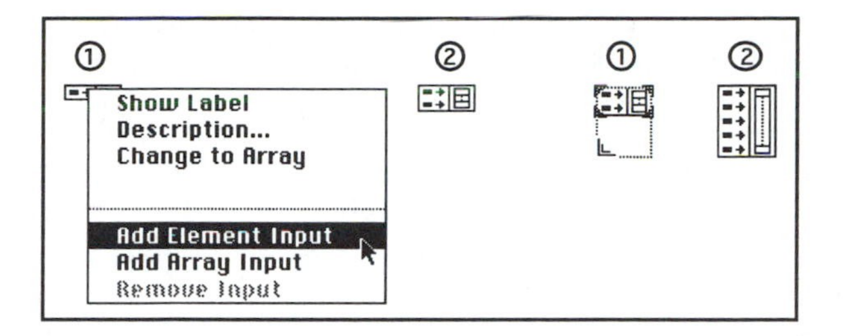

Polymorphism

Most of the array functions accept *n*-dimensional arrays of any type, although the wiring diagrams in the function descriptions show numeric arrays as the default data type.

The **Bundle** and **Unbundle** functions do not show the data type for their individual input or output terminals until you wire objects to these terminals. When you wire them, these terminals look similar to the data types of the corresponding front panel control or indicator terminals.

Out-of-Range Index Values

Attempting to index an array beyond its bounds results in a default value determined by the array element type.

Array Functions

Array Max & Min

Searches for the maximum and minimum values in **numeric array**. This function also returns the indices where it finds the maximum and minimum values.

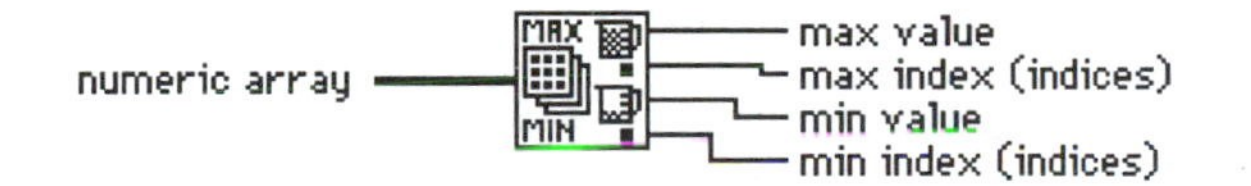

numeric array can have any number of dimensions.

max value and **min value** are of the same data type as the elements in **numeric array**.

max index and **min index**. If **numeric array** is multidimensional, then **max index** and **min index** are arrays, each of whose elements is the index for the corresponding dimension in **numeric array**.

The function compares each data type according to the rules discussed in the *Comparison Functions* section.

Array Size

Returns the number of elements in each dimension of **array**.

array — size(s)

array can be an *n*-dimensional array of any type.

size(s). If **array** is one-dimensional, the size is a scalar number. If **array** is multidimensional, the returned value is a one-dimensional array in which each element represents the number of elements in the corresponding dimension of **array**. For example, a 3D array with two rows, three columns, and four pages has the following output array.

4	2	3
Pages	Rows	Columns

Array Subset

Returns a portion of **array** starting at **index** and containing **length** elements.

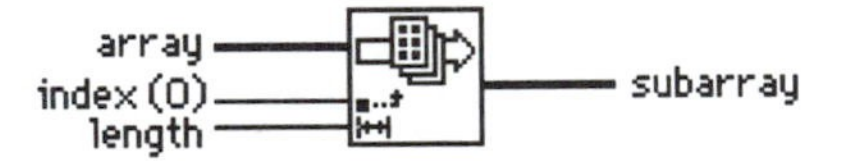

array can be an *n*-dimensional array of any type. If **array** is multidimensional, you must add a pair of **index** and **length** terminals for each dimension by resizing or adding a dimension.

index must be a scalar number. If **index** is less than 0, the function sets it to 0. If **index** is greater than or equal to the array size, the function returns an empty array.

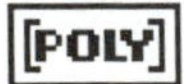

length must be a scalar number. If **index** plus **length** is larger than the size of the array, the function returns only as much data as is available.

subarray is of the same type as the elements of **array**.

Build Array

Appends all inputs in top-to-bottom order to create **array with appended element**.

array
element
array with appended element

array. The elements of **array** can be of any data type.

element must be of the same data type as the elements of **array**.

array with appended element.

Initially, all the inputs of this function are elements. To change an element input to an array input, pop up on the input and select **Change to Array** from the pop-up menu. To create a 1D array, connect scalar values to the element inputs and 1D arrays to the array inputs. To build a 2D array, connect 1D arrays to element inputs and 2D arrays to the array inputs. In general, to build an array of *n* dimensions, each input **array** must be of the same dimension, *n*, and each element must have *n*–1 dimensions. Thus, wiring 2D arrays to a series of element inputs produces a 3D array.

Index Array

Returns the **element** of **array** at **index**. If **array** is multidimensional, you must add additional **index** terminals for each dimension of **array** by resizing or popping up and adding elements.

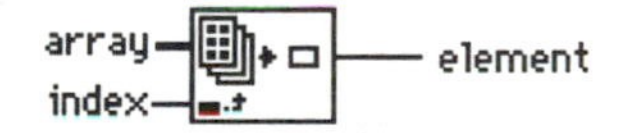

array can be an *n*-dimensional array of any type (except array).

index must be a scalar number.

element has the same type as the elements of **array**.

In addition to extracting an element of the array, you can *slice* out a higher-dimensional component by disabling one or more of the index terminals.

Initialize Array

Returns an *n*-dimensional array in which every element is initialized to the specified value.

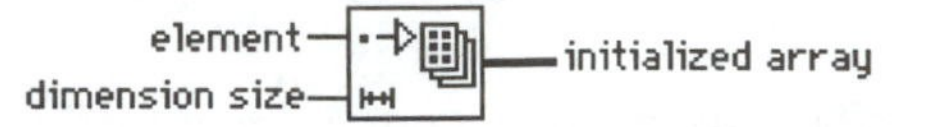

element determines the type of the array and the initial value of each array element. **element** cannot be an array.

dimension size. You must add a dimension size terminal for each dimension of the initialized array.

[POLY] **initialized array** is an array of the same type as the type you wire to **element.**

Replace Array Element

Replaces with **new element** the element in **array** at **index.**

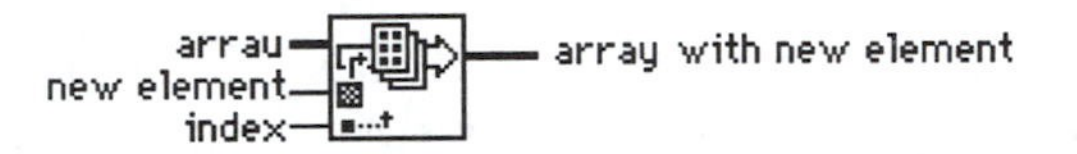

array can be an *n*-dimensional array of any type.

new element must be of the same type as the other elements in **array.**

NUM **index** must be scalar. If **array** is multidimensional, you must wire an **index** terminal for each dimension.

[POLY] **array with new element** is of the same composition as **array.**

Reshape Array

Changes the dimension of an array according to the value of **dimension size.**

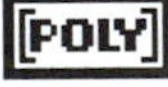

n-dim array can be an *n*-dimensional array of any type except Boolean.

dimension size. You must add *m* **dimension size** terminals for *m* dimensions. **dimension size** must be a scalar number.

m-dim array. If the product of the dimension sizes is greater than the number of elements in the input array, the function pads the new array with default values. If the product of the dimension sizes is less than the number of elements in the input array, the function truncates the new array.

Sort 1D Array

Returns a sorted version of **array** in which the elements are arranged in ascending order. The rules for comparing each data type are described in the *Comparison Functions* section.

1D array can be a one-dimensional array of any type except Boolean.

sorted array.

Split 1D Array

Divides **array** at **index** and returns the two portions.

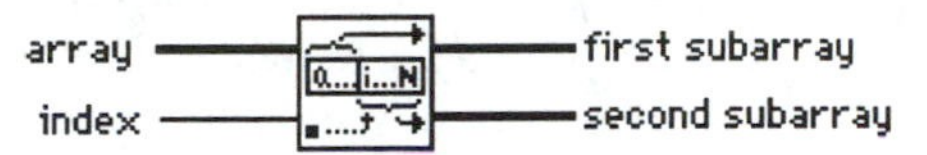

array can be a one-dimensional array of any type.

index must be scalar. If **index** is negative or 0, **first subarray** is empty. If **index** is greater than the size of **array**, **second subarray** is empty.

first subarray and **second subarray** contain elements of the same data type as the elements of **array**. **first subarray** contains **array**[0] through **array**[**index**–1], and **second subarray** contains the rest of **array**.

Transpose 2D Array

Rearranges the elements of **2D array** such that **2D array**[i,j] becomes **transposed array**[j,i].

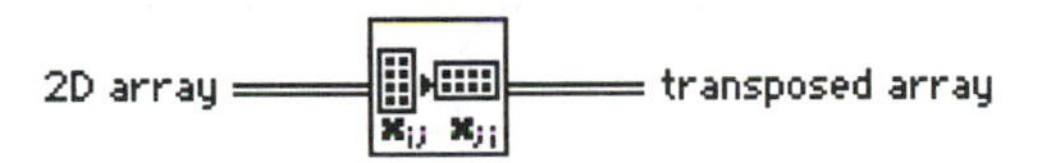

2D array can be a two-dimensional array of any type.

transposed array.

Cluster Functions

The cluster functions described in the following sections are expandable. You can use the Positioning tool to stretch them to contain the desired number of input and output terminals.

Bundle

Assembles all the individual input components into a single cluster or replaces elements in a cluster.

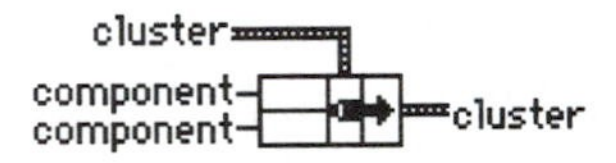

component, **component**, and **cluster**. The input parameters of this function are polymorphic—they can be of any data type. If you wire **cluster**, only those components you want to change must be wired. The unwired components remain unchanged. If you do not wire **cluster**, you must wire all the components.

cluster.

Unbundle

Splits a cluster into its individual components.

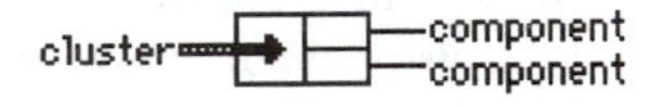

cluster.

component and **component**. The output parameters of this function are polymorphic. The function arranges the components from top to bottom in cluster order, the order in which you initially added them to the cluster.

Time & Dialog Functions

This chapter describes the one- and two-button dialog box functions and the timing functions, which you can use to measure elapsed time or to suspend an operation for a specific period of time.

You access the dialog and timing function palette by selecting **Time & Dialog** from the **Functions** menu.

User Dialog Function Descriptions

One Button Dialog

Displays a dialog box that contains a **message** and a single button. The **button name** is the name displayed on the dialog box button.

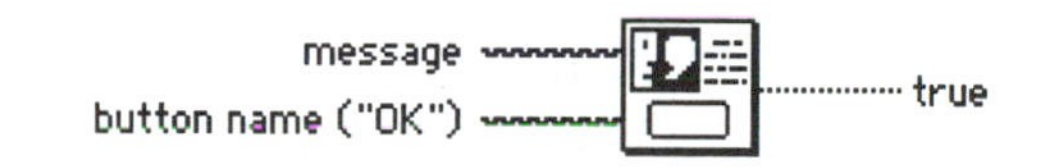

message.

button name. If you do not wire this input, **button name** defaults to OK.

true. When you click on the button, **true** contains a value of TRUE.

Two Button Dialog

Displays a dialog box that contains a **message** and two buttons. **T button name** and **F button name** are the names displayed on the buttons of the dialog.

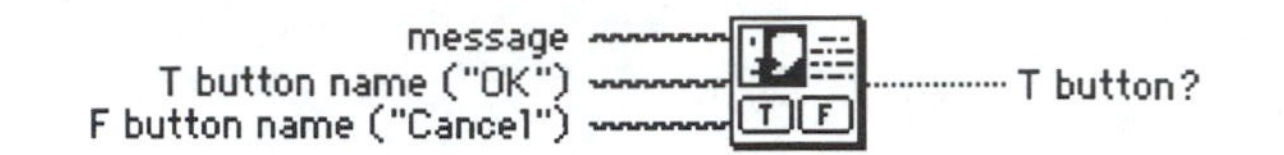

message.

T button name. If you do not wire this input, the name defaults to OK.

F button name. If you do not wire this input, the name defaults to Cancel.

T button?. If the user clicks on the front panel control whose terminal you wire to **T button name, T button?** returns a value of TRUE. If the user clicks on the front panel control whose terminal you wire to **F button name, T button?** returns a value of FALSE.

Timing Function Descriptions

Get Date/Time String

Outputs in string format the date and time specified by the number of **seconds** that have elapsed since 12:00 A.M., Friday, January 1, 1904 GMT or its equivalent in the local time zone.

date format (0)
seconds (now)
want seconds? (F)
date string
time string

NUM

date format selects the appearance of the **date string.** The following date formats are available.

0: 1/1/04

1: Friday, January 1, 1904

2: Fri Jan 1 04

Notice that these formats depend on your system resource file. Thus, if you do not have a U.S. system resource file, these formats are different.

seconds. If you leave **seconds** unwired, LabVIEW defaults to "now" and uses the current date and time as output. Otherwise, the function returns the date and time that correspond with the number of **seconds** since 12:00 A.M., January 1, 1904.

want seconds? controls the display of seconds in the **time string.**

date string.

time string is of the form: 15:24.

Tick Count (ms)

Returns the value of the millisecond timer. The base reference time (millisecond zero) is undefined; that is, you cannot convert **millisecond timer value** to a real-world time or date. Be careful when you use this function in comparisons because the value of the millisecond timer wraps from $2^{32}–1$ to 0. To safeguard against wrapping occurring without your knowledge, you can compare the current reading with the previous reading. If the current value is less than the previous value, the millisecond timer has wrapped around.

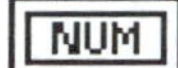

millisecond timer value.

Note

The timing resolution is different from platform to platform and is limited by the internal clock. On the Mac, one tick of the internal clock is about 17 ms. Under Windows, one tick is usually 55 ms.

Wait (ms)

Waits the specified number of milliseconds. The function returns the value of the millisecond timer when the wait expires.

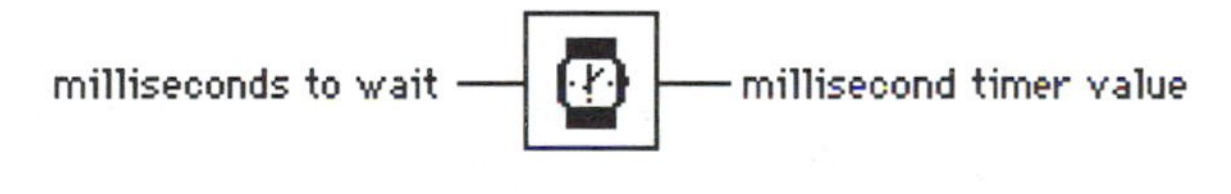

milliseconds to wait.

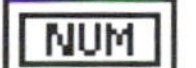

millisecond timer value.

millisecond timer value.

Note

The timing resolution is different from platform to platform and is limited by the internal clock. On the Mac, one tick of the internal clock is about 17 ms. Under Windows, one tick is usually 55 ms.

File & Error Functions

This chapter describes the VIs that perform high-level file I/O and error handling operations. The **File & Error** palette can be accessed from the **Functions** menu.

These VIs use three terminal label styles in the VI help window to distinguish the importance of each input and output. Labels in **bold** font are all you need to use in many applications. You use labels in plain text less often, and labels enclosed in brackets [ ] least of all.

File VI Descriptions

Write Characters To File

Writes a character string to a new byte stream file or appends the string to an existing file. The VI opens or creates the file beforehand and closes it afterward.

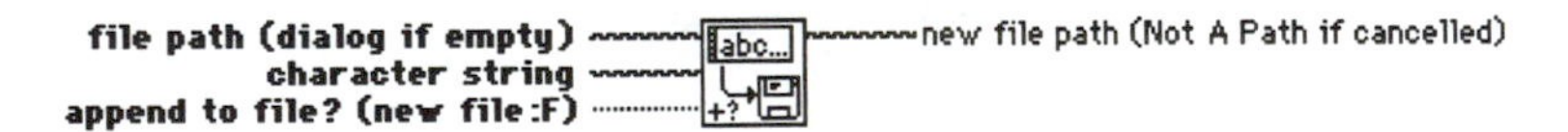

file path is the path name of the file. If **file path** is empty (default value) or is Not A Path, the VI displays a File dialog box from which you can select a file. Error 43 occurs if the user cancels the dialog.

character string is the data the VI writes to the file.

append to file. Set to TRUE if you want to append the data to a existing file; you can also set it TRUE to write to a new file. Set to FALSE (default value) if you want to write the data to a new file or to replace an existing file.

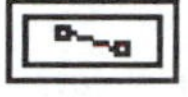

new file path is the path of the file to which the VI wrote data. You can use this output to determine the path of a file that you open using a dialog box. **new file path** returns Not A Path if the user selects Cancel from the dialog box.

Read Characters From File

Reads a specified number of characters from a byte stream file beginning at a specified character offset. The VI opens the file beforehand and closes it afterward.

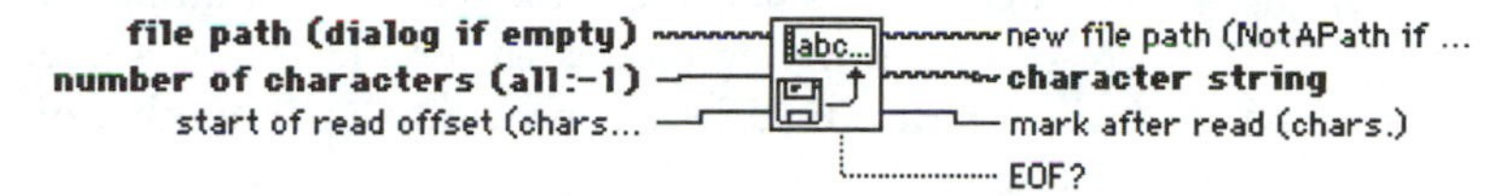

file path is the path name of the file. If **file path** is empty (default value) or is Not A Path, the VI displays a File dialog box from which you can select a file. Error 43 occurs if the user cancels the dialog.

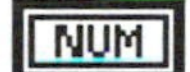

number of characters is the maximum number of characters the VI reads. The VI reads fewer characters if it reaches the EOF first. If **number of characters** <0, the VI reads the entire file. The default value is –1.

start of read offset is the position in the file, measured in characters (or bytes), at which the VI begins reading.

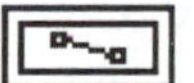

new file path is the path of the file from which the VI read data. You can use this output to determine the path of a file that you open using a dialog box. **new file path** returns Not A Path if the user selects Cancel from the dialog box.

character string is the data read from the file.

mark after read is the location of the file mark after the read; it points to the character in the file following the last character read.

EOF? is TRUE if you attempt to read past the end of file.

Read Lines From File

Reads a specified number of lines from a byte stream file beginning at a specified character offset. The VI opens the file beforehand and closes it afterward.

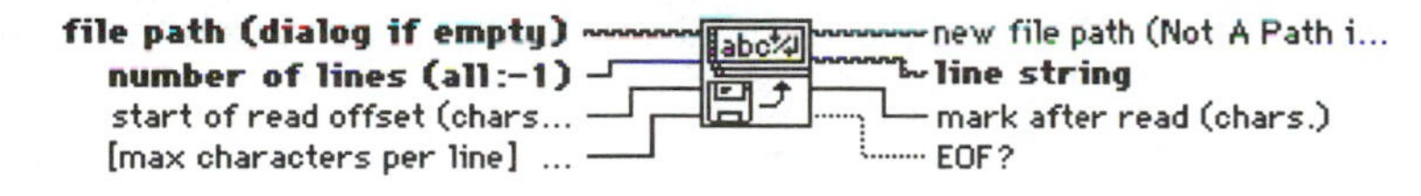

file path is the path name of the file. If file path is empty (default value) or is Not A Path, the VI displays a File dialog box from which you can select a file. Error 43 occurs if the user cancels the dialog.

NUM

number of lines is the maximum number of lines the VI reads. If **number of lines**<0, the VI reads the entire file. The default value is –1. For this VI, a line is a character string ending with a carriage return, line feed, or a carriage return followed by a line feed; a string ending at the EOF; or a string that has the maximum line length specified by the max characters per line input.

start of read offset is the position in the file, measured in characters (or bytes), at which the VI begins reading.

NUM

max characters per line is the maximum number of characters the VI reads before ending the search for the end of a line. The default is 0, which means that there is no limit to the number of characters the VI reads.

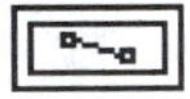

new file path is the path of the file from which the VI read data. You can use this output to determine the path of a file that you open using a dialog box. **new file path** returns Not A Path if the user selects Cancel from the dialog box.

line string is the data read from the file.

mark after read is the location of the file mark after the read; it points to the character in the file following the last character read.

EOF? is TRUE if you attempt to read past the end of the file.

Write To Spreadsheet File

Converts a 2D or 1D array of numbers to a text string and writes the string to a new byte stream file or appends the string to an existing file. You can optionally transpose the data. This VI opens or creates the file beforehand and closes it afterwards. You can use this VI to create a text file readable by most spreadsheet applications.

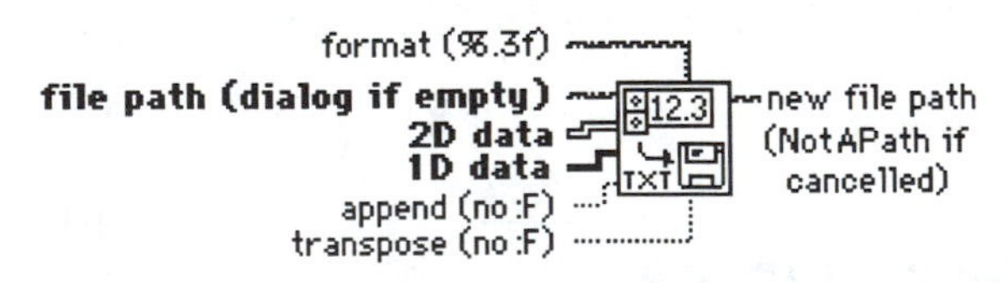

file path is the path name of the file. If file path is empty (default value) or is Not A Path, the VI displays a File dialog box from which you can select a file. Error 43 occurs if the user cancels the dialog.

2D data contains the single-precision numbers the VI writes to the file if 1D data is not wired or is empty.

1D data contains the single-precision numbers the VI writes to the file if this input is not empty. The VI converts the 1D array into a 2D array before transposing it and converting it to a string and writing it to the file. If transpose? is FALSE, each call to this VI creates a new line or row in the file.

append to file? Set to TRUE if you want to append the data to a existing file; you can also set it TRUE to write to a new file. Set to FALSE (default value) if you want to write the data to a new file or to replace an existing file.

transpose? Set TRUE to transpose the data before converting it to a string. The default value is FALSE.

format specifies how to convert the numbers to characters. If the format string is %.3f (default), the VI creates a string long enough to contain the number, with three digits to the right of the decimal point. If the format is %d, the VI converts the data to integer form using as many characters as necessary to contain the entire number. Refer to the discussion of format strings and the **Array To Spreadsheet String** function in the *String Functions* section.

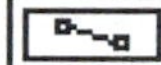

new file path is the path of the file to which the VI wrote data. You can use this output to determine the path of a file that you open using dialog. **new file path** returns Not A Path if the user selects Cancel from the dialog box.

Read From Spreadsheet File

Reads a specified number of lines or rows from a numeric text file beginning at a specified character offset and converts the data to a 2D array of numbers. You can optionally transpose the array. The VI opens the file beforehand and closes it afterward. You can use this VI to read spreadsheet file saved in text format.

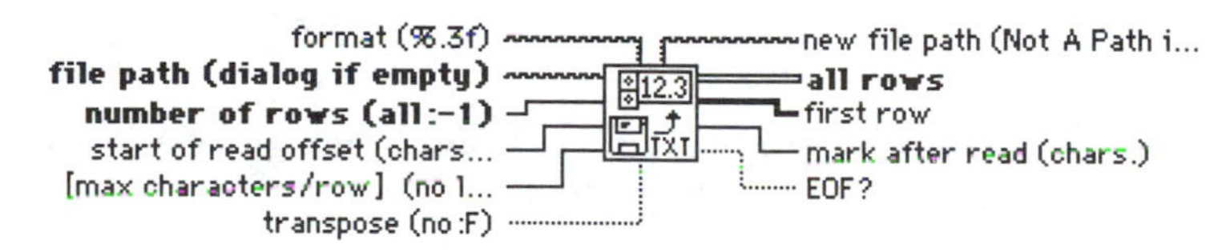

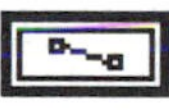

file path is the path name of the file. If file path is empty (default value) or is Not A Path, the VI displays a File dialog box from which you can select a file. Error 43 occurs if the user cancels the dialog.

number of rows is the maximum number of rows or lines the VI reads. If **number of rows** <0, the VI reads the entire file. The default value is –1. For this VI, a row is a character string ending with a carriage return, line feed, or a carriage return followed by a line feed; a string ending at the EOF; or a string that has the maximum line length specified by the max characters per row input.

start of read offset is the position in the file, measured in characters (or bytes), at which the VI begins reading.

max characters per row is the maximum number of characters the VI reads before ending the search for the end of a row or line. The default is 0, which means that there is no limit to the number of characters the VI reads.

transpose? Set TRUE to transpose the data after converting it from a string. The default value is FALSE.

format specifies how to convert the characters to numbers; the default is %.3f. Refer to the discussion of format strings and the **Spreadsheet String To Array** function in the *String Functions* section.

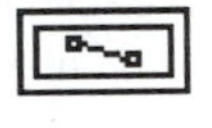

new file path is the path of the file from which the VI read data. You can use this output to determine the path of a file that you open using dialog. new file path returns Not A Path if the user selects Cancel from the dialog box.

all rows is the data read from the file in the form of a 2D array of numbers.

first row is the first row of the **all rows** array in the form of a 1D array of numbers. You can use this output when you want to read one row into a 1D array.

mark after read is the location of the file mark after the read; it points to the character in the file following the last character read.

EOF? is TRUE if you attempt to read past the end of the file.

Linking LabVIEW to HiQ

LabVIEW has two VIs you can use to read numeric text files created by HiQ and write to numeric text files readable by HiQ.

Write to HiQ Text File

Converts a 1D or 2D array of numbers to a text string and writes the string to a new file or appends the string to an existing file. You can optionally transpose the data. This VI opens or creates the file beforehand and closes it afterward. You can use this VI to create a text file that HiQ can read.

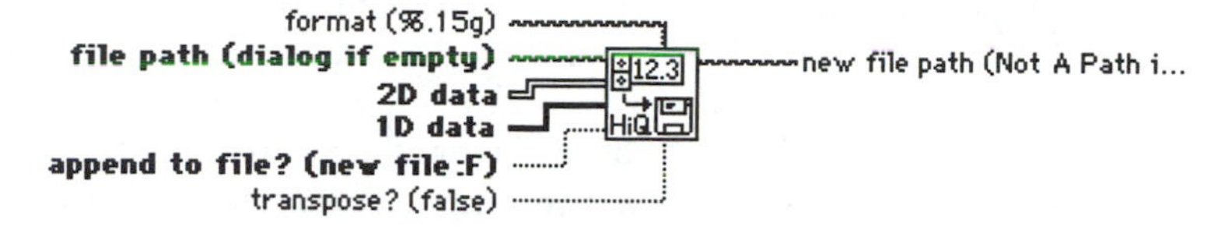

file path is the path name of the file. If file path is empty (default value) or is Not A Path, the VI displays a file dialog box from which you can select a file. Error 43 occurs if the user cancels the dialog.

2D data contains the numbers the VI writes to the file if 1D data is not wired or is empty.

1D data contains the numbers the VI writes to the file if this input is not empty. The VI converts the 1D array into a 2D array before optionally transposing it, converting it to a string and writing it to the file. If **transpose?** is FALSE, each call to this VI creates a new line or row in the file.

append to file? Set to TRUE if you want to append the data to an existing file. Set to FALSE (default value) if you want to write the data to a new file or to replace an existing file.

transpose? Set to TRUE to transpose the data before converting it to a string. The default value is FALSE.

format specifies how to convert the numbers to characters. If the format string is %.15g (default), the VI creates a string long enough to contain the number, with 15 digits to the right of the decimal point. If the format is %d, the VI converts the data to integer form using as many characters as necessary to contain the entire number. Refer to the discussion of format strings and the **Format & Append** function in the *String Functions* section.

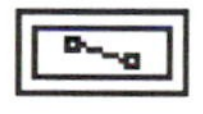

new file path is the path of the file to which the VI wrote data. Not A Path is returned if the user selects Cancel from the dialog box.

Read from HiQ Text File

Reads a specified number of rows from a HiQ numeric text file beginning at a specified character offset and converts the data to a 2D array of numbers. You can optionally transpose the array. The VI opens the file beforehand and closes it afterward.

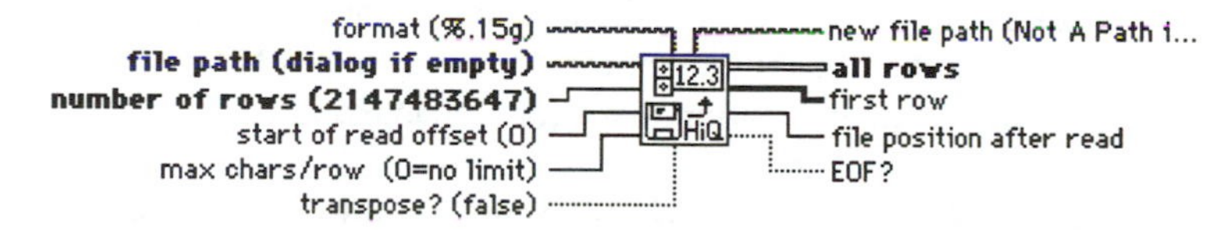

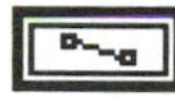

file path is the path name of the file. If file path is empty (default value) or is Not A Path, the VI displays a file dialog box from which you can select a file. Error 43 occurs if the user cancels the dialog.

number of rows is the maximum number of rows or lines the VI reads. For this VI, a row is a character string ending with a carriage return, line feed, or a carriage return followed by a line feed; a string ending with end of file; or a string that has the maximum line length specified by the **max characters per row** input. If **number of rows** <0, the VI reads the entire file. The default value is –1.

start of read offset is the position in the file, measured in characters, at which the VI begins reading.

max chars/row is the maximum number of characters the VI reads before ending the search for the end of a row or line. The default is 0, which means there is no limit to the number of characters the VI reads.

transpose? Set to TRUE to transpose the data after converting it from a string. The default value is FALSE.

format specifies how to convert the characters to numbers; the default is %.15g. Refer to the discussion of format strings and the **Format & Append** function in the *String Functions* section.

new file path is the path of the file from which the VI read data. Not A Path is returned if the user selects Cancel from the dialog box.

all rows is the data read from the file in the form of a 2D array of numbers.

first row is the first row of the **all rows** array in the form of a 1D array of numbers. You can use this output when you want to read one row into a 1D array.

file position after read is the location of the file mark after the read; it points to the character in the file following the last character read.

EOF? is TRUE if you attempt to read past the end of the file.

Error Handling

When designing applications, you should consider the possibility of unexpected situations and decide how you are going to handle them. For instance, when writing to a file you should consider the possibility that the operation might fail due to a lack of disk space. If you are designing an application that needs reliable operation, you should check for errors and be prepared to handle them.

LabVIEW I/O operations check for errors and return error codes. Error handling is different for every application, so LabVIEW doesn't perform automatic handling of errors. For instance, you may want to shut the application down after receiving an error, or you may just want to notify the user so that he or she can correct the problem.

The error handler VI has both an error input and output. If the input indicates an error, then the VI either does nothing or shuts down any I/O operation that it is managing. One advantage of this error input/output design is that you can connect several I/O operations together. If an error occurs, LabVIEW will not perform subsequent unnecessary VIs, giving the user the opportunity to handle the error.

Another advantage of this error input/output design is that it allows you to establish the order of a set of I/O operations, even though there may not be any data dependency between the operations. You could establish execution order with a Sequence structure, but the error input/output design allows you to establish the order with all of the operations on the same level of the block diagram.

Error Input/Output Cluster

The error input/output used by many of the high-level I/O operations is a cluster containing a Boolean indicating when an error has been detected, the error code, and a descriptive string that usually contains the name of the source of the error. This cluster is shown in the following illustration.

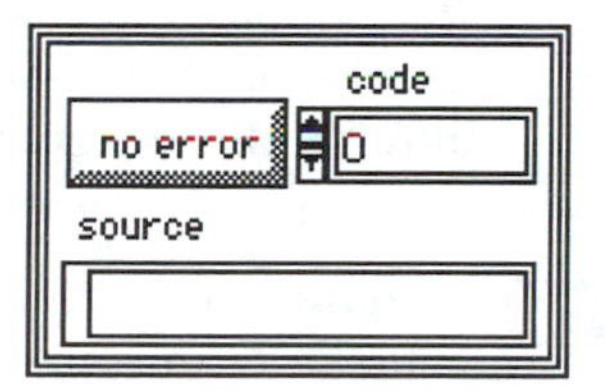

Simple Error Handler VI Overview

Simple Error Handler

Ascertains whether an error has occurred. If it finds an error, this VI creates a description of the error and optionally displays a dialog box.

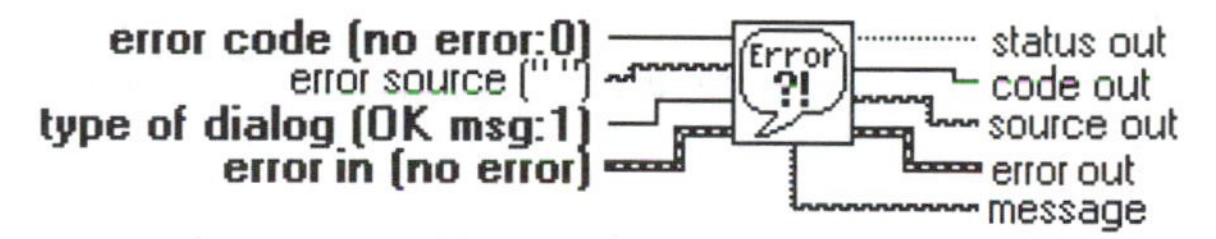

Simple Error Handler will take care of the error handling needed by most applications. You can pass the error information to this VI either as the error cluster or as an error code with an optional descriptive string. If **error cluster in** indicates an error, the VI reports that error as output. If not, or if **error cluster in** is not wired, the VI checks **error code** to see if it is nonzero. If it is, the VI converts the error code to a description of the error, if possible. You can also have the VI display a dialog box with the error description and an OK button, or a dialog box with the error description and buttons that allow the user to stop or continue execution.

type of dialog determines what type of dialog box to display, if any. Regardless of its value, the VI outputs the error information and **message** describing the error.

0: displays no dialog box. This is useful if you want to have programmatic control over how the error is handled.

1: (the default value) displays a dialog box with a single OK button. After the user acknowledges the dialog box, the VI returns control to the main VI.

2: displays a dialog box with buttons allowing the user to either continue or stop. If the user cancels, the VI calls the Stop function to halt execution.

error code is a numeric error code. The VI ignores this value if **error in** indicates an error. Otherwise, this value is tested. A nonzero value signifies an error.

error source is an optional string that you can use to describe the source of **error code**. If **error code** indicates an error, the VI uses this string in the **message** that this VI returns and possibly displays.

error in describes an error that you want to check. If you leave **error in** unwired, this VI checks **error code** for errors. It has the following parameters:

status is TRUE if an error occurred. If this value is FALSE, the VI assumes that no error occurred according to the **error in**, and then checks **error code**.

code is the error code associated with an error.

source. In case of an error, most VIs that use the error in and error out clusters set **source** to the name of the VI or function that produced the error.

status out is TRUE if the VI found an error.

code out is the error code indicated by **error in** or **error code**.

source out indicates the source of the error.

error out is a cluster containing the same information as in **status out**, **code out**, and **source out**. It has the same structure as **error in**.

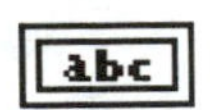

message describes the error that occurred and the source of the error.

Analysis VIs

This chapter discusses the analysis VIs, their uses, their parameters, and the algorithms they use. The analysis VIs fall into three submenus:

- **DSP (Digital Signal Processing)** contains VIs that perform frequency domain transformations, frequency domain analysis, time domain analysis, IIR digital filtering functions, data windowing, and generate digital waveforms.
- **Numeric** contains VIs that perform numerical analysis and algebraic functions on vectors and matrices.
- **Statistics** contains VIs that perform descriptive statistics functions, such as identifying the mean or the standard deviation of a set of data, as well as inferential statistics functions for probability, interpolation, and curve-fitting functions

Error Reporting

Each analysis VI has an error output parameter, which returns a nonzero number when invalid input conditions arise. Refer to the list of error codes in *Appendix C* for the error condition that corresponds to the code.

In general, if the VI cannot resolve the error conflicts, or if it cannot complete the operation, the VI sets output arrays to empty arrays and floating-point output scalars to undefined, which it displays as *not a number* (NaN). The following figure shows the front panel of the **Dot Product** VI, which returns NaN and an error code of –20003, which means the input arrays must not be empty, when you try to calculate the dot product of X and Y.

Dot Product

X Vector

0 0.00

Y Vector

0 0.00

X * Y

NaN

error

-20003

Notation and Naming Conventions

To help you identify the type of parameters and operations, this document uses the following notation and naming conventions unless otherwise specified in a VI description. Although there are a few scalar functions and operations, most of the analysis VIs process large blocks of data in the form of one-dimensional arrays (or vectors) and two-dimensional arrays (or matrices).

Normal lowercase letters represent scalars or constants. For example,

$$a\,,$$
$$\pi\,,$$
$$b = 1.234.$$

Capital letters represent arrays. For example,

$$X\,,$$
$$A\,,$$
$$Y = a\,X + b.$$

In general, X and Y denote 1D arrays, and A, B, and C represent matrices.

Array indexes in LabVIEW are zerobased. The index of the first element in the array, regardless of its dimension, is zero. The following sequence of numbers represents a 1D array X containing n elements.

$$X = \{x_0, x_1, x_2, \ldots, x_{n-1}\}$$

The following scalar quantity represents the ith element of the sequence X.

$$x_i, \quad 0 \le i < n$$

The first element in the sequence is x_0 and the last element in the sequence is x_{n-1}, for a total of n elements.

The following sequence of numbers represents a 2D array containing n rows and m columns.

$$A = \begin{bmatrix} a_{00} & a_{01} & a_{02} & \cdots & a_{0m-1} \\ a_{10} & a_{11} & a_{12} & \cdots & a_{1m-1} \\ a_{20} & a_{21} & a_{22} & \cdots & a_{2m-1} \\ \vdots & \vdots & \vdots & \vdots & \vdots \\ a_{n-10} & a_{n-11} & a_{n-2} & \cdots & a_{n-1m-1} \end{bmatrix}$$

The total number of elements in the 2D array is the product of n and m. The first index corresponds to the row number, and the second index corresponds to the column number. The following scalar quantity represents the element located on the ith row and the jth column.

$$a_{ij}, \quad 0 \leq i < n \text{ and } 0 \leq j < m$$

The first element in A is $a_{0\,0}$ and the last element is $a_{n-1\ m-1}$.

Unless otherwise specified, this manual uses the following simplified array operation notations.

Setting the elements of an array to a scalar constant is represented by

$$X = a\,,$$

which corresponds to the sequence

$$X = \{a, a, a, \ldots, a\}$$

and is used instead of

$$x_i = a, \text{ for } i = 0, 1, 2, \ldots, n\text{–}1.$$

Multiplying the elements of an array by a scalar constant is represented by

$$Y = a\,X,$$

which corresponds to the sequence

$$Y = \{ax_0, ax_1, ax_2, \ldots, ax_{n-1}\}$$

and is used instead of

$$y_i = ax_i, \text{ for } i = 0, 1, 2, \ldots, n\text{–}1.$$

Similarly, multiplying a 2D array by a scalar constant is represented by

$$B = k\,A,$$

which corresponds to the sequence

$$B = \begin{bmatrix} ka_{00} & ka_{01} & ka_{02} & \cdots & ka_{0m-1} \\ ka_{10} & ka_{11} & ka_{12} & \cdots & ka_{1m-1} \\ ka_{20} & ka_{21} & ka_{22} & \cdots & ka_{2m-1} \\ \vdots & \vdots & \vdots & \vdots & \vdots \\ ka_{n-10} & ka_{n-11} & ka_{n-12} & \cdots & ka_{n-1m-1} \end{bmatrix}$$

and is used instead of

$$b_{ij} = ka_{ij}, \text{ for } i = 0, 1, 2, \ldots, n{-}1 \text{ and } j = 0, 1, 2, \ldots, m{-}1.$$

Empty arrays are possible in LabVIEW. An array with no elements is an empty array and is represented by

$$\text{Empty} = \text{NULL} = \varnothing = \{\ \}.$$

In general, operations on empty arrays result in empty output arrays or undefined results.

Sampling Signals

To use digital signal processing techniques, you must convert an analog signal into its digital representation. This section includes only a brief discussion of the notation that represents a digital signal. This section does not discuss the mathematical background or problems associated with sampling techniques.

Consider an analog signal $x(t)$ and the sampling interval Δt. The signal $x(t)$ can be represented by the discrete sequence of samples

$$\{x(0), x(\Delta t), x(2\Delta t), x(3\Delta t), \ldots, x(k\Delta t), \ldots\}.$$

Because Δt establishes only the sampling rate and has no bearing on the actual sampled (digitized) value, the sample at

$$t = i\Delta t, \text{ for } i = 0, 1, 2, \ldots$$

corresponds to the ith element in the sequence. Thus,

$$x_i = x(i\Delta t)$$

and $x(t)$ can be represented by the sequence X whose values are

$$X = \{x_0, x_1, x_2, x_3, \ldots, x_k, \ldots\}.$$

If n samples are obtained from the signal $x(t)$, then the sequence

$$X = \{x_0, x_1, x_2, x_3, \ldots, x_{n-1}\}$$

is the digital representation or the sampled version of $x(t)$.

DSP VIs

This section describes the VIs that process and analyze an acquired or simulated signal, found in the **Analysis-DSP** palette. The signal processing VIs perform frequency domain transformations, frequency domain analysis, time domain analysis, and other transform.

The Fast Fourier Transform (FFT)

The Fourier transform establishes the relationship between a signal and its representation in the frequency domain. The Fourier transform is a powerful analysis tool for spectral analysis, applied mechanics, acoustics, medical imaging, numerical analysis, instrumentation, and telecommunications. The definition of the Fourier transform of a signal $x(t)$ is

$$X(f) = F(x\{t\}) = \int_{-\infty}^{\infty} x(t)e^{-j2\pi ft}dt \quad , \qquad \textit{(Equation 1)}$$

and the inverse Fourier transform of a signal $X(f)$ is

$$x(t) = F^{-1}\{X(f)\} = \int_{-\infty}^{\infty} X(f)e^{j2\pi ft}dt \quad . \qquad \textit{(Equation 2)}$$

A notation often used to indicate that the signals $x(t)$ and $X(f)$ are a Fourier transform pair and are related via the Fourier transform is

$$x(t) \Leftrightarrow X(f). \qquad \textit{(Equation 3)}$$

You can derive the discrete representation of the Fourier transform equations, equations 1 and 2, by sampling the Fourier transform pair in equation 3 using the following sampling relationships:

$$\Delta t = \frac{1}{f_s} \qquad \Delta f = \frac{f_s}{n}$$

where Δt is the sampling interval, Δf is the frequency resolution, f_s is the sampling frequency, and n is the number of samples in both the time and frequency domain.

Thus, the discrete transform pair

$$x_i \Leftrightarrow X_k \qquad \textit{(Equation 4)}$$

is obtained and the discrete Fourier transform is given by

$$X_k = \sum_{i=0}^{n-1} x_i e^{-j2\pi ik/n}\Delta t \qquad \textit{(Equation 5)}$$

and the inverse by

$$x_i = \sum_{i=0}^{n-1} X_k e^{j2\pi ik/n}\Delta f \quad . \qquad \textit{(Equation 6)}$$

X_k in equation 5 represents an amplitude spectral density. By multiplying the right-hand side of equation 5 by the frequency resolution Δf, we arrive at the amplitude spectrum. This amplitude spectrum is the final form of the Discrete Fourier Transform (DFT) and inverse DFT, given by equations 7 and 8, respectively. Notice that the DFT is independent of the sampling rate.

$$X_k = \sum_{i=0}^{n-1} x_i e^{-j2\pi ik/n} \qquad \text{for } k = 0, 1, 2, \ldots, n-1 \qquad \textit{(Equation 7)}$$

$$x_i = \frac{1}{n}\sum_{k=0}^{n-1} X_k e^{j2\pi ik/n} \qquad \text{for } i = 0, 1, 2, \ldots, n-1. \qquad \textit{(Equation 8)}$$

Direct implementation of the DFT requires approximately n^2 complex operations, and until recently, it was a time-consuming process. However, when the size of the sequence is

$$n = 2^m \text{ for } m = 1, 2, 3, \ldots$$

you can implement the computation of the DFT with approximately $n \log_2(n)$ operations. DSP literature refers to these algorithms as fast Fourier transforms (FFTs). Furthermore, with the aid of the FFT, you can find the DFT of any size sequence in approximately $3n \log_2(n)$ operations where n is the next power of 2 that accommodates intermediate results. You can find a more detailed explanation of FFT theory in most introductory texts on DSP.

The algorithm implemented in the LabVIEW analysis VIs is known as the Split-Radix algorithm. This algorithm has a form similar to the Radix-4 algorithms with the efficiency of Radix-8 algorithms. The Split-Radix algorithm requires the least number of multiplications among the Radix-2, Radix-4, and Mixed-Radix algorithms.

This manual uses the following notation to denote the discrete Fourier transform of a sequence x

$$X = F\{x\},$$

and

$$x = F^{-1}\{X\}$$

to denote the discrete inverse Fourier transform. The Fourier transform always results in a complex output sequence, and the input sequence can be either *real* or *complex*. Unless otherwise specified, the complex sequences are represented by two real sequences. If X is a complex sequence, then

$$X_{\text{Re}} = \text{Re}\{X\}$$

represents the real part of the complex sequence X,

$$X_{\text{Im}} = \text{Im}\{X\}$$

represents the imaginary part of the complex sequence X, and

$$X = X_{\text{Re}} + j\, X_{\text{Im}} = \text{Re}\{X\} + j\, \text{Im}\{X\}.$$

Transform VIs

FFT

Computes the Fast Fourier Transform (FFT) or the Discrete Fourier Transform (DFT) of the input sequence **X**.

Note

The LabVIEW Student Edition does not contain a Power Spectrum VI. You can calculate the power spectrum by squaring the magnitude of the FFT.

The input sequence has real and imaginary components.

The **FFT** VI executes FFT routines if the size of the input sequence is a valid power of two:

$$\text{size} = 2^m,\ m = 1, 2, \ldots, 23.$$

If the size of the input sequence is not a power of two, the **FFT** VI calls an efficient DFT routine.

The output sequence Y = FFT[**X**] is complex and is returned in two arrays, one containing the real and the other containing the imaginary components.

$$Y = Y_{\text{Re}} + jY_{\text{Im}}$$

X(RE) is the real input sequence.

X(IM) is the imaginary input sequence.

FFT{X}(RE) is the real component of the FFT of **X**.

FFT{X}(IM) is the imaginary component of the FFT of **X**.

error. See the *Analysis Error Codes* in *Appendix C* for a description of the error.

Inverse FFT

Computes the Inverse Fast Fourier Transform (FFT) or the Inverse Discrete Fourier Transform (DFT) of the input sequence **FFT{X}**.

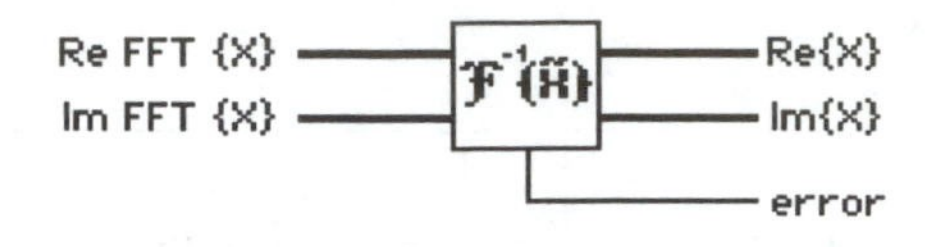

The input sequence has real and imaginary components.

The **Inverse FFT** VI executes Inverse FFT routines if the size of the input sequence is a valid power of two:

$$\text{size} = 2^m,\ m = 1, 2, ..., 23.$$

If the size of the input sequence is not a power of two, the **Inverse FFT** VI calls an efficient Inverse DFT routine.

The output sequence **X** = Inverse FFT[**FFT{X}**] is returned in two arrays, one containing the real and the other containing the imaginary components.

[NUM] **Re FFT{X}** is the real input sequence.

[NUM] **Im FFT{X}** is the imaginary input sequence.

[NUM] **Re{X}** is the Inverse Real FFT of **FFT{X}**.

[NUM] **Im{ X}** is the Inverse Imaginary FFT of **FFT{X}**.

NUM **error**. See the *Analysis Error Codes* in *Appendix C* for a description of the error.

Time Series Analysis VIs

Convolution

Computes the convolution of the input sequences **X** and **Y**.

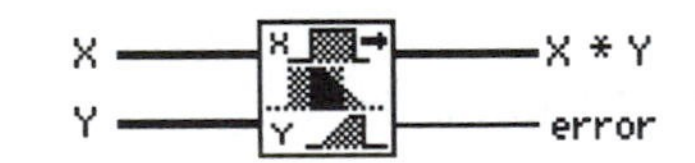

The convolution $h(t)$, of the signals $x(t)$ and $y(t)$ is defined as

$$h(t) = x(t)^* y(t) = \int_{-\infty}^{\infty} x(\tau)y(t-\tau)d\tau$$

where the symbol * denotes convolution.

[NUM] **X**.

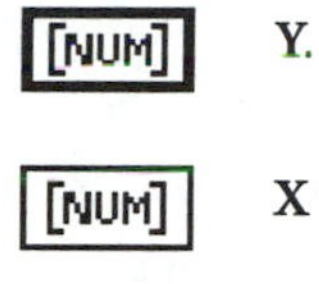

Y.

X * Y. The convolution of **X** with **Y.**

error. See the *Analysis Error Codes* in *Appendix C* for a description of the error.

For the discrete implementation of the convolution, let H represent the output sequence **X * Y**, n be the number of elements in the input sequence **X**, and m be the number of elements in the input sequence **Y**. Assuming that indexed elements of **X** and **Y** that lie outside their range are zero,

$$x_i = 0, \quad i < 0 \text{ or } i \geq n$$

and

$$y_j = 0, \quad j < 0 \text{ or } j \geq m,$$

then you obtain the elements of H using

$$h_i = \sum_{k=0}^{n-1} x_k y_{i-k} \qquad \text{for } i = 0, 1, 2, \ldots, \text{size}-1 \text{ ,}$$

$$\text{size} = n + m - 1,$$

where size is the total number of elements in the output sequence **X * Y.**

Note

This is not a circular convolution. Because $x(t) * y(t) \Leftrightarrow X(f)Y(f)$ ***is a Fourier transform pair, you can create a circular version of the convolution using a diagram similar to the following diagram.***

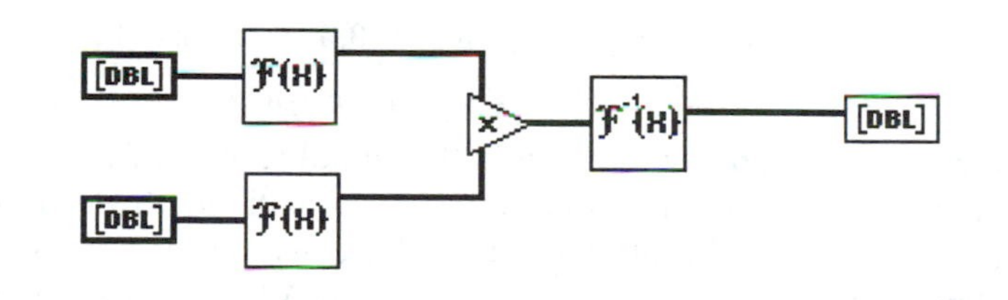

Correlation

Computes the cross correlation of the input sequences **X** and **Y.**

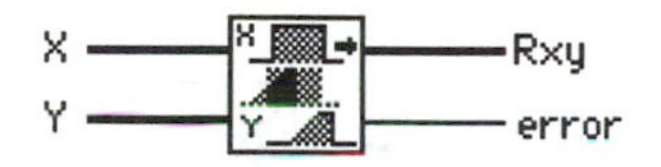

The cross correlation $R_{xy}(t)$ of the signals $x(t)$ and $y(t)$ is defined as

$$R_{xy}(t) = x(t) \otimes y(t) = \int_{-\infty}^{\infty} x(\tau)y(t+\tau)d\tau \quad ,$$

where the symbol ⊗ denotes correlation.

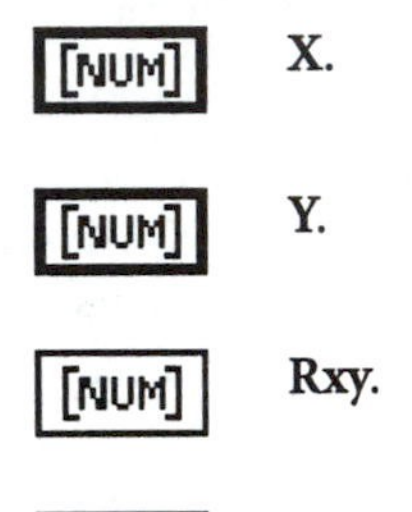

X.

Y.

Rxy.

error. See the *Analysis Error Codes* in *Appendix C* for a description of the error.

The discrete implementation of this VI is as follows. Let H represent a sequence whose indexing can be negative, let n be the number of elements in the input sequence **X**, let m be the number of elements in the sequence **Y**, and assume that the indexed elements of **X** and **Y** that lie outside their range are equal to zero,

$$x_j = 0, \; j < 0 \text{ or } j \geq n,$$

and

$$y_j = 0, \; j < 0 \text{ or } j \geq m$$

Then the VI obtains the elements of H using

$$h_j = \sum_{k=0}^{n-1} x_k x_{j+k} \qquad \text{for } j = -(n-1), -(n-2), \ldots, -2, -1, 0, 1, 2, \ldots, m-1 \,.$$

The elements of the output sequence **Rxy** are related to the elements in the sequence H by

$$rxy_i = h_{i-(n-1)} \text{ for } i = 0, 1, 2, \ldots, \text{size}-1 \,,$$

$$\text{size} = n + m - 1$$

where size is the number of elements in the output sequence **Rxy**.

Because you cannot index LabVIEW arrays with negative numbers, the corresponding cross correlation value at $t = 0$ is the nth element of the output sequence **Rxy**. Therefore, **Rxy** represents the correlation values that the VI shifted n times in index.

Unwrap Phase

Unwraps the **Phase** array by eliminating discontinuities whose absolute values exceed π.

Phase — Unwrap — Unwrapped Phase, error

Phase is expressed in radians.

Unwrapped Phase is expressed in radians.

error. See the *Analysis Error Codes* in *Appendix C* for a description of the error.

Integral x(t)

Performs the discrete integration of the sampled signal **X**.

The integral $F(t)$ of a function $f(t)$ is defined as

$$F(t) = \int f(t)dt$$

For an example of how to use this VI, look for **Integral & Derivative Example** in ANALYSIS.LLB in the EXAMPLES folder or directory.

X is the sampled signal.

initial condition defaults to 0.0.

final condition defaults to 0.0.

dt is the sampling interval, and must be greater than zero. If **dt** is less than or equal to zero, the VI sets **Integral X** to an empty array and returns an error. **dt** defaults to 1.0.

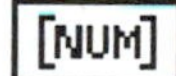

Integral X is the sampled output sequence.

error. See the *Analysis Error Codes* in *Appendix C* for a description of the error.

Let Y represent the sampled output sequence **Integral X**. The VI obtains the elements of Y using

$$y_i = \frac{1}{6}\sum_{j=0}^{i}(x_{j-1} + 4x_j + x_{j+1})\,dt \qquad \text{for } i = 0, 1, 2, \ldots, n-1,$$

where n is the number of elements in **X**, x_{-1} is specified by **initial condition** when $i = 0$, and x_n is specified by **final condition** when $i = n-1$.

The **initial condition** and **final condition** minimize the overall error by increasing the accuracy at the boundaries, especially when the number of samples is small. Determining boundary conditions before the fact enhances accuracy.

Derivative x(t)

Performs a discrete differentiation of the sampled signal **X**.

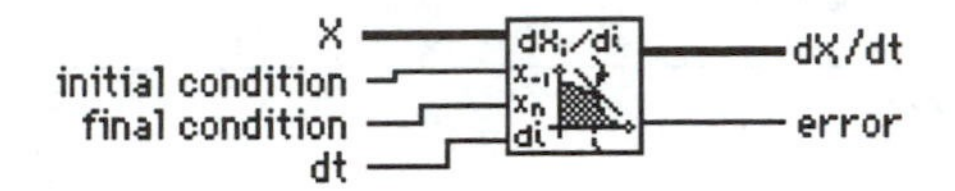

The differentiation $f(t)$ of a function $F(t)$ is defined as

$$f(t) = \frac{d}{dt}F(t) \quad .$$

For an example of how to use this VI, look for **Integral & Derivative Example** in ANALYSIS.LLB in the EXAMPLES folder or directory.

X is the sampled signal.

Initial Condition defaults to 0.0.

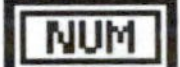

Final Condition defaults to 0.0.

dt is the sampling interval and must be greater than zero. If **dt** is less than or equal to zero, the VI sets **d/dt X** to an empty array and returns an error. **dt** defaults to 1.0.

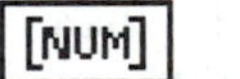

dX/dt.

[NUM]

error. See the *Analysis Error Codes* in *Appendix C* for a description of the error.

Let Y represent the sampled output sequence **d/dt X**. The discrete implementation is given by

$$y_i = \frac{1}{2dt}(x_{i+1} - x_{i-1}) \qquad \text{for } i = 0, 1, 2, \ldots, n-1 \,,$$

where n is the number of samples in $\boldsymbol{x(t)}$, x_{-1} is specified by **initial condition** when $i = 0$, and x_n is specified by **final condition** when $i = n-1$.

Initial condition and **final condition** minimize the error at the boundaries.

Peak Detector

Analyzes the input sequence **X** for valid peaks and keeps a **count** of the number of peaks encountered and a record of **Indices**, which locates the points that exceed the

threshold in a valid peak. A peak is valid when the number of consecutive elements of **X** that exceed the **threshold** is at least equal to **width**.

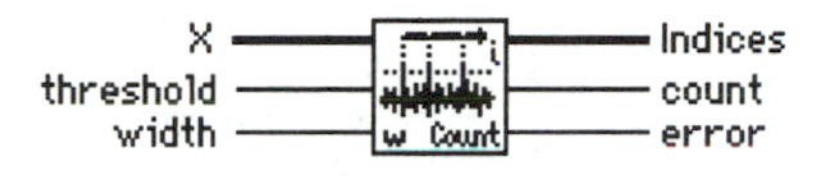

X. The number of samples in **X** must be greater than the specified **width**. If **X** is less than or equal to **width**, the VI sets **count** to zero and returns an error.

threshold defaults to 0.0.

width must be greater than zero. If **width** is less than or equal to zero, the VI sets **count** to zero and returns an error. **width** defaults to 1.

Indices.

count.

error. See the *Analysis Error Codes* in *Appendix C* for a description of the error.

Rectangular To Polar

Converts the rectangular coordinates x and y into the polar coordinates magnitude and phase, according to the following formulas:

$$\text{magnitude} = \sqrt{x^2 + y^2}$$

$$\text{phase} = \tan^{-1}\left(\frac{y}{x}\right).$$

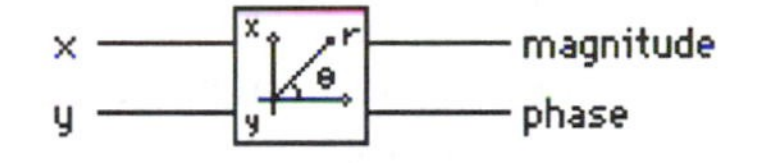

x is the first rectangular coordinate input.

y is the second rectangular coordinate input.

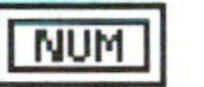

magnitude is the output magnitude.

phase is the output phase expressed in radians.

Polar To Rectangular

Converts the polar coordinates magnitude and phase into rectangular coordinates x and y to the following formulas:

$$x = a \cos$$

$$x = \textbf{magnitude} \cos(\textbf{phase})$$

$$y = \textbf{magnitude} \sin(\textbf{phase})$$

magnitude — x
phase — y

NUM **magnitude** is the magnitude.

NUM **phase** must be expressed in radians.

NUM **x** is the first rectangular coordinate output.

NUM **y** is the second rectangular coordinate output.

Amplitude and Phase Spectrum

Computes the single-sided, scaled amplitude spectrum magnitude and phase of a real time-domain signal.

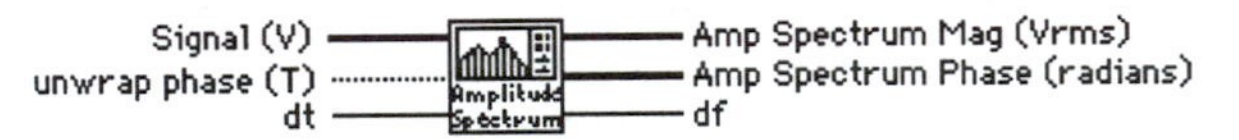

The VI computes the amplitude spectrum as

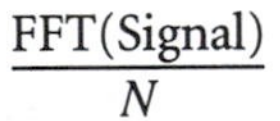

$$\frac{\mathrm{FFT(Signal)}}{N}$$

where N is the number of points in the signal array. The VI then converts the amplitude spectrum to single-sided rms magnitude and phase spectra.

For an example of how to use this VI, look for **Amplitude Spectrum Example** in ANALYSIS.LLB in the EXAMPLES folder or directory.

[NUM] **Signal (V)** is the array containing the time-domain signal.

unwrap phase (T). Set to TRUE to enable phase unwrapping on the output phase Amp Spectrum Phase (radians). If you set **unwrap phase (T)** to FALSE, the VI does not perform unwrapping. The default setting is TRUE.

dt is the sample period of the time-domain signal, usually in seconds. **dt** is also $1/f_s$ where *fs* is the sampling frequency of the time-domain signal.

Amp Spectrum Mag (Vrms) is the single-sided amplitude spectrum magnitude in volts rms if the input signal is in volts. If the input signal is not in volts, the results are in input signal units rms.

Amp Spectrum Phase (radians) is the single-sided amplitude spectrum phase in radians.

df is the line frequency interval of the power spectrum, in hertz, if **dt** is in seconds.

Time Domain Window

Applies the selected window to the real time-domain signal. The VI scales the result so that when the power or amplitude spectrum of the windowed waveform is computed, all windows provide the same level within the accuracy constraints of the window. This VI also returns important window constants for the selected window.

For an example of how to use this VI, look in ANALYSIS.LLB in the EXAMPLES folder or directory and open **Amplitude Spectrum Example.**

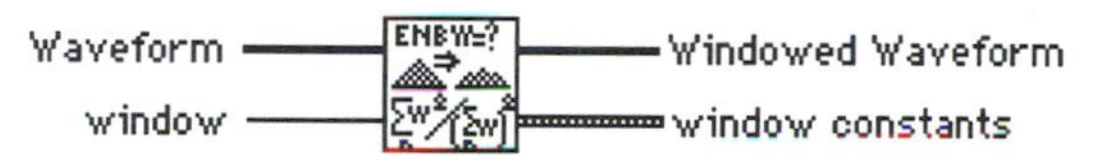

Waveform is the real time-domain signal.

window is the time-domain window to be used.

0: Triangle/Bartlett
1: Hanning
2: Hamming

Windowed Waveform is the time-domain signal multiplied by the scaled window.

window constants contains the following important constants for the selected window.

eq noise BW is the equivalent noise bandwidth of the selected window. You can use this value to divide a sum of individual power spectra of the power spectrum or to compute the power in a given frequency span.

coherent gain is the inverse of the scaling factor that was applied to the window.

Signal Generator

This VI generates an array containing a sine, square, sawtooth, or triangle wave.

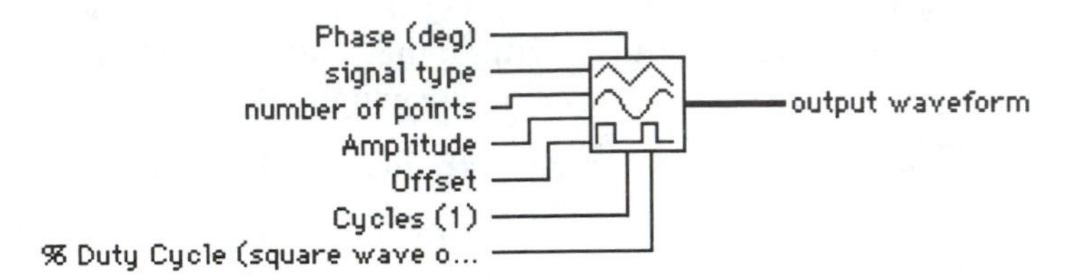

If the sequence *Y* represents **Output Waveform**, the VI generates the pattern according to the following formula:

$$y_i = amp * \text{function}(\text{phase}[i]), \text{ for } i = 0, 1, 2, \ldots, n\text{–}1,$$

where *function* is sine, square, sawtooth, or triangle, *amp* is the **amplitude**, and phase[*i*] = initial_phase + **f***360.0* *i*,

For examples of how to use this VI, look in ANALYSIS.LLB in the EXAMPLES folder or directory.

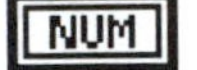

Phase is the initial phase, in degrees, of **Output Waveform**.

Signal Type is the type of waveform generated.

0: Sine

1: Square

2: Sawtooth

3. Triangle

number of points is the number of samples of the **Output Waveform**.

amplitude is the amplitude of the **Output Waveform**.

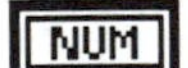

offset is the DC component of the **Output Waveform**.

cycles is the number of cycles of the **Output Waveform**. It defaults to one cycle. Note that the output wave will always have the number of points specified by the number of points input regardless of how many cycles it includes.

duty cycle is the duty cycle, in percent. Applies only to the **Square Wave**.

Output Waveform is the output wave.

Noise Generator

Generates a Uniform White Noise or a Gaussian White Noise pattern.

For an example of how to use this VI, look in ANALYSIS.LLB in the EXAMPLES folder or directory and open **Curve Fit Example.**

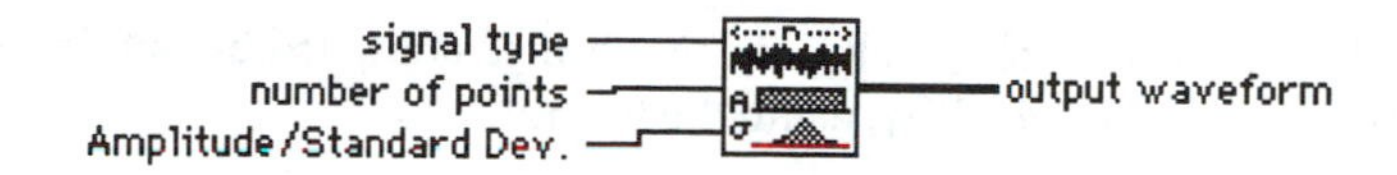

Uniform White Noise

Generates a uniformly distributed pseudo random pattern whose values are in the range [–a:a], where a is the absolute value of **amplitude.** The VI generates the pseudo random sequence using a modified version of the Very-Long-Cycle random number generator algorithm.

Given that the probability density function, $f(x)$, of the uniformly distributed **Uniform White Noise** is

$$f(x) = \begin{cases} \frac{1}{2a} & \text{if } -a \le x \le a \\ 0 & \text{elsewhere} \end{cases},$$

where a is the absolute value of the specified **amplitude**, and that you can compute the expected values, $E\{\cdot\}$, using the formula

$$E(x) = \int_{-\infty}^{\infty} x(f(x))\,dx \quad ,$$

then the expected mean value, μ, and the expected standard deviation value, σ, of the pseudorandom sequence are

$$\mu = E\{x\} = 0 ,$$

$$\sigma = \left[E\{(x-\mu)^2\}\right]^{1/2} = \frac{a}{\sqrt{3}} \approx 0.57735a \quad .$$

The pseudo random sequence produces approximately 2^{90} samples before the pattern repeats itself.

Gaussian White Noise

Generates a Gaussian distributed pseudo random pattern whose statistical profile is $(\mu, \sigma) = (0, s)$, where s is the absolute value of the specified **standard deviation.**

The VI generates the Gaussian distributed pseudo random sequence using a modified version of the Very-Long-Cycle random number generator algorithm based upon the

Central Limit Theorem. Given that the probability density function, $f(x)$, of the Gaussian distributed **Gaussian Noise Pattern** is

$$f(x) = \frac{1}{\sqrt{2\pi}s} e^{\left(-\frac{1}{2}\right)\left(\frac{x}{s}\right)^2} ,$$

where s is the absolute value of the specified **standard deviation** and that you can compute the expected values, E{•}, using the formula

$$E(x) = \int_{-\infty}^{\infty} x(f(x))\,dx ,$$

then the expected mean value, μ, and the expected standard deviation value, σ, of the pseudo random sequence are

$$\mu = E\{x\} = 0 ,$$

$$\sigma = \left[E\{(x-\mu)^2\}\right]^{1/2} = s .$$

The pseudo random sequence produces approximately 2^{90} samples before the pattern repeats itself.

signal type specifies Uniform or Gaussian White Noise patterns.

0: Uniform White Noise

1: Gaussian White Noise

number of points must be greater than or equal to 0. If **samples** is less than zero, the VI sets **output waveform** to an empty array but does *not* return an error. **number of points** defaults to 128.

amplitude/standard deviation defaults to 1.0. The Uniform White Noise function uses this input as the **amplitude.** The Gaussian White Noise function uses this input as the **standard deviation.**

output waveform is the noise pattern.

Digital Filtering Functions

This section contains a brief discussion of digital filter theory and describes VI that implements the IIR filters.

Analog filter design is one of the most important areas of electronic design. Although analog filter design books featuring simple, well-tested filter designs exist, filter design is often reserved for specialists because it requires advanced mathematical knowledge and understanding of the processes involved in the system affecting the filter.

Modern sampling and digital signal processing tools have made it possible to replace analog filters with digital filters in applications that require flexibility and programmability. These applications include audio, telecommunications, geophysics, and medical monitoring.

Digital filters have the following advantages over their analog counterparts.

- They are software programmable.
- They are stable and predictable.
- They do not drift with temperature or humidity or require precision components.
- They have a superior performance-to-cost ratio.

You can use digital filters in LabVIEW to control parameters such as filter order, cutoff frequencies, amount of ripple, and stopband attenuation.

The digital filter VI described in this section follows the virtual instrument philosophy. The VI handles all the design issues, computations, memory management, and actual data filtering internally, transparently to the user. You do not have to be an expert in digital filters or digital filter theory to process the data.

The following discussion of sampling theory is intended to give you a better understanding of the filter parameters and how they relate to the input parameters.

The sampling theorem states that you can reconstruct a continuous-time signal from discrete, equally spaced samples if the sampling frequency is at least twice that of the highest frequency in the time signal. Assume you can sample the time signal of interest at Δt equally spaced intervals without losing information. The Δt parameter is the sampling interval.

You can obtain the sampling rate or sampling frequency f_s from the sampling interval

$$f_s = \frac{1}{\Delta t} ,$$

which means that, according to the sampling theorem, the highest frequency that the digital system can process is

$$f_{Nyq} = \frac{f_s}{2} .$$

The highest frequency the system can process is known as the Nyquist frequency. This also applies to digital filters. For example, if your sampling interval is

$$\Delta t = 0.001 \text{ sec} ,$$

then the sampling frequency is

$$f_s = 1000 \text{ Hz} ,$$

and the highest frequency that the system can process is

$$f_{Nyq} = 500 \text{ Hz} .$$

Infinite Impulse Response Filters

Infinite impulse response filters (IIR) filters are digital filters whose impulse response can theoretically be infinite in duration. The general block diagram for IIR filters is shown in the following figure.

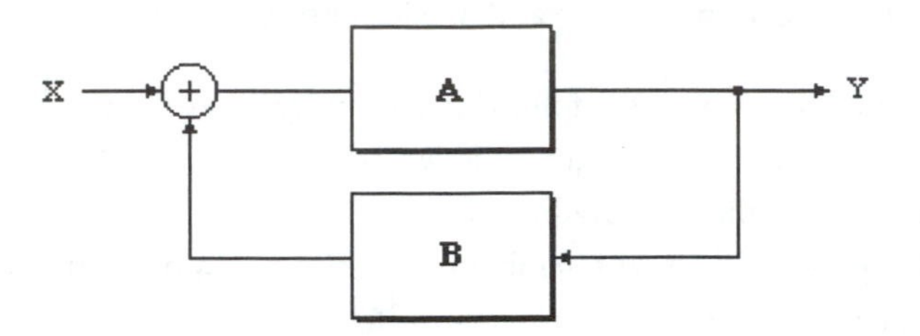

The digital characterization between the input sequence X and the output sequence Y is

$$y_i = \frac{1}{b_0}\left(\sum_{j=0}^{n} a_j x_{i-j} - \sum_{k=1}^{m} b_k y_{i-k} \right) ,$$

where *n* is the polynomial order of the forward branch labeled A, and *m* is the polynomial order of the feedback branch labeled B.

The delays associated with the feedback branch of the filter configuration cause the infinite duration of the impulse response. Even when no more external samples feed into the system, the feedback portion continues to send samples to the filter structure. In practical filter applications, the input sequence is finite, and the output sequence is truncated to a finite number of samples.

IIR filters in LabVIEW have the following properties.

- They check for the validity of the Nyquist criterion.
- Negative indices resulting from the preceding equation are assumed to be zero the first time you call the VI.
- Because the initial filter state is assumed to be zero (negative indices), a transient proportional to the filter order occurs before the filter reaches steady state. The duration of the transient response, or delay, for lowpass and highpass filters is equal to the filter order

delay = order.

The duration of the transient response for bandpass and bandstop filters is twice the filter order

delay = 2 * order.

You can eliminate this transient on successive calls by enabling state memory. To enable state memory, set the **mode** parameter of the VI to TRUE.

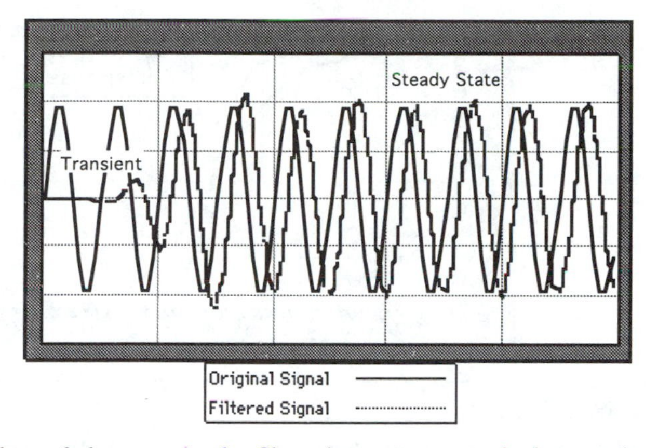

- The number of elements in the filtered sequence equals the number of elements in the input sequence.
- The filter retains the remaining forward and feedback values when the filtering is completed.

The advantage of digital IIR filters over finite impulse response (FIR) filters is that IIR filters require fewer coefficients to perform the same filtering operation. Thus, IIR filters execute much faster and do not require extra memory because they execute in place.

The disadvantage of IIR filters is that the phase response is nonlinear. If the application does not require phase information, such as simple signal monitoring, IIR filters are well suited. You must generally use FIR filters to process applications requiring linear phase response.

IIR filters are also known as recursive filters or autoregressive moving-average (ARMA) filters.

Butterworth Filters

A smooth response at all frequencies and a monotonic decrease from the specified cutoff frequencies characterize the frequency response of Butterworth filters. Butterworth filters are maximally flat—the ideal response of unity in the passband and zero in the stopband. The half power frequency or the 3-dB down frequency corresponds to the specified cutoff frequencies.

The following illustration shows the response of a lowpass Butterworth filter. The advantage of Butterworth filters is a smooth, monotonically decreasing frequency response. Once the cutoff frequency is set, the *steepness* of the transition is proportional to the filter order. Higher-order Butterworth filters approach the ideal lowpass filter response.

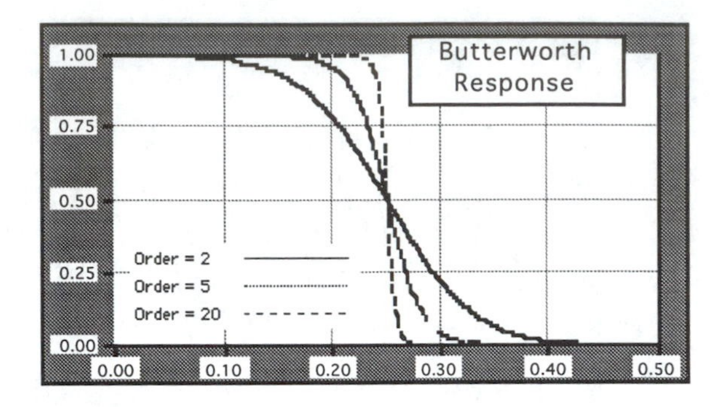

Chebyshev Filters

Butterworth filters do not always provide a good approximation of the ideal filter response because of the slow rolloff between the passband (the portion of interest in the spectrum) and the stopband (the unwanted portion of the spectrum).

Chebyshev filters minimize peak error in the passband by accounting for the maximum absolute value of the difference between the ideal filter and the filter response you want (the maximum tolerable error in the passband). The frequency response characteristics of Chebyshev filters have an equi-ripple magnitude response in the passband, monotonically decreasing magnitude response in the stopband, and a sharper rolloff than Butterworth filters.

The following graph shows the response of a lowpass Chebyshev filter. Notice that the equi-ripple response in the passband is constrained by the maximum tolerable ripple error and that the sharp rolloff appears in the stopband. The advantage of Chebyshev filters over Butterworth filters is that Chebyshev filters have a sharper transition between the passband and the stopband with a lower order filter. This produces a smaller absolute error and higher execution speeds.

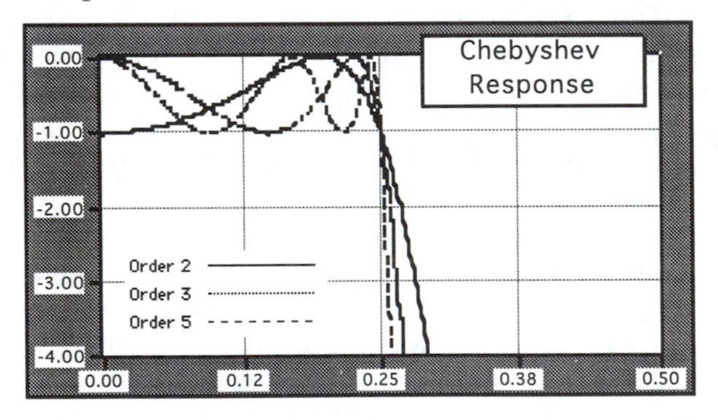

Chebyshev II or Inverse Chebyshev Filters

Chebyshev II, also known as inverse Chebyshev or Type II Chebyshev filters, are similar to Chebyshev filters, except that Chebyshev II filters distribute the error over the stopband (as opposed to the passband), and Chebyshev II filters are maximally flat in the passband (as opposed to the stopband).

Chebyshev II filters minimize peak error in the stopband by accounting for the maximum absolute value of the difference between the ideal filter and the filter response you want. The frequency response characteristics of Chebyshev II filters are equi-ripple magnitude response in the stopband, monotonically decreasing magnitude response in the passband, and a rolloff sharper than Butterworth filters.

The following graph plots the response of a lowpass Chebyshev II filter. Notice that the equi-ripple response in the stopband is constrained by the maximum tolerable error and that the smooth monotonic rolloff appears in the stopband. The advantage of Chebyshev II filters over Butterworth filters is that Chebyshev II filters give a sharper transition between the passband and the stopband with a lower-order filter. This difference corresponds to a smaller absolute error and higher execution speed. One advantage of Chebyshev II filters over regular Chebyshev filters is that for Chebyshev II filters, the error is distributed in the stopband instead of the passband.

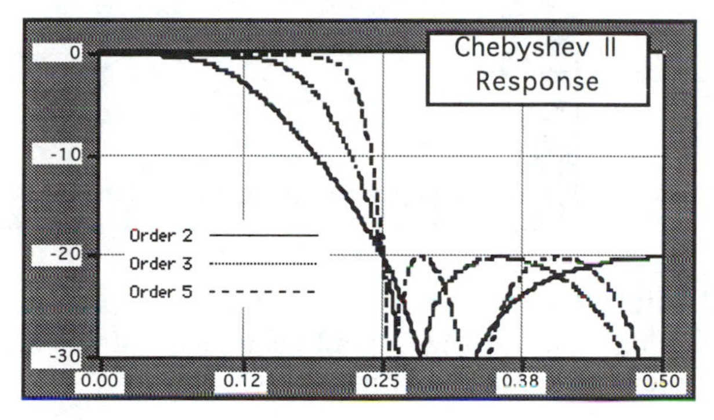

Bessel Filters

You can use Bessel filters to preserve linear phase. IIR filters have a nonlinear phase response. In higher-order filters and those with a steeper rolloff, this condition is more pronounced, especially in the transition regions of the filters. Bessel filters have maximally flat response in both magnitude and phase. Furthermore, the phase response in the passband of Bessel filters, which is the region of interest, is nearly linear. Like Butterworth filters, Bessel filters require high-order filters to minimize the error and, for this reason, are not widely used. You can also obtain linear phase response using FIR filter designs.

The following graphs plot the response of a lowpass Bessel filter. Notice that the response is smooth at all frequencies, as well as monotonically decreasing in both magnitude and phase. Also notice that the phase in the passband is nearly linear.

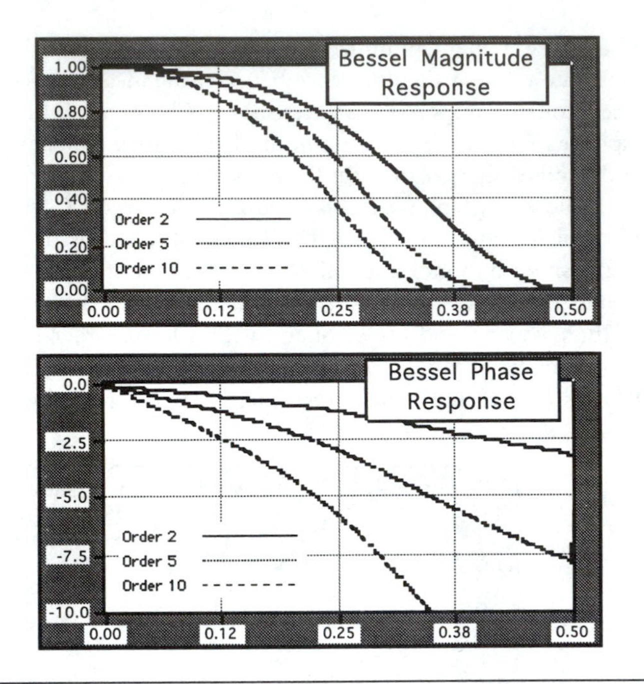

All IIR Filters

Filters the input sequence **Input Data** using the IIR filter specified by **Filter Design** and by **Reverse Coefficients** and **Forward Coefficients.**

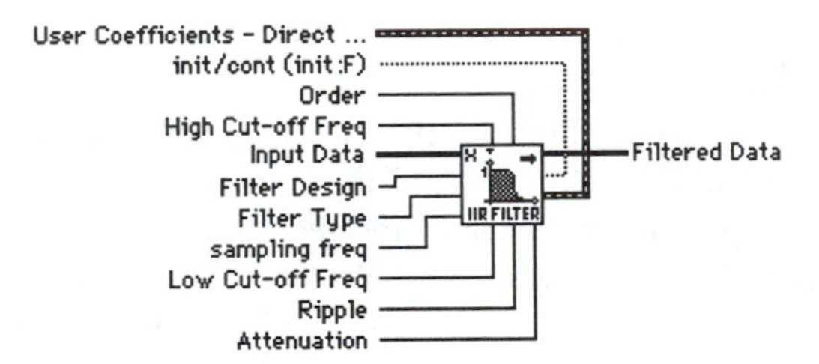

If Y represents the output sequence **Filtered Data,** the VI obtains the elements of Y using

$$y_i = \frac{1}{b_0}\left(\sum_{j=0}^{n-1} a_j x_{i-j} - \sum_{k=1}^{m-1} b_k y_{i-k}\right)$$

where n is the number of **Forward Coefficients** (represented by a_j), and m is the number of **Reverse Coefficients** (represented by b_k).

For an example of how to use this VI, look for **IIR Filter Design** in ANALYSIS.LLB in the EXAMPLES folder or directory.

User Coefficients—Direct Form contains the following parameters.

Forward Coefficients. This VI does not place any restrictions on the coefficient arrays. If both coefficient arrays are empty, the VI performs no filtering and sets **Filtered Data** to the value of **Input Data.**

Reverse Coefficients.

Note

You can use the IIR Filter VI to perform FIR filtering by passing an empty array into Reverse Coefficients.

---

init/cont (init:F) controls the initialization of the internal filter states. When **init/cont** is FALSE (default), the internal states are initialized to zero. When **init/cont** is TRUE, the internal filter states are initialized to the *final* filter states from the previous call to this instance of this VI. To filter a large data sequence that has been split into smaller blocks, set this control to FALSE for the first block, and to TRUE for continuous filtering of all remaining blocks.

order must be greater than zero. If **order** is less than or equal to zero, the VI sets **Filtered X** to an empty array and returns an error. **order** defaults to 2.

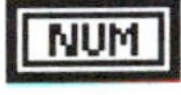

high cutoff freq : fh is the high cutoff frequency. The VI ignores this parameter when **filter type** is 0 (lowpass) or 1 (highpass)

Input Data.

Filter Design specfies the type of filter used.

0: Butterworth

1: Chebyshev

2: Chebyshev II

3: Bessel

4: Use Input Coefficients

filter type specifies the passband of the filter according to the following values.

sampling freq : fs is the sampling frequency and must be greater than zero. If it is less than or equal to zero, the VI sets **Filtered Data** to an empty array and returns an error. **sampling freq : fs** defaults to 1.0.

low cutoff freq : fl is the low cutoff frequency and must observe the Nyquist criterion.

ripple must be greater than zero (for Chebyshev only—other filters do not use this parameter), and you must express it in decibels. If **ripple** is less than or equal to zero, the VI sets **Filtered Data** to an empty array and returns an error. **ripple** defaults to 0.1.

attenuation must be greater than zero (for Chebyshev II only—other filters do not use this parameter). If **attenuation** is less than or equal to zero, the VI sets **Filtered X** to an empty array and returns an error. **attenuation** defaults to 60.0.

Filtered Data.

Numeric VIs

This section describes the VIs in the **Analysis- Numeric** palette that perform algebraic functions on vectors and matrices.

1D Polynomial Evaluation

Performs a polynomial evaluation of **X** using **Coefficients : a.**

The output array **Y** is given by

$$Y = \sum_{n=0}^{m} a_n X^n \quad ,$$

where m is the polynomial order.

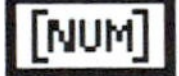

X is the input data to be used in the polynomial evaluation.

Coefficients : a. The total number of elements in **Coefficients : a** is the polynomial order plus one.

Y is the output data.

error. See the *Analysis Error Codes* in *Appendix C* for a description of the error.

Dot Product

Computes the dot product of **X Vector** and **Y Vector.**

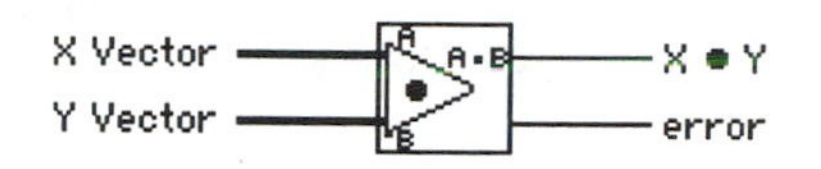

Let X represent the input sequence **X Vector** and Y represent the input sequence **Y Vector**. The VI obtains the dot product **X•Y** using the formula

$$X \bullet Y = \sum_{i=0}^{n-1} x_i y_i \quad ,$$

where n is the number of data points. Notice that the output value **X•Y** is a scalar value.

X Vector. If the number of elements in **X Vector** is different from the number of elements in **Y Vector**, the VI computes the dot product based on the sequence that contains the fewest elements. If **X Vector** is an empty array, the dot product is NaN.

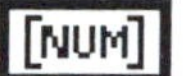

Y Vector. If **Y Vector** is an empty array, the dot product is NaN.

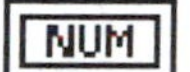

X • Y is the dot product.

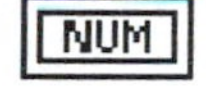

error. See the *Analysis Error Codes* in *Appendix C* for a description of the error.

Outer Product

Computes the outer product of **X Vector** and **Y Vector**.

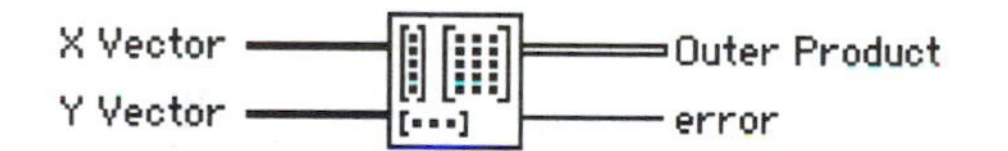

Let X represent the input sequence **X Vector** and Y represent the input sequence **Y Vector**. The VI obtains **Outer Product** using the formula

$$A_{ij} = x_i y_j \text{, for } \begin{cases} i = 0, 1, 2, \ldots, n-1 \\ j = 0, 1, 2, \ldots, m-1 \end{cases} \quad ,$$

where A represents the 2D output sequence **Outer Product**, n is the number of elements in the input sequence **X Vector**, and m is the number of elements in the input sequence **Y Vector**.

X Vector is the first input vector.

Y Vector is the second input vector.

Outer Product. If one of the input sequences is an empty array, **Outer Product** is an empty array.

error. See the *Analysis Error Codes* in *Appendix C* for a description of the error.

Unit Vector

Finds the **norm** of the input **Input Vector** and obtains its corresponding **Unit Vector** by normalizing the original **Input Vector** with its **norm**.

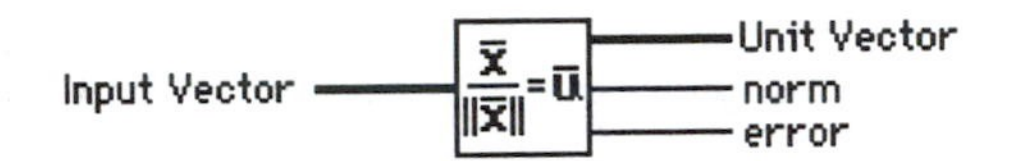

Let X represent the input **Input Vector**, then **norm** is given by

$$\|X\| = \sqrt{x_0^2 + x_1^2 + \ldots + x_{n-1}^2} \quad ,$$

where $\|X\|$ is **norm**, and the VI calculates **Unit Vector,** U, using

$$U = \frac{X}{\|X\|} \quad .$$

Input Vector. If **Input Vector** is an empty array, **Unit Vector** is also an empty array, and **norm** is NaN.

Unit Vector is the output normalized vector.

norm is the norm of **Input Vector.**

error. See the *Analysis Error Codes* in *Appendix C* for a description of the error.

Normalize Vector

Normalizes the input **Vector** using its statistical profile (μ,σ), where μ is the **mean** and σ is the **standard deviation,** to obtain a **Normalized Vector** whose statistical profile is (0,1).

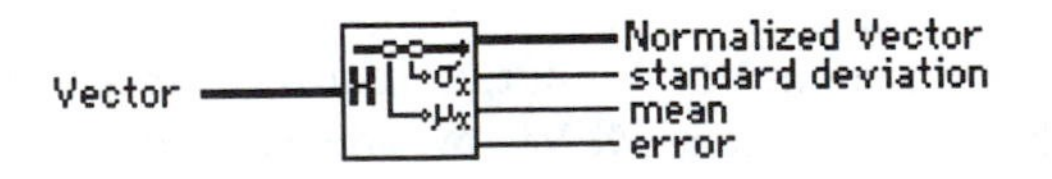

The VI obtains **Normalized Vector** using

$$Y = \frac{X - \mu}{\sigma} \quad ,$$

$$\mu = \frac{\sum_{i=0}^{n-1} x_i}{n}$$

$$\sigma = \sqrt{\frac{\sum_{i=0}^{n-1} (x_i^2 - \mu^2)}{n}}$$

where Y represents the output sequence **Normalized Vector**, and X represents the input sequence **Vector** of length n, and X_i is the ith element of X.

Vector. If **Vector** is an empty array, **Normalized Vector** is also an empty array, and **mean** and **standard deviation** are NaN.

Normalized Vector is the output normalized vector.

standard deviation is the standard deviation of **Vector.**

mean is the mean of **Vector.**

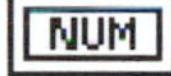

error. See the *Analysis Error Codes* in *Appendix C* for a description of the error.

A x B

Performs the matrix multiplication of two input matrices.

If A is an n-by-k matrix and B is a k-by-m matrix, the matrix multiplication of A and B $C = AB$ results in a matrix C whose dimensions are n by m. Let A represent the 2D input array **A** matrix, B represent the 2D input array **B** matrix, and C represent the 2D output array **A x B.** The VI obtains the elements of C using the formula

$$c_{ij} = \sum_{l=0}^{k-1} a_{il} b_{lj} \quad \text{for} \begin{cases} i = 0, 1, 2, \ldots, n-1 \\ j = 0, 1, 2, \ldots, m-1 \end{cases},$$

where n is the number of rows in **A** matrix, k is the number of columns in **A** matrix and the number of rows in **B** matrix, and m is the number of columns in **B** matrix.

Note

The A x B VI performs a strict matrix multiplication and not an element-by-element 2D multiplication. To perform an element-by-element multiplication you must use the LabVIEW Multiply function. In general, $AB \neq BA$.

A. The number of columns in **A** must match the number of rows in **B** and must be greater than zero: $k > 0$. If the number of columns in **A** does not match the number of rows in **B**, the VI sets **AxB** to an empty array and returns an error.

B is the second matrix. If the number of rows in *B* does not match the number of columns in *A*, the VI sets **AxB** to an empty array and returns an error.

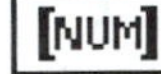

A x B is the matrix containing the result of the matrix multiplication *A x B*.

error. See the *Analysis Error Codes* in *Appendix C* for a description of the error.

A x Vector

Performs the multiplication of an input matrix and an input vector.

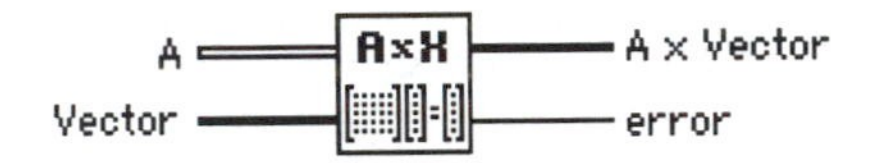

If *A* is an *n*-by-*k* matrix, and **X** is a vector with *k* elements, the multiplication of **A** and **X** $Y = AX$ results in a vector *Y* with n elements. Let *Y* represent the output **AxX.** The VI obtains the elements of *Y* using the formula

$$y_i = \sum_{j=0}^{k-1} a_{ij} x_j \quad \text{, for } i = 0, 1, 2, \ldots, n-1,$$

where n is the number of rows in **A**, and k is the number of columns in **A** and the number of elements in **X**.

A. The number of columns in **A** must match the number of elements in **X** and must be greater than zero: $k > 0$. If the number of columns in **A** does not match the number of elements in **X**, the VI sets **A x X** to an empty array and returns an error.

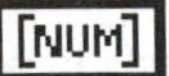

Vector is the input vector.

A x Vector is the output vector containing the result of *A* x Vector.

error. See the *Analysis Error Codes* in *Appendix C* for a description of the error.

Inverse Matrix

Finds the **Inverse Matrix** of the **Input Matrix**.

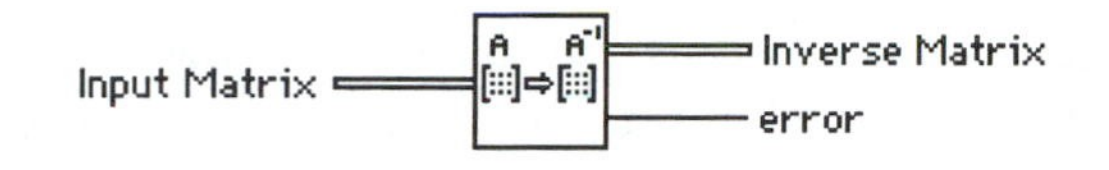

Let A be the **Input Matrix** and I be the identity matrix. **Inverse Matrix** is obtained by solving the system $AB = I$ for B.

If A is a nonsingular matrix, you can show that the solution to the preceding system is unique and that it corresponds to the inverse matrix of A

$$B = A^{-1},$$

and B is therefore **Inverse Matrix**. A nonsingular matrix is a matrix in which no row or column is a linear combination of any other row or column, respectively.

Note

You cannot always determine beforehand whether the matrix is singular, especially with large systems. The Inverse Matrix VI detects singular matrices and returns an error, so you do not need to verify whether you have a valid system before using this VI.

[NUM] **Input Matrix** must be nonsingular and must have as many rows as columns. If **Input Matrix** is singular or is not square, the VI sets **Inverse Matrix** to an empty array and returns an error.

Note

The numerical implementation of the matrix inversion is not only numerically intensive but, because of its recursive nature, is also highly sensitive to round-off error introduced by the floating-point numeric coprocessor. Although the computations use the maximum possible accuracy, the VI cannot always solve the system.

Inverse Matrix is the inverse matrix of **Input Matrix**.

error. See the *Analysis Error Codes* in *Appendix C* for a description of the error.

Determinant

Finds the **determinant** of **X**.

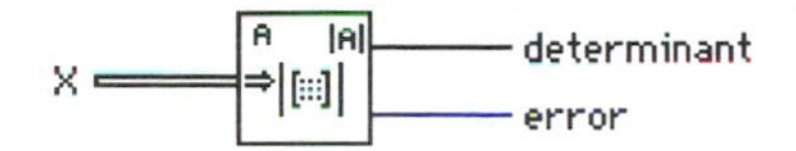

Let A be a square matrix that represents **X** and let L and U be the lower and upper triangular matrices, respectively, of A such that

$$A = LU,$$

where the main diagonal elements of the lower triangular matrix L are arbitrarily set to one. The VI finds the determinant of A by the product of the main diagonal elements of the upper triangular matrix U

$$|A| = \prod_{i=0}^{n-1} u_{ii} \quad ,$$

where $|A|$ is the **determinant** of **X**, and n is the dimension of **X**.

[NUM] **X** must have as many rows as columns, and its dimensions must be greater than zero: $n > 0$. If **X** is an empty array or is not square, the VI sets **determinant** to NaN and returns an error.

determinant. The **determinant** of a matrix is a scalar value.

Special Case: The **determinant** of a singular matrix is zero. This is a valid result and is *not* an error.

$|A| = 0.0$ if A is singular.

error. See the *Analysis Error Codes* in *Appendix C* for a description of the error.

Trace

Finds the **trace** of **Input Matrix**.

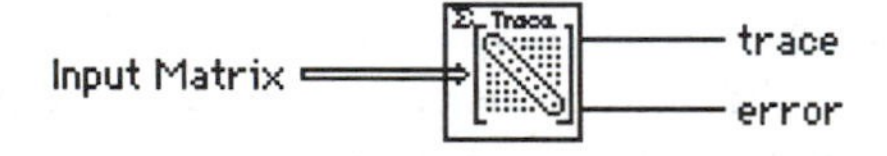

Let A be a square matrix that represents **Input Matrix** and $\mathrm{tr}(A)$ be **trace.** The **trace** of A is the sum of the main diagonal elements of A

$$\mathrm{tr}(A) = \sum_{i=0}^{n-1} a_{ii} \quad ,$$

where n is the dimension of **Input Matrix.**

[NUM] **Input Matrix** must have as many rows as columns, and its dimensions must be greater than zero: $n > 0$. If **Input Matrix** is an empty array or is not square, the VI sets **trace** to NaN and returns an error.

trace. The **trace** of a matrix is a scalar value.

error. See the *Analysis Error Codes* in *Appendix C* for a description of the error.

Linear Equations

Solves a linear system of simultaneous equations.

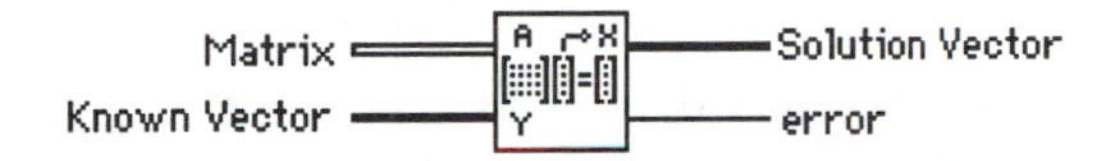

Let A be an n-by-n square matrix that represents the input **Matrix**, Y be the set of n coefficients in **Solution Vector**, and X be the set of n coefficients that solves the system

$$AX = Y.$$

If A is a nonsingular matrix—no row or column is a linear combination of any other row or column, respectively—then you can solve the system for X by decomposing the input matrix A into its lower and upper triangular matrices, L and U, such that

$$AX = LZ = Y,$$

and

$$Z = UX$$

can be an alternate representation of the original system. Notice that Z is also an n element vector.

Triangular systems are easy to solve using recursive techniques. Consequently, when you obtain the L and U matrices from A, you can find Z from the $LZ = Y$ system and X from the $UX = Z$ system.

Note

You cannot always determine beforehand whether the matrix is singular, especially with large systems. The Inverse Matrix VI detects singular matrices and returns an error, so you do not need to verify whether you have a valid system before using this VI.

The numerical implementation of the matrix inversion is numerically intensive and, because of its recursive nature, is also highly sensitive to round-off error. Although the computations use the maximum possible accuracy, the VI cannot always solve the system.

Matrix must be nonsingular and must be square. If **Matrix** is singular or is not square, the VI sets **Solution Vector** to an empty array and returns an error.

Known Vector. The number of elements in **Known Vector** must match the dimension size of the input **Matrix**. If the number of elements in **Known Vector** does not match the size of **Matrix**, the VI sets **Known Vector** to an empty array and returns an error.

Solution Vector is the solution X to $AX = Y$.

error. See the *Analysis Error Codes* in *Appendix C* for a description of the error.

Statistics VIs

This section describes the VIs, located in the **Analysis-Statistics** palette, that perform descriptive statistics, probability, analysis of variance, and interpolation functions.

Descriptive Statistics

Standard Deviation/Mean

Computes the **mean** value and the **standard deviation** of the values in the input sequence **X.**

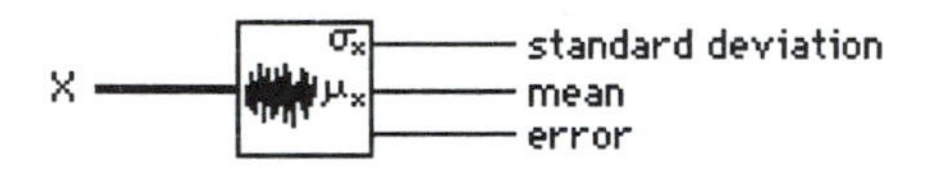

The VI computes **standard deviation** (σ) and **mean** (μ) using the following formula:

$$\sigma_x = \sqrt{\frac{1}{n}\sum_{i=0}^{n-1}(x_i - \mu)^2} ,$$

where

$$\mu = \frac{1}{n}\sum_{i=0}^{n-1} x_i ,$$

and n is the number of elements in **X**.

X. If the input sequence **X** is empty, **standard deviation** and **mean** are NaN.

standard deviation.

mean.

error. See the *Analysis Error Codes* in *Appendix C* for a description of the error.

Moment About Mean

Computes the moment about the mean of the input sequence **X** using the specified **order.** Note that you can find the **variance** of a function by taking the second moment about the mean.

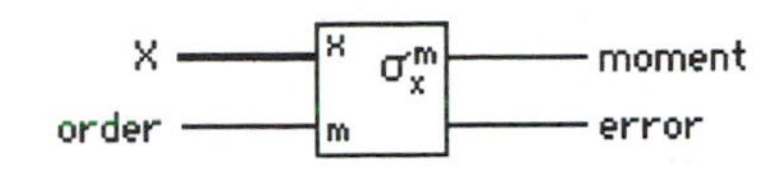

Let m be the desired **order**. The VI computes the mth order **moment** using the formula

$$\sigma_x^m = \frac{1}{n}\sum_{i=0}^{n-1}(x_i - \mu)^m \quad ,$$

where $\sigma_x{}^m$ is the mth-order **moment**, and n is the number of elements in the input sequence **X**.

X. If the input sequence **X** is empty, **moment** is NaN.

order must be greater than zero. If **order** is less than or equal to zero, the VI sets **moment** to NaN and returns an error. **order** defaults to 2.

moment.

error. See the *Analysis Error Codes* in *Appendix C* for a description of the error.

Median

Finds the median value of the input sequence **X** by sorting the values of **X** and selecting the middle element(s) of the sorted array.

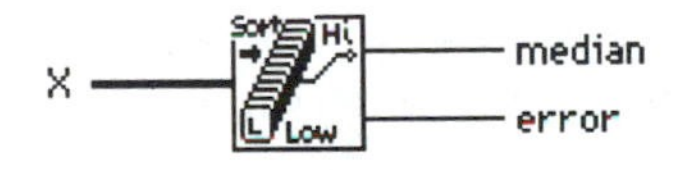

Let n be the number of elements in the input sequence **X,** and let S be the sorted sequence of X. The VI finds **median** using the following identity:

$$\text{median} = \begin{cases} s_i & \text{if } n \text{ is odd} \\ 0.5\,(s_{k-1} + s_k) & \text{if } n \text{ is even} \end{cases}$$

where

$$i = \frac{n-1}{2} \quad ,$$

and

$$k = \frac{n}{2} \quad .$$

X. If the input sequence **X** is empty, **median** is NaN.

median.

error. See the *Analysis Error Codes* in *Appendix C* for a description of the error.

Mode

Finds the **mode** of the input sequence **X** by computing the histogram of the values of **X** and selecting the largest bin. The **mode** is the center value of the bin with the largest count. Mode calls the **Histogram** VI.

X must contain at least one sample. If the **X** is empty, the histogram is undefined, the error is returned via the **Histogram** VI, and the **Mode** VI sets **mode** to NaN.

Special Case: If the input sequence has a constant value, the **Mode** VI ignores the number of intervals and sets **mode** to the constant value in the input sequence

$$\text{if } \mathbf{X} = a \Rightarrow \mathbf{mode} = a.$$

intervals. The number of **intervals** must be greater than zero. If the number of **intervals** is less than or equal to zero, the histogram is undefined, the error is returned via the **Histogram** VI, and the **Mode** VI sets **mode** to NaN. **intervals** defaults to 1.

mode is the value that occurs most often in a sequence of values. For example, if the input sequence is

$$\mathbf{X} = \{0, 1, 3, 3, 4, 4, 4, 5, 5, 7\},$$

then the **mode** of **X** is 4 because that is the value that most often occurs in **X**.

Because the VI finds **mode** with the aid of a histogram, you should read the **Histogram** VI description. The VI obtains **mode** as follows. The VI generates a discrete histogram $h(x)$ with the specified number of **intervals** of the input sequence **X** and then scans $h(x)$ for the interval Δ_i that has the maximum count. Once the VI identifies the interval, the VI selects the center value of the interval as the **mode** of the input sequence **X**

$$h(\text{mode}) = \max[h(x)].$$

error. See the *Analysis Error Codes* in *Appendix C* for a description of the error.

Histogram

Finds the discrete histogram of the input sequence **X**. The histogram is a frequency count of the number of times that a specified interval occurs in the input sequence.

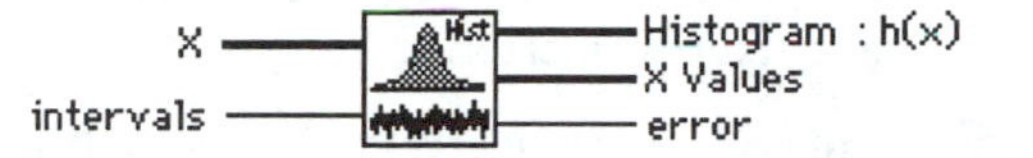

If the input sequence is

$$\mathbf{X} = \{0, 1, 3, 3, 4, 4, 4, 5, 5, 8\},$$

then the **Histogram h(x)** of **X** for eight intervals is

$$h(X) = \{h_0, h_1, h_2, h_3, h_4, h_5, h_6, h_7\} = \{1, 1, 0, 2, 3, 2, 0, 1\}.$$

Notice that the histogram of the input sequence **X** is a function of **X**.

The VI obtains **Histogram : h(x)** as follows. The VI scans the input sequence **X** to determine the range of values in it. Then the VI establishes the interval width, Δx, according to the specified number of **intervals**,

$$\Delta x = \frac{\max - \min}{m},$$

where max is the maximum value found in the input sequence **X**, min is the minimum value found in the input sequence **X**, and m is the specified number of **intervals**.

Let χ represent the output sequence **X Values**, because the histogram is a function of **X**. The VI evaluates elements of χ using

$$\chi_i = \min + 0.5\Delta x + i\Delta x \qquad \text{for } i = 0, 1, 2, \ldots, m-1.$$

The VI defines the ith interval Δ_i to be the range of values from $\chi_i - 0.5\,\Delta x$ up to but not including $\chi_i + 0.5\,\Delta x$,

$$\Delta_i = [\chi_i - 0.5\,\Delta x : \chi_i + 0.5\,\Delta x), \text{ for } i = 0, 1, 2, \ldots, m-1,$$

and defines the function $y_i(x)$ to be

$$y_i(x) = \begin{cases} 1 & \text{if } x \in\supseteq \Delta i \\ 0 & \text{elsewhere} \end{cases}$$

The function has unity value if the value of x falls within the specified interval. Otherwise it is zero. Notice that the interval Δ_i is centered about χ_i, and its width is Δx.

The last interval, Δ_{m-1}, is defined as $[\chi_i - 0.5\Delta x : \chi_i + 0.5\Delta x]$. In other words, if a value is equal to max, it is counted as belonging to the last interval.

Finally, the VI evaluates the histogram sequence H using

$$h_i = \sum_{j=0}^{n-1} y_i(x_j) \qquad \text{for } i = 0, 1, 2, \ldots, m-1,$$

where h_i represents the elements of the output sequence **Histogram : h(x)**, and n is the number of elements in the input sequence **X**.

[NUM] **X** must contain at least one sample. If **X** is empty, the histogram is undefined, and the VI sets **Histogram : h(x)** and **X Values** to empty arrays and returns an error.

NUM **intervals** must be greater than zero. If **intervals** is less than or equal to zero, the histogram is undefined, and the VI sets **Histogram : h(X)** and **X Values** to empty arrays and returns an error. **intervals** defaults to 0.

[NUM] **Histogram : h(x)**.

[NUM] **X Values.**

NUM **error.** See the *Analysis Error Codes* in *Appendix C* for a description of the error.

Probability

Normal Distribution

Computes the one-sided **probability**, p_1, of the Normally distributed random variable, **x**,

$$p_1 = \text{Prob}\{X \leq \mathbf{x}\},$$

where X is standard Normally distributed, p is the **probability**, and **x** is the value.

x — Normal Dist — probability

NUM **x.**

NUM **probability** is greater than or equal to zero and less than or equal to one: $0.0 \leq p \leq 1.0$.

This function computes only the one-sided probability. You can obtain the two-sided probability (Prob$\{-x \leq X \leq x\}$), p_1, using the following formula:

$$p_1 = 1-2\,(1-p) = 2p-1 = 2\ \text{Prob}\{X \leq \mathbf{x}\} -1.$$

Inv Normal Distribution

Computes the value of **x** such that the condition

$$p = \text{Prob}\{X \leq \mathbf{x}\}$$

is satisfied, given the **probability** value, p, of a Normally distributed random variable, X.

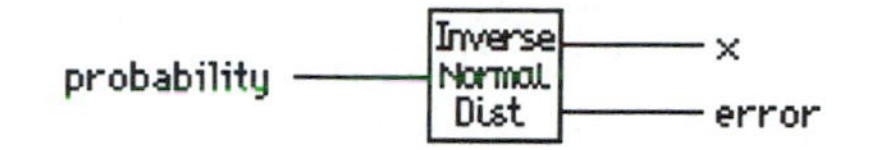

probability must meet the following condition: $0.0 < p < 1.0$. If **probability** is out of range, the VI sets **x** to NaN and returns an error.

x.

error. See the *Analysis Error Codes* in *Appendix C* for a description of the error.

---

erf(x)

Evaluates the error function at the input value. Note that

$$\mathbf{erfc}(x) = 1 - \mathrm{erf}(x).$$

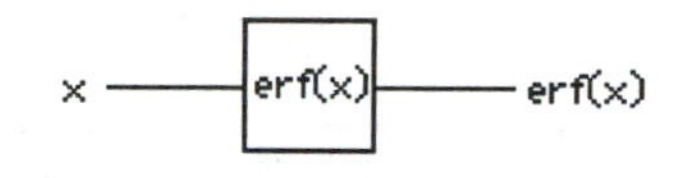

x.

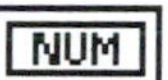

erf(x) is accurate to 15 decimal places.

$$\mathrm{erf}(x) = \frac{2}{\sqrt{\pi}} \int_0^x \exp(-t^2)\, dt$$

Interpolation

Polynomial Interpolation

Interpolates the function f at x, given a set of n points (x_i, y_i), where $f(x_i) = y_i$, f is any function, and given a number x in the range of the X values. The VI calculates output **interpolation value** $P_{n-1}(x)$, where P_{n-1} is the unique polynomial of degree $n-1$ that passes through the n points (x_i, y_i).

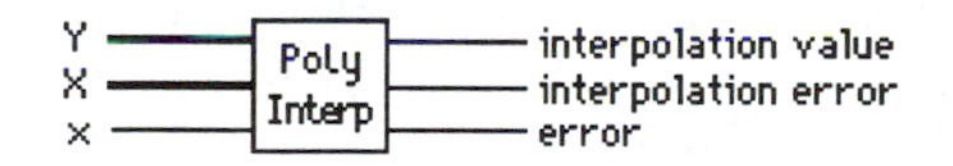

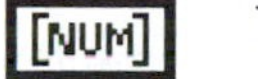

Y. Array.

X. If the number of elements in **X** is different from the number of elements in *Y*, the VI sets the output **interpolation value** and **interpolation error** to NaN and returns an error.

x.The value of *x* should be in the range of **X.** Otherwise, there will be a large interpolation error.

interpolation value.

interpolation error is an estimate of the error in the interpolation.

error. See the *Analysis Error Codes* in *Appendix C* for a description of the error.

Curve Fit

Finds the curve values and the set of coefficients that describe the curve that best represents the input data set according to the specified fit type.

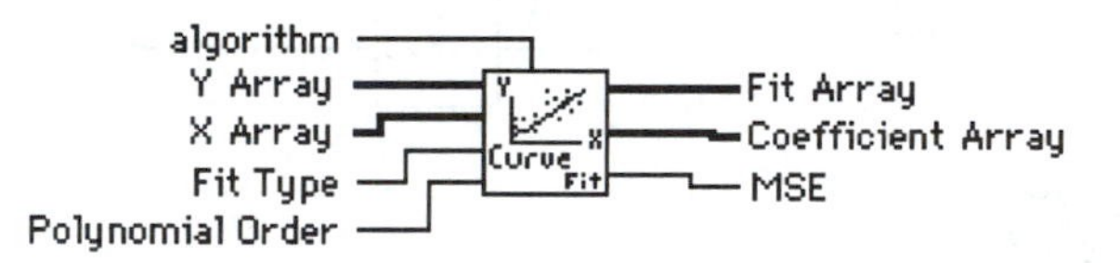

The general form of the fit is given by

$$F = \sum_{j=0}^{m} a_j X^j$$

where *F* represents the output sequence **FitArray,**

X represents the input sequence **X Values,**

A represents the **Coefficient Array,** and

m is the **Polynomial Order.**

The VI obtains **mse** using the formula

$$\text{mse} = \frac{1}{n} \sum_{j=0}^{n-1} (f_i - y_i)^2,$$

where *Y* represents the input sequence **Y Values,** and *n* is the number of data points.

For an example of how to use this VI, look at **Curve Fit Example** in ANALYSIS.LLB in the EXAMPLES folder or directory.

Y Values. The number of sample points in Y Values must be greater than **Polynomial Order.** If the number of sample points is less than or equal to the Polynomial Order, the VI sets **Fit Coefficients** to an empty array and returns an error.

X Values. The number of sample points in X Values must be greater than **Polynomial Order**. If the number of sample points is less than or equal to **Polynomial Order**, the VI sets **Fit Coefficients** to an empty array and returns an error.

Fit Type is the type of fit used:

0: Linear

1: Exponential

2: Polynomial

Polynomial Order must be greater than or equal to zero: $0 \leq m < n - 1$, where n is the number of sample points, and m is the **Polynomial Order.** If **Polynomial Order** is less than zero, the VI sets **Coefficient Array** to an empty array and returns an error. **Polynomial Order** defaults to 2.

algorithm has six selections:

0: SVD

1: Givens

2: Givens2

3: Householder

4: LU decomposition

5: Cholesky

It defaults to SVD.

Fit Array.

Coefficient Array. The total number of elements in **Polynomial Fit Coefficients** is $m + 1$.

mse is the mean squared error between the raw and fitted data.

error. See the *Analysis Error Codes* in *Appendix C* for a description of the error.

DAQ VIs

This chapter contains background information about data acquisition and reference descriptions of the DAQ VIs. You can run these VIs from their front panels or use them as subVIs in very basic applications.

You need only one DAQ VI to perform each basic I/O operation. These VIs automatically alert you to errors with a dialog box that gives you the option to halt the execution of the VI or to ignore the error. You can see software examples of how to use the DAQ VIs in `EXAMPLES\DAQ.LLB`. DAQ error codes are listed in *Appendix C.*

Note
To operate the DAQ VIs, you need only wire the terminals highlighted in boldface. Parentheses () indicate default parameters that LabVIEW automatically selects when you leave the terminal unwired.

The *LabVIEW Student Edition* will support only the National Instruments Lab-NB, Lab-LC, NB-DIO-24, and all NB-MIO boards on the Macintosh. It supports the Lab-PC, Lab-PC+, PC-LPM-16, PC-DIO-24, DAQcard-700, DAQpad-1200, and all AT-MIO boards on ISA computers running Windows.

Analog Input

When measuring analog signals with a DAQ board, you must consider the following factors that affect the digitized signal quality: single-ended and differential inputs, range, resolution, sampling rate, accuracy, and noise.

Single-ended inputs are all referenced to a common ground point. You use these inputs when the input signals are high level (greater than 1 V), the leads from the signal source to the analog input hardware are short (less than 15 ft), and all input signals share a common ground reference. If the signals do not meet these criteria, you should use *differential* inputs. With differential inputs, each input has its own ground reference. Differential inputs also reduce noise errors because they cancel out common-mode noise picked up by the leads.

Resolution is the number of bits that the analog-to-digital converter (ADC) uses to represent the analog signal. The higher the resolution, the higher the number of divisions into which the range is broken, and therefore, the smaller the detectable voltage change. The following illustration shows a sine wave and its corresponding digital image that a 3-bit and a 16-bit ADC obtains. A 3-bit converter (which is seldom used but makes a convenient example) divides the range into 2^3 or 8 divisions. A binary code between 000 and 111 represents each division. Clearly, the digital signal is not a good representation of the original signal because information has been lost in the conversion. By increasing the resolution to 16 bits, however, the ADC's number of codes increases from 8 to 65,536 (2^{16}), and it can therefore obtain an extremely accurate representation of the analog signal.

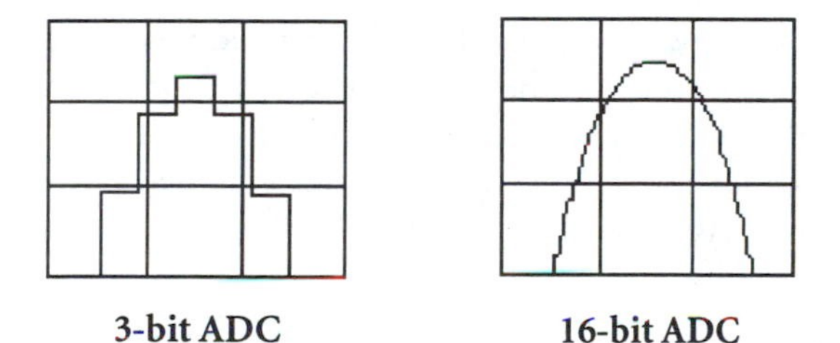

3-bit ADC **16-bit ADC**

Range refers to the minimum and maximum voltage levels that the ADC can quantize. DAQ boards offer selectable ranges (typically 0 to 10 V or –10 to 10 V), so you can match the signal range to that of the ADC to take best advantage of the resolution available to accurately measure the signal. The range, resolution, and gain available on a DAQ board determine the smallest detectable change in the voltage. This change in voltage represents 1 least significant bit (LSB) of the digital value and is often called the *code width*. The smallest detectable change is calculated as voltage range/(gain × $2^{\text{resolution in bits}}$). For example, a 12-bit DAQ board with a 0 to 10 V input range and a gain of 1 detects a 2.4 mV change, while the same board with a –10 to 10 V input range would detect only a change of 4.8 mV.

$$\frac{10}{1 \times 2^{12}} = 2.4 \text{ mV} \qquad \frac{20}{1 \times 2^{12}} = 4.8 \text{ mV}$$

Sampling rate determines how often the conversions take place. A fast sampling rate acquires more points in a given time and therefore can often form a better representation of the original signal than a slow sampling rate. All input signals must be sampled at a sufficiently fast rate to faithfully reproduce the analog signal. According to the Nyquist Sampling Theorem, you must sample at least twice the rate of the maximum frequency component you want to detect to properly digitize the signal. For example, audio signals converted to electrical signals often have frequency components up to 20 kHz; therefore, you need a board with a sampling rate greater than 40 kHz to properly acquire the signal. On the other hand, temperature transducers usually do not require a high sampling rate because temperature does not change rapidly in most applications; a board with a slower sampling rate can acquire temperature signals properly. The following illustration contrasts the results of adequately sampling a signal with the outcome of undersampling.

Adequately sampled

Aliased due to undersampling

Averaging. Unwanted noise distorts the analog signal before it is converted to a digital signal. The source of this noise may be external or internal to the computer. You can limit external noise error by using proper signal conditioning. You also can minimize the effects of this noise by oversampling the signal and then averaging the oversampled points. The level of noise is reduced by a factor of

$$1 / \sqrt{\text{number of points averaged}}$$

For example, if you average 100 points, you reduce the effect of the noise in the signal by a factor of $1/10$.

DAQ VI Descriptions

AI Sample Channel

Measures the signal attached to the specified channel and returns the measured voltage.

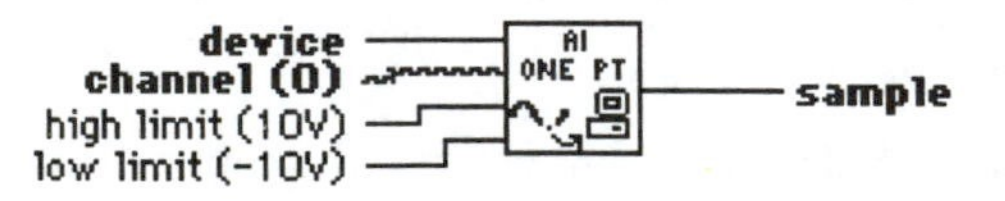

You use the **AI Sample Channel** VI to perform a single, untimed measurement of a channel. If an error occurs, a dialog box appears and gives you the option to abort the operation or continue execution.

device is the device number you assigned to the plug-in data acquisition board when you configured that board on computers running Windows. On Macs, **device** is the slot number your board is in. The default input is 1.

channel identifies the analog input channel you want to measure. The default input is onboard channel 0.

low limit specifies the minimum voltage the board measures. The default input is –10 V (–5 V for the Lab-NB and Lab-LC).

high limit specifies the maximum voltage the board measures. The default input is 10 V (5 V for the Lab-NB and Lab-LC).

sample contains the scaled analog input data for the specified channel.

AI Acquire Waveform

Acquires the specified number of samples at the specified sample rate from a single input channel and returns the acquired data.

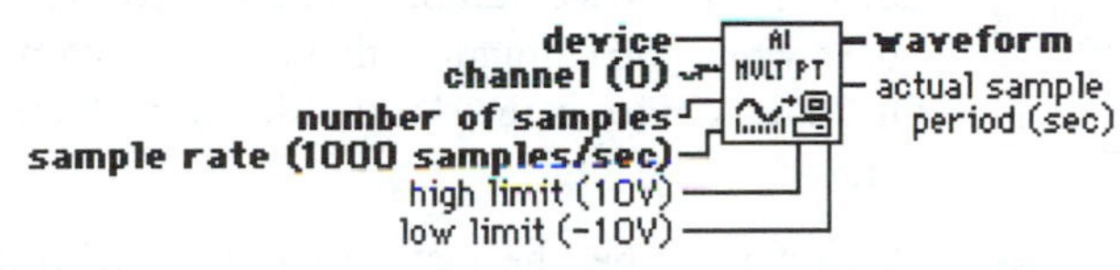

You use the **AI Acquire Waveform** VI to perform a timed measurement of a waveform (multiple voltage readings at a specified sampling rate) on a specified analog input channel. If an error occurs, a dialog box appears and gives you the option to abort the operation or continue execution.

device is the device number you assigned to the plug-in data acquisition board when you configured that board on computers running Windows. On Macs, **device** is the slot number your board is in. The default input is 1.

channel identifies the analog input channel you want to measure. The default input is onboard channel 0.

number of samples is the number of single-channel samples the VI acquires before the acquisition is complete.

sample rate is the number of samples per second the VI acquires from the specified channel. The default rate is 1,000 samples/sec.

low limit specifies the minimum voltage the board measures. The default input is –10 V (–5 V for the Lab-NB and Lab-LC).

high limit specifies the maximum voltage the board measures. The default input is 10 V (5 V for the Lab-NB and Lab-LC).

[NUM]

waveform is a one-dimensional array that contains scaled analog input data in volts.

NUM

actual sample period is the time between samples, which is the inverse of the sample rate the VI used to acquire the data. The **actual sample period** may differ slightly from the requested sample rate, depending on the capabilities of your hardware.

AI Sample Channels

Performs a single voltage reading from each of the specified channels.

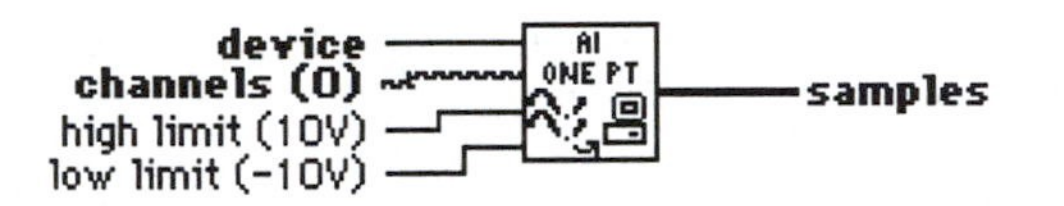

You use the **AI Sample Channels** VI to measure a single voltage from each of the specified analog input channels. If an error occurs, a dialog box appears and gives you the option to abort the operation or continue execution.

device is the device number you assigned to the plug-in data acquisition board when you configured that board on computers running Windows. On Macs, **device** is the slot number your board is in. The default input is 1.

abc

channel identifies the analog input channel you want to measure. If x, y, and z refer to channels, you can specify a list of channels by separating the individual channels by commas—for example, x,y,z. If x refers to the first channel in a consecutive channel range and y refers to the last channel, you can specify the range by separating the first and last channels by a colon—for example, x:y.

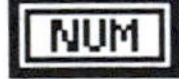

low limit specifies the minimum voltage the board measures. The default input is –10 V (–5 V for the Lab-NB and Lab-LC).

high limit specifies the maximum voltage the board measures. The default input is 10 V (5 V for the Lab-NB and Lab-LC).

[NUM]

samples is a one-dimensional array that contains scaled analog input data in volts.

AI Acquire Waveforms

Acquires data from the specified channels and scans the list at the specified scan rate.

device
channels (0)
number of samples/ch
scan rate (1000 scans/sec)
high limit (10V)
low limit (-10V)
AI
MULT PT
waveforms
actual scan period (sec)

You use the **AI Acquire Waveforms** VI to perform a timed measurement of multiple waveforms on the specified analog input channels. If an error occurs, a dialog box appears and gives you the option to abort the operation or continue execution.

device is the device number you assigned to the plug-in data acquisition board when you configured that board on computers running Windows. On Macs, **device** is the slot number your board is in. The default input is 1.

channel identifies the analog input channel you want to measure. If x, y, and z refer to channels, you can specify a list of channels by separating the individual channels by commas—for example, x,y,z. If x refers to the first channel in a consecutive channel range and y refers to the last channel, you can specify the range by separating the first and last channels by a colon—for example, x:y.

number of samples/ch is the number of samples per channel the VI acquires before the acquisition is complete.

scan rate is the number of scans per second the VI acquires. The default rate is 1,000 scans/sec.

low limit specifies the minimum voltage the board measures. The default input is –10 V (–5 V for the Lab-NB and Lab-LC).

high limit specifies the maximum voltage the board measures. The default input is 10 V (5 V for the Lab-NB and Lab-LC).

waveforms is a two-dimensional array that contains scaled analog input data in volts.

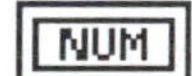

actual scan period is the time between scans, which is the inverse of the scan rate the VI used to acquire the data. This rate may differ slightly from the scan rate you specified. LabVIEW attempts to get as close as possible to the specified scan rate, given the limitations of your particular data acquisition board.

AO Update Channel

Writes a specified voltage value to an analog output channel.

You use the **AO Update Channel** VI to write a single update to an analog output channel. If an error occurs, a dialog box appears and gives you the option to abort the operation or continue execution.

device is the device number you assigned to the plug-in data acquisition board when you configured that board on computers running Windows.

On Macs, **device** is the slot number your board is in. The default input is 1.

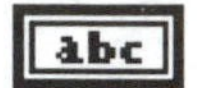

channel identifies the analog output channel you want to use. The default input is 0.

voltage contains the voltage value to be written to the specified analog output channel. You must specify **voltage**.

AO Generate Waveform

Generates a voltage waveform on an analog output channel at the specified update rate.

You use the **AO Generate Waveform** VI to generate a multipoint voltage waveform on a specified analog output channel. If an error occurs, a dialog box appears and gives you the option to abort the operation or continue execution. Note that on the Macintosh, you need a plug-in DMA board such as the NB-DMA2800 in order to perform waveform generation.

device is the device number you assigned to the plug-in data acquisition board when you configured that board on computers running Windows. On Macs, **device** is the slot number your board is in. The default input is 1.

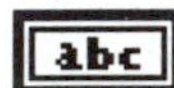

channel identifies the analog output channel you want to use.

update rate is the number of voltage updates to generate per second. The default rate is 1,000 updates/sec.

waveform is a one-dimensional array that contains data in volts to be written to the analog output channel. You must specify **waveform**.

AO Update Channels

Writes voltage values to each of the specified analog output channels.

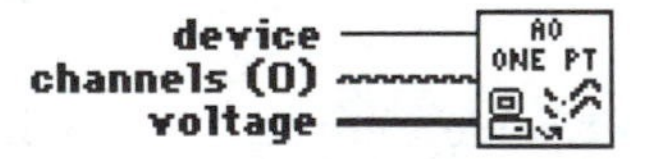

You use the **AO Update Channels** VI to update multiple analog output channels with single voltage values. If an error occurs, a dialog box appears and gives you the option to abort the operation or continue execution.

device is the device number you assigned to the plug-in data acquisition board when you configured that board on computers running Windows. On Macs, **device** is the slot number your board is in. The default input is 1.

channel identifies the analog output channel you want to use. If x, y, and z refer to channels, you can specify a list of channels by separating the individual channels by commas—for example, x,y,z. If x refers to the first channel in a consecutive channel range and y refers to the last channel, you can specify the range by separating the first and last channels by a colon—for example, x:y.

voltage is a one-dimensional array that contains the analog output data in volts. You must specify **voltage**.

AO Generate Waveforms

Generates multiple voltage waveforms on the specified analog output channels at the specified update rate.

You use the **AO Generate Waveforms** VI to generate multiple timed voltage updates on the specified analog output channels. If an error occurs, a dialog box appears and gives you the option to abort the operation or continue execution. Note that on the Macintosh, only the NB-MIO-16X board in conjunction with a plug-in DMA board has the capability to generate more than one waveform at a time.

device is the device number you assigned to the plug-in data acquisition board when you configured that board on computers running Windows. On Macs, **device** is the slot number your board is in. The default input is 1.

channels identifies the analog output channel you want to use. If x, y, and z refer to channels, you can specify a list of channels by separating the individual channels by commas—for example, x,y,z. If x refers to the first channel in a consecutive channel range and y refers to the last channel, you can specify the range by separating the first and last channels by a colon—for example, x:y.

update rate is the number of voltage updates to generate per second. The default rate is 1,000 updates/sec. All channels in the **channels** string update simultaneously.

waveforms is a two-dimensional array that contains data to be written to the analog output channels in volts. The array must be in column-major order—that is, the first dimension must be data, and the second dimension must be the channel. The channel order of the data must be identical to the channel order you specify in **channels**. You must specify

waveforms, where the row is the update number and the column is the channel number.

Read From Digital Port

Reads a digital port that you configure.

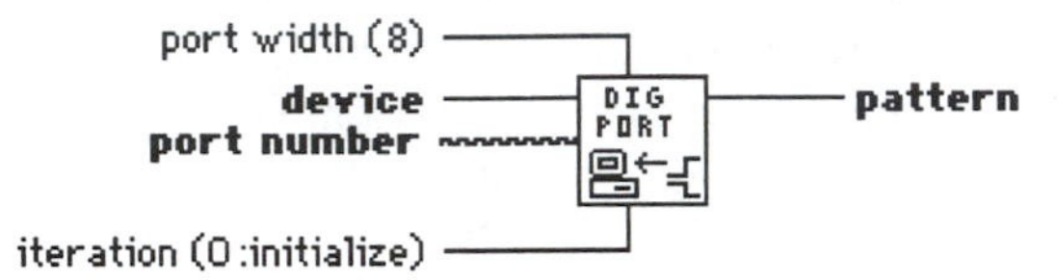

NUM **device** is the device number you assigned to the plug-in data acquisition board when you configured that board on computers running Windows. On Macs, **device** is the slot number your board is in. The default input is 1.

abc **port number** specifies the port this VI configures. A **port number** value of 0 signifies port 0, a **port number** of 1 signifies port 1, and so on. If you use an SCXI-1160, SCXI-1161, SCXI-1162, or SCXI-1163 module, use the `SCx!MDy!0` syntax, where `x` is the chassis ID and `y` is the module device number, to specify the port on a module.

NUM **port width** is the total width in bits or the number of lines of the port. For example, you can combine two 8-bit ports into a 16-bit port on a Lab PC+ board by setting **port width** to 16.

When **port width** is greater than the physical port width of a digital port, the following restrictions apply. The **port width** must be an integral multiple of the physical port width, and the port numbers in the combined port must begin with the port named by **port number** and must increase consecutively. For example, if **port number** is 3 and **port width** is 24 (bits), the VI uses ports 3, 4, and 5.

iteration defaults to 0 so that you can optimize operation when you execute this VI in a loop.

NUM **pattern** is the data the VI read from the port.

Read From Digital Line

Reads the logical state of a digital line on a port that you configure.

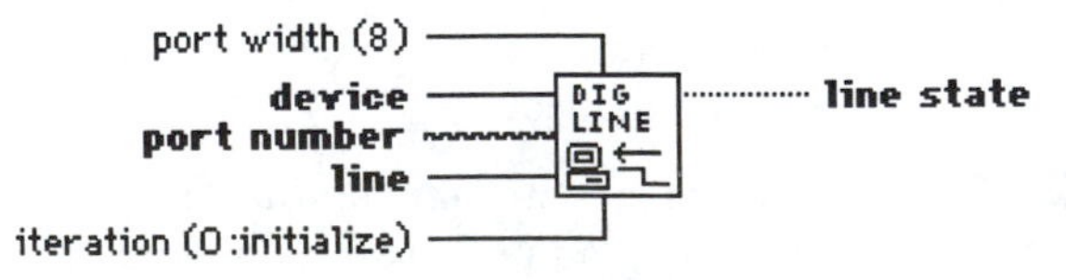

NUM **device** is the device number you assigned to the plug-in data acquisition board when you configured that board on computers running Windows.

On Macs, **device** is the slot number your board is in. The default input is 1.

port number specifies the port this VI configures. A **port number** value of 0 signifies port 0, a **port number** of 1 signifies port 1, and so on. If you use an SCXI-1160, SCXI-1161, SCXI-1162, or SCXI-1163 module, use the `SCx!MDy!0` syntax, where `x` is the chassis ID and `y` is the module device number, to specify the port on a module.

line specifies the individual bit or line within the port to be used for I/O.

port width is the total width in bits of the port. For example, you can combine two 8-bit ports into a 16-bit port on a Lab PC+ board by setting **port width** to 16.

When **port width** is greater than the physical port width of a digital port, the following restrictions apply. The **port width** must be an integral multiple of the physical port width, and the port numbers in the combined port must begin with the port named by **port number** and must increase consecutively. For example, if **port number** is 3 and **port width** is 24 (bits), the VI uses ports 3, 4, and 5.

iteration defaults to 0 so that you can optimize operation when you execute this VI in a loop.

line state indicates whether the logic of the value read from the line is high or low.

Write To Digital Port

Outputs a decimal pattern to a digital port that you specify.

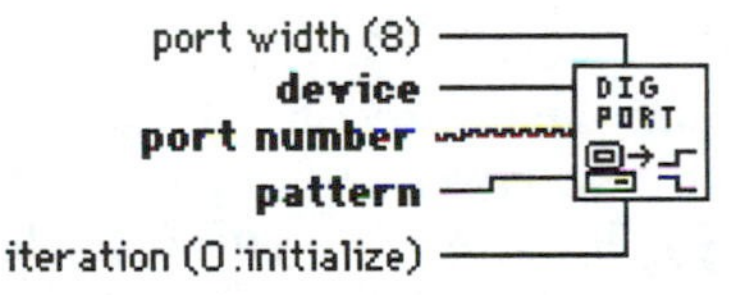

device is the device number you assigned to the plug-in data acquisition board when you configured that board on computers running Windows. On Macs, **device** is the slot number your board is in. The default input is 1.

port number specifies the port this VI configures. A **port number** value of 0 signifies port 0, a **port number** of 1 signifies port 1, and so on. If you use an SCXI-1160, SCXI-1161, SCXI-1162, or SCXI-1163 module, use the `SCx!MDy!0` syntax, where `x` is the chassis ID and `y` is the module device number, to specify the port on a module.

port width is the total width in bits of the port. For example, you can combine two 8-bit ports into a 16-bit port on a Lab PC+ board by setting **port width** to 16.

When **port width** is greater than the physical port width of a digital port, the following restrictions apply. The **port width** must be an integral multiple of the physical port width, and the port numbers in the combined port must begin with the port named by **port number** and must increase consecutively. For example, if **port number** is 3 and **port width** is 24 (bits), the VI uses ports 3, 4, and 5.

pattern specifies the new state of the lines in the port.

iteration defaults to 0 so that you can optimize operation when you execute this VI in a loop.

Write To Digital Line

Sets the output logic state of a digital line on a digital port that you specify to high or low.

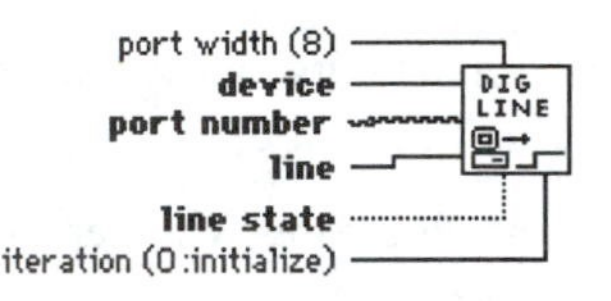

device is the device number you assigned to the plug-in data acquisition board when you configured that board on computers running Windows. On Macs, **device** is the slot number your board is in. The default input is 1.

port number specifies the port this VI configures. A **port number** value of 0 signifies port 0, a **port number** of 1 signifies port 1, and so on. If you use an SCXI-1160, SCXI-1161, SCXI-1162, or SCXI-1163 module, use the `SCx!MDy!0` syntax, where `x` is the chassis ID and `y` is the module device number, to specify the port on a module.

port width is the total width in bits of the port. For example, you can combine two 8-bit ports into a 16-bit port on a Lab PC+ board by setting **port width** to 16.

When **port width** is greater than the physical port width of a digital port, the following restrictions apply. The **port width** must be an integral multiple of the physical port width, and the port numbers in the combined port must begin with the port named by **port number** and must increase consecutively. For example, if **port number** is 3 and **port width** is 24 (bits), the VI uses ports 3, 4, and 5.

line specifies the individual bit or line within the port to be used for I/O.

line state determines whether the VI sets the logic of the line to high or low.

iteration defaults to 0 so that you can optimize operation when you execute this VI in a loop.

Counter VIs

This section describes five VIs for programming counters on MIO boards with the AM9513 or DAQ-STC counter/timer chip**: Count Time or Events, Generate Delayed Pulse, Generate Pulse Train, Measure Frequency,** and **Measure Pulse Width or Period.** These VIs can generate a single delayed TTL pulse, a finite or continuous train of pulses, and measure the frequency, pulse width, or period of a TTL signal. *These VIs do not work with Lab boards, DAQCards, and other boards that have the 8253 chip.*

This section also describes the **ICTR Control** VI, which you can use with boards that have the 8253 chip.

Some of these VIs use other counters in addition to the one specified. In this case, a logically adjacent counter is chosen, which is referred to as counter+1 when it is the adjacent, logically higher counter and counter -1 when it is the adjacent, logically lower counter. For a device with the AM9513 chip, the counter is 1. Then counter+1 is counter 2 and counter-1 is counter 5.

Count Events or Time

Configures one or two counters to count external events or elapsed time. An external event is a high or low signal transition on the specified SOURCE pin of the counter. This VI works on boards with the AM9513 or DAQ-STC counter/timer chip, such as the MIO series. It does not work on boards with the 8253 chip, such as the Lab series.

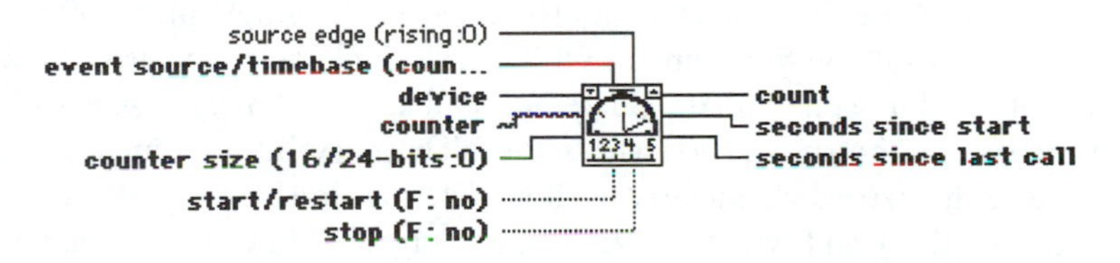

To count events, set the event source/timebase to 0.0 and connect the signal you want to count to the SOURCE pin of the counter. To count time, set this control to the timebase frequency you want to use.

event source/timebase (Hz) is set to the frequency of the internal signal whose cycles are counted, or it is set to ≤ 0.0 (default) to count the signal on the SOURCE pin of the counter.

device is the device number you assigned to the DAQ device during configuration on computers running Windows. On Macs, **device** is the slot number your board is in. The default input is 1.

counter is the ASCII number of the counter to use for the operation.

counter size is set to 0 (default) to use a single 16-bit Am9513 counter or 24-bit DAQ-STC counter, or is set to 1 to use two Am9513 counters as a 32-bit counter. Leave this input set to 0 for a DAQ-STC counter.

start/restart is set TRUE to configure and start the counter(s).

stop is set TRUE to stop the counter(s).

source edge is the edge of the counter clock signal on which it decrements.

0: count on low to high transition.

1: count on high to low transition.

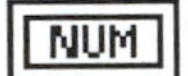

count is the value of the counter at the time it is read. If there are two Am9513 counters assigned to the tasked, the value of the higher-order counter is multiplied by 10000 hex, shifting it to 16 bits. The higher-order counter is then added to the value of the lower counter. The count will be incorrect if the **taskID** is for two concatenated DAQ-STC counters.

seconds since start is the amount of time that has elapsed since counting began. This indicator is not used for event counting.

seconds till overflow is the amount the time remaining until the counter reaches TC and overflows. This indicator is not used for event counting.

Generate Delayed Pulse

Configures and starts a counter to generate a single pulse with the specified delay and pulse width on the counter's OUT pin. A single pulse consists of a delay phase (phase 1), followed by a pulse phase (phase 2), and than a return to the phase 1 level. If an internal timebase is chosen, the VI selects the highest-resolution timebase for the counter to achieve the desired characteristics. If an external timebase signal is chosen, the user indicates the delay and width as cycles of that signal. You can optionally gate or trigger the pulse with a signal on the counter's GATE pin. This VI works on boards with the AM9513 or DAQ-STC counter/timer chip, such as the MIO series. It does not work on boards with the 8253 chip, such as the Lab series.

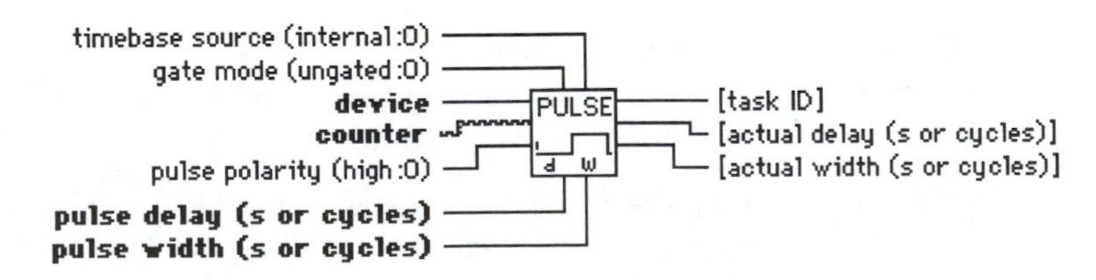

device is the device number you assigned to the DAQ device during configuration on computers running Windows. On Macs, **device** is the slot number your board is in. The default input is 1.

counter is the counter number on the device to use for the operation.

pulse delay (s or cycles) is the desired duration of the first phase of the pulse, phase 1. The unit is seconds if **timebase source** is 0 (internal) and cycles if **timebase source** is 1 (external). If **pulse delay**=0.0 and **timebase source** is internal, the VI selects a minimum delay of three cycles of the timebase used.

pulse width (s or cycles) is the desired duration of the second phase of the pulse, phase 2. The unit is seconds if **timebase source** is 0 (internal) and cycles if **timebase source** is 1 (external). If **pulse width**=0.0 and **timebase source** is internal, the VI selects a minimum width of three cycles of the timebase used.

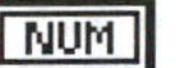

timebase source is the signal that counts the counter.

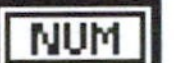

gate mode specifies how the counter's GATE signal is used.

0: ungated/software start: ignore the gate source and start when the VI is called. (default)

1: count while the gate signal is TTL high.

2: count while the gate signal is TTL low.

3: start counting on the rising edge of the TTL gate signal.

4: start counting on the falling edge of the TTL gate signal.

5: restart counting on each rising edge of the TTL gate signal.

6: restart counting on each falling edge of the TTL gate signal

Use **gate mode** 3 or 4 to generate one delayed pulse on the first gate edge after starting. Use **gate mode** 5 or 6 to generate a delayed pulse for each gate edge (i.e., retriggerable one-shot behavior).

pulse polarity is the polarity of the second phase (phase two) of each delayed pulse.

0: high pulse: the pulse starts at a low TTL level and ends at a high level (default)

1: low pulse: the pulse starts at a high TTL level and ends at a low level.

taskID is the reference number assigned for the device and counter(s) reserved for this operation.

actual delay (s or cycles) is the achieved delay. It may differ from the desired delay because the hardware has limited resolution and range.

actual width (s or cycles) is the achieved pulse width. It may differ from the desired width because the hardware has limited resolution and range.

Helpful VI Tips

- The delayed pulse appears on counter's OUT pin.
- If pulse delay or pulse width =0.0, a minimum delay or width is computed.
- If gate mode >0, connect your gating signal to counter's GATE pin.

Generate Pulse Train

Configures the specified counter to generate a continuous pulse train on the counter's OUT pin, or to generate a finite-length pulse train using the specified counter and an adjacent counter. The signal has the prescribed frequency, duty cycle, and polarity. Each cycle of the pulse train consist of a delay phase (phase 1) followed by a pulse phase (phase 2).

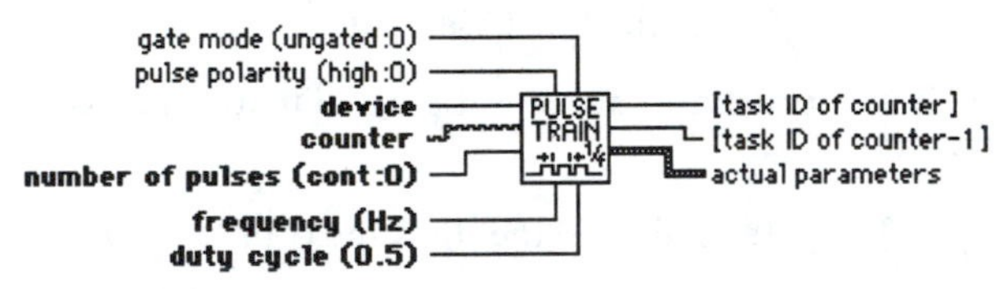

This VI uses only the specified **counter** to generate a continuous pulse. For a finite-length pulse, the VI also uses counter-1 to generate a minimum-delayed pulse to gate **counter**. To stop a continuous pulse train, execute this VI again with a –1 wired to **number of pulses**. You must externally wire counter-1's OUT pin to **counter's** GATE pin for a finite-length pulse train. You can optionally gate or trigger the start of the train with a signal on the counter-1's GATE pin. This VI works on boards with the AM9513 or DAQ-STC counter/timer chip, such as the MIO series. It does not work on boards with the 8253 chip, such as the Lab series.

Note

A pulse train consists of a series of delayed pulses, where phase 1 or the first phase of each pulse is the inactive state of the output (low for a high pulse) and the phase 2 of the second phase is the pulse itself. Refer to the following illustration of a high-polarity pulse train.

Helpful VI Tips

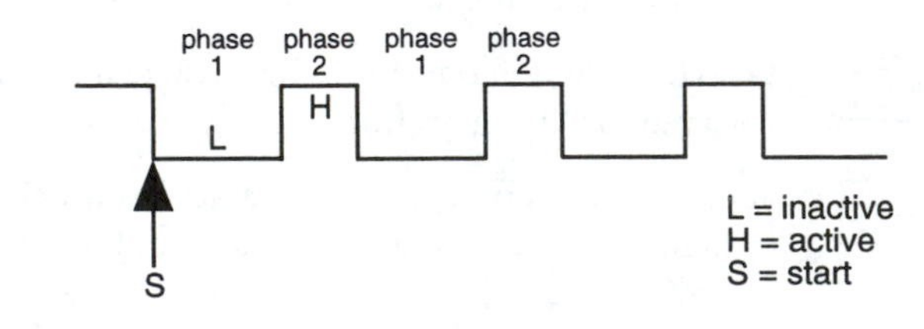

The pulse train might end in either the active or inactive states. Under certain circumstances you can accommodate for this behavior by changing pulse polarity and duty cycle.

- The pulse train appears on counter's OUT pin.
- If number of pulses >0, connect counter's GATE pin to counter-1's OUT pin.
- If gate mode >0, connect your gating signal to counter's GATE pin if number of pulses =0 (continuous) or to counter-1's GATE pin if number of pulses >0.

device is the device number you assigned to the DAQ device during configuration on computers running Windows. On Macs, **device** is the slot number your board is in. The default input is 1.

counter is the counter number on the device to use for the operation.

number of pulses is the number of pulses you want in the pulse train. If the value is = 0 (default), the VI generates a continuous pulse train. To stop a continuous pulse train, execute this VI again with a –1 wired to **number of pulses**.

frequency (Hz) is the desired repetition rate of the continuous pulse train.

duty cycle is the desired ratio of the second phase (phase two) of the pulse to the period of one cycle (1/frequency); default is 0.5. If **duty cycle**=0.0 or 1.0, the VI computes the closest achievable duty cycle using a minimum period of three timebase cycles. A duty cycle very close to 0.0 or 1.0 may not be possible.

gate mode specifies how the signal on the counter's GATE pin is used for a continuous pulse train or how the signal on counter-1's GATE pin is used for a finite pulse train.

0: ungated/software start: ignore the gate source and start when the VI is called. (default)

1: count while the gate signal is TTL high.

2: count while the gate signal is TTL low.

3: start one pulse train on the rising edge of the TTL gate signal.

4: start one pulse train on the falling edge of the TTL gate signal.

If **number of pulses**=0 (continuous pulse train), **gate mode** 3 or 4 generates one pulse per gate edge, which is the behavior of a retriggerable one shot. If **number of pulses** >0 and you want to retriggered the pulse train, you must restart **counter** between the end of one train and the next edge of the gate signal.

pulse polarity is the polarity of second phase (phase 2) of each delayed pulse.

0: high pulse: the pulse starts at a low TTL level and ends at a high level (default)

1: low pulse: the pulse starts at a high TTL level and ends at a low level.

taskID of gate counter is the taskID of counter-1, the one used to gate the continuous pulses.

taskID of continuous pulse counter is the taskID of counter, the one used to generate the continuous pulse.

actual parameters are the achieved parameters.These parameters may differ from the desired parameters because the hardware has limited resolution and range.

frequency (Hz) is the achieved frequency.

duty cycle is the achieved duty cycled.

pulse delay is the achieved minimum delay to the gating pulse.

pulse width is the achieved width of the gating pulse.

Measure Frequency

Measures the frequency of a TTL signal on the specified counter's SOURCE pin. In addition to connecting the input signal, you must connect the counter's GATE pin to counter-1's OUT pin. This VI is useful for relatively high frequency signals. Use the **Measure Pulse Width or Period** VI for relatively low frequency signals. Keep in mind that period(s) = 1/frequency (Hz). You can optionally gate or trigger the operation with a signal on counter-1's GATE pin.

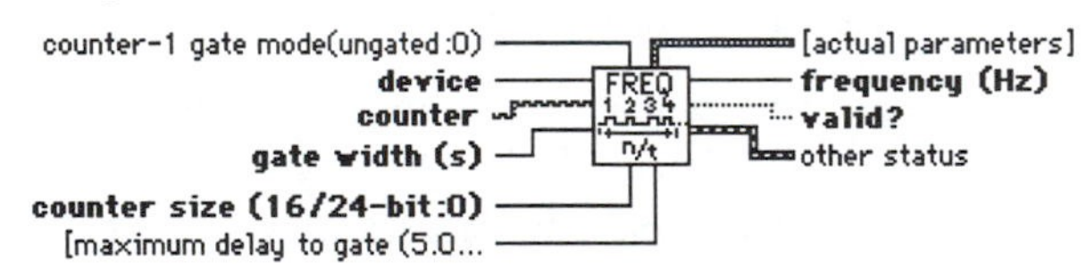

This VI configures the specified **counter** and counter+1 (optional) as event counters to count rising edges of the signal on counter's SOURCE pin. The VI also configures counter-1 to generate a minimum-delayed pulse to gate the event counter, starts the event counter and then the gate counter, waits the expected gate period, and then reads the gate counter until its output state is low. Next the VI reads the event counter and computes the signal frequency (**number of events/actual gate pulse width**) and stops the counters. You can optionally gate or trigger the operation with a signal on counter-1's GATE pin. This VI works on boards with the AM9513 or DAQ-STC counter/timer chip, such as the MIO series. It does not work on boards with the 8253 chip, such as the Lab series.

Helpful VI Tips

- Connect the signal you want to measure to counter's SOURCE pin.

- Connect counter's GATE pin to counter-1's OUT pin.
- If counter-1 gate mode >0, connect the gating signal to counter-1's GATE pin and set **maximum delay to gate** to the expected maximum delay.
- On a counter overflow, use 32-bit counter size or decrease **gate width**; if the number of rising edges is low, increase the gate width.

device is the device number you assigned to the DAQ device during configuration on computers running Windows. On Macs, **device** is the slot number your board is in. The default input is 1.

counter is the ASCII number of the base counter to use to count events.

gate width (s) is the desired length of the pulse used to gate the signal; the lower the signal frequency, the longer the width must be.

counter size is set to zero (default) to use the specified 16-bit Am9513 counter or 24-bit DAQ-STC counter. It is set to 1 to use the two Am9513 counters as a 32-bit counter. Leave this input 0 for a DAQ-STC counter.

frequency (Hz) is the frequency of the signal, computed as **number of events/actual gate pulse width.**

valid? is TRUE if the measurement completes without a counter overflow.

counter-1 gate mode specifies how the counter-1's GATE signal is used.

0: ungated/software start: ignore the gate source and start when the VI is called. (default)

1: count while the gate signal is TTL high.

2: count while the gate signal is TTL low.

3: start counting on the rising edge of the TTL gate signal.

4: start counting on the falling edge of the TTL gate signal.

maximum delay to gate is the maximum expected delay between the time the VI is called and the start of the gating pulse. If counter-1 gate mode>0 and the gate does not start in this time, a timeout occurs.

actual parameters is a cluster of lesser parameters.

number of rising edges is the number of rising edges of the signal counted during the gate.

gate pulse (s) width is the length of the gating pulse that is used.

other status is a cluster of additional status information.

counter overflow? is TRUE if the event counter reaches terminal count.

timeout? is TRUE if the time limit expires during the call. A timeout and valid measurement may occur at the same time. A **timeout** does not produce an error.

Measure Pulse Width or Period

Measures the pulse width (length of time a signal is high or low) or period (length of time between adjacent rising or falling edges) of a TTL signal connected to the counter's GATE pin. The method used gates an internal timebase clock with the signal being measured. This VI is useful in measuring the period or frequency (1/period) of relatively low frequency signals, when many timebase cycles occur during the gate. Use the **Measure Frequency** VI to measure the period or frequency of relatively high frequency signals.

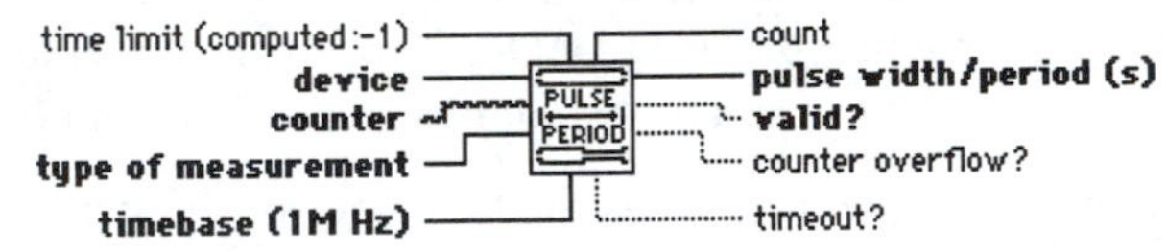

The VI iterates until a valid measurement, timeout, or counter overflow occurs. A valid measurement exists when **count** ≥ 4 without a counter overflow. If counter overflow occurs, lower the timebase. If you start a pulse width measurement during the phase you want to measure, you get an incorrect low measurement. Therefore, make sure the pulse does not occur until after the counter is started. This restriction does not apply to period measurements. This VI works on boards with the AM9513 or DAQ-STC counter/timer chip, such as the MIO series. It does not work on boards with the 8253 chip, such as the Lab series.

Helpful VI Tips

- Connect the signal you want to measure to counter's GATE pin.
- To measure pulse width, the pulse must occur after the counter is started.

device is the device number you assigned to the DAQ device during configuration on computers running Windows. On Macs, **device** is the slot number your board is in. The default input is 1.

counter is the ASCII number of the counter to use for operation.

type of measurement identifies the type of pulse width or period measurement to make. The following illustration demonstrates the various values for **type of measurement**.

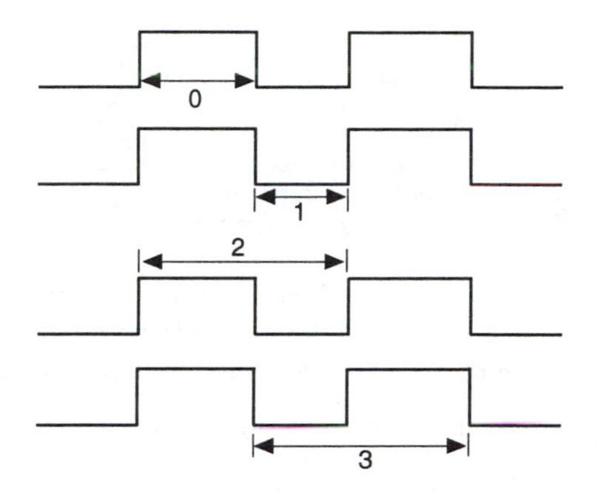

0: measure high pulse width from rising to falling edge.

1: measure low pulse width from falling to rising edge.

2: measure period between adjacent rising edges. (default)

3: measure period between adjacent falling edges.

timebase is the internal clock signal to use (default 1 MHz). If the counter overflows because **timebase** is too high, lower it until a valid reading occurs or until the lowest timebase (100 Hz) is used and a time-out occurs.

pulse width/period (s) is the measured pulse width or period; it equals count divided by the timebase and may be valid or invalid.

valid is TRUE if counter has not underflowed (if count >=4) or overflowed.

time limit is the period to wait for a valid measurement. If **time limit** = –1.0 (default), the time limit is set to five seconds or four times the range of the counter at the selected **timebase** (4*65,536/**timebase** in seconds for the Am9513 and 4*16,777,216/**timebase** for the DAQ-STC), whichever is larger.

count is the counter reading; it may be valid or invalid. For best accuracy, choose a timebase frequency that maximizes the count without overflowing it.

counter overflow is TRUE if counter reaches terminal count. Overflow does not produce an error.

timeout is TRUE if a valid reading is not within the prescribed or computed time limit. The **timeout** parameter does not produce an error.

ICTR Control

Controls counters on devices that use the 8253 chip such as Lab boards and DAQcard700s, SCXI-1200, DAQPad-1200, PC-LPM-16, and DAQCard 700.

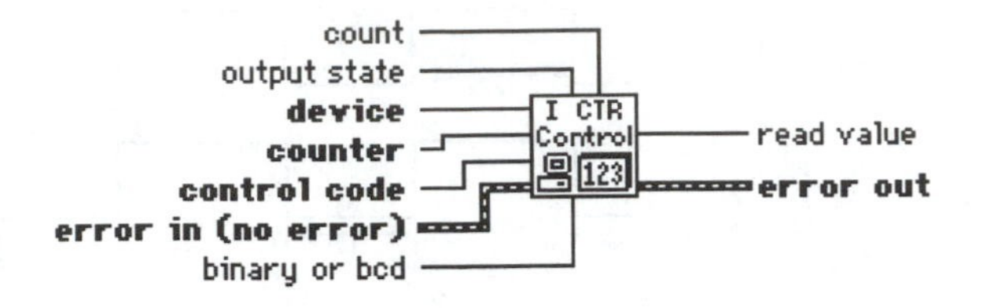

device is the device number you assigned to the DAQ device during configuration on computers running Windows. On Macs, **device** is the slot number your board is in. The default input is 1.

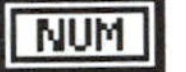

counter is the counter this VI controls, ranging from 0 through 2.

control code determines the counter's operating mode.

0: Setup mode 0 – Toggle output from low to high on TC (default).
1: Setup mode 1 – Programmable one-shot.
2: Setup mode 2 – Rate generator.
3: Setup mode 3 – Square wave rate generator.
4: Setup mode 4 – Software-triggered strobe.
5: Setup mode 5 – Hardware-triggered strobe.
6: Read.
7: Reset.

In setup mode 0, as shown in the following illustration, the output becomes low after the mode set operation, and the counter begins to count down while the gate input is high. The output becomes high when counter reaches the TC (that is, when the counter decreases to 0) and stays high until you set the selected counter to a different mode.

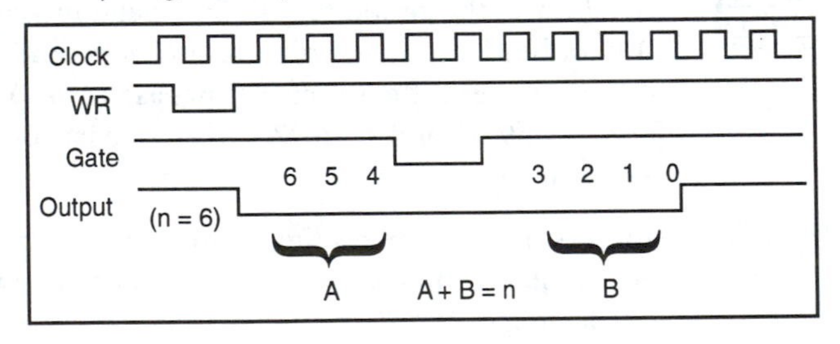

Mode Timing Diagram

In setup mode 1, as shown in the following illustration, the output becomes low on the count following the leading edge of the gate input and becomes high on TC.

Mode 1 Timing Diagram

In setup mode 2, as shown in the following illustration, the output becomes low for one period of the clock input. The **count** indicates the period between output pulses.

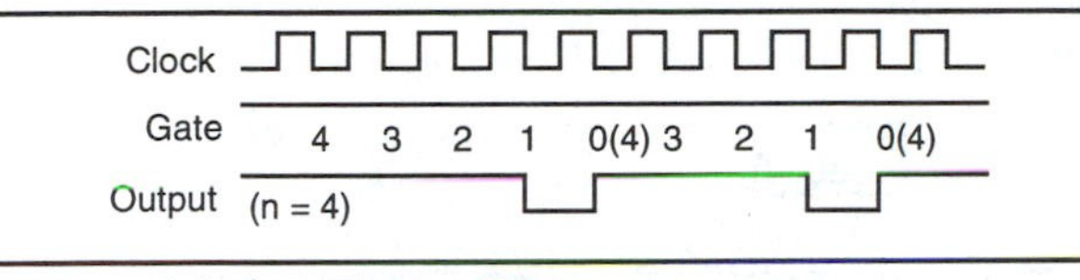

Mode 2 Timing Diagram

In setup mode 3, the output stays high for one-half of the **count** clock pulses and stays low for the other half. Refer to the next figure.

Mode 3 Timing Diagram

In setup mode 4, as in the next illustration, the output is initially high, and the counter begins to count down while the gate input is high. On terminal count, the output becomes low for one clock pulse, then becomes high again.

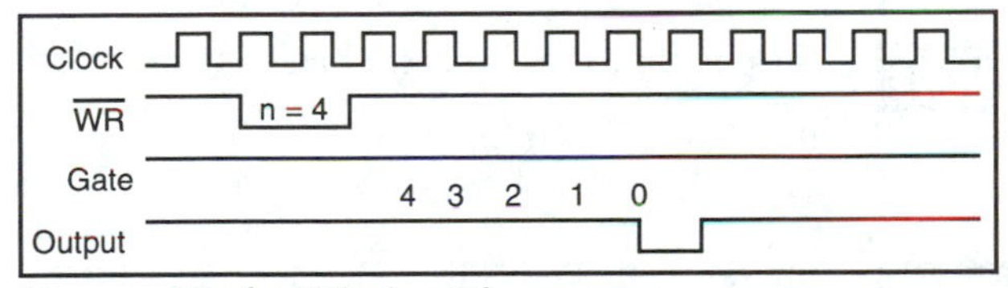

Mode 4 Timing Diagram

Setup mode 5 is similar to mode 4, except that the gate input triggers the count to start. See the next figure for an illustration of mode 5.

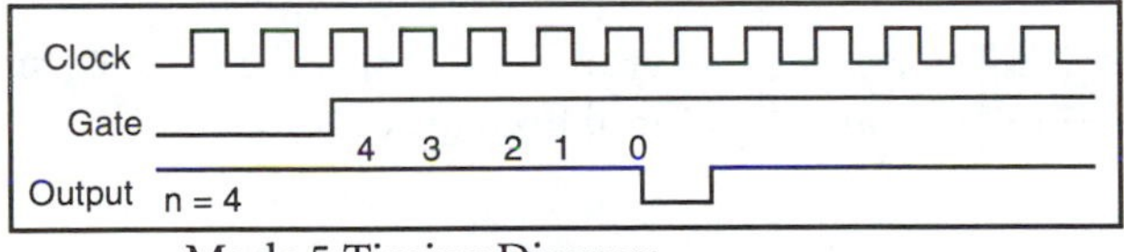

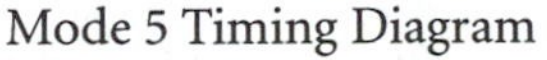
Mode 5 Timing Diagram

See the 8253 Programmable Interval Timer data sheet in your Lab board user manual for details on these modes and their associated timing diagrams.

error in describes any error conditions prior to this VI's execution. This cluster defaults to no error. The **error in** cluster contains the following parameters.

status is TRUE if an error occurred. If **status** is TRUE, this VI does not do the configuration.

code is the error code number identifying an error. A value of 0 generally means no error, a negative value means a fatal error, and a positive value is a warning. The error codes are listed in *Appendix C.*

source identifies where an error occurred. The **source** string is usually the name of the VI that produced the error.

error out contains error information. If the **error in** cluster indicated an error, the **error out** cluster contains the same information. Otherwise, **error out** describes the error status of this VI.

count is the period between output pulses. If **control code** is 0, 1, 4, or 5, **count** can be 0 through 65,535 in binary counter operation and 0 through 9,999 in binary-coded decimal (BCD) counter operation. If **control code** is 2 or 3, **count** can be 2 through 65,535 and 0 in binary counter operation and 2 through 9,999 and 0 in BCD counter operation.

Note

0 is equivalent to 65,536 in binary counter operation and 10,000 in BCD counter operation.

output state is only valid when **control code** = 7 (reset).

0: Low (default input).

1: High.

binary or BCD controls whether the counter operates as a 16-bit binary counter or as a 4-decade BCD counter.

0: 4-decade BCD counter.

1: 16-bit binary counter.

read value. When you set **control code** to 6 (read) **read value** returns the value the VI read from the counter.

GPIB and Serial Port VIs

This chapter contains an overview of the LabVIEW GPIB and serial port VIs, a history of the GPIB, and an explanation of GPIB improvements and standards. If you want more information on GPIB operation and configuration, see *Appendix A* or look in the manuals that came with your GPIB board.

GPIB VI Overview

Introduction to the GPIB

The General Purpose Interface Bus (GPIB) is a link, or interface system, through which interconnected electronic devices communicate.

History of the GPIB

Hewlett-Packard designed the GPIB (originally called the HP-IB) to interconnect and control its line of programmable instruments. The GPIB was soon applied to other applications such as intercomputer communication and peripheral control because of its 1 Mbytes/sec maximum data transfer rates. It was later accepted as IEEE Standard 488-1975 and has since evolved into ANSI/IEEE Standard 488.1-1987. The versatility of the system prompted the name General Purpose Interface Bus.

National Instruments brought the GPIB to users of non-Hewlett-Packard computers and devices, specializing in both high-performance, high-speed hardware interfaces and comprehensive, full-function software. The GPIB VIs for LabVIEW follow the IEEE 488.1 specification.

Compatible GPIB Hardware

The following National Instruments GPIB hardware products are compatible with the *LabVIEW Student Edition.*

LabVIEW for Windows

• AT-GPIB	IEEE 488.2 interface board for the IBM PC AT and compatible computers that have 16-bit plug-in slots
• MC-GPIB	IEEE 488.2 interface board for the IBM PS/2 and compatible computers that have MicroChannel plug-in slots
• GPIB-PCII/IIA	IEEE 488.2 interface board for the IBM PC/XT/AT
• GD-GPIB	IEEE 488.2 interface board for the GRiDCASE 1500 series computer
• GPIB-232CT-A	External RS-232/IEEE 488.2 Controller

LabVIEW for Macintosh

• NB-GPIB	IEEE 488.2 interface board for Macintosh computers with full-size NuBus plug-in slots
• NB-GPIB-P	IEEE 488.2 interface board for Macintosh computers with half-size NuBus plug-in slots
• LC-GPIB	IEEE 488.2 interface board for the Macintosh LC series computers
• NB-DMA2800	IEEE 488 DMA interface board for Macintosh NuBus computers
• NB-DMA-8-G	IEEE 488 DMA interface board for Macintosh NuBus computers
• GPIB-SCSI-A	External SCSI-to-IEEE 488.2 Controller
• GPIB-SCSI	External SCSI-to-IEEE 488 Controller
• GPIB-422CT	External RS-422/IEEE 488 Controller

Operation of the GPIB

This section describes basic concepts you need to understand to operate the GPIB. It also contains a description of the physical and electrical characteristics of the GPIB and configuration requirements of the GPIB.

Types of Messages

The GPIB carries device-dependent messages and interface messages.

- Device-dependent messages, often called *data* or *data messages*, contain device-specific information such as programming instructions, measurement results, machine status, and data files.
- Interface messages manage the bus itself. They are usually called *commands* or *command messages.* Interface messages perform such tasks as initializing the bus, addressing and unaddressing devices, and setting device modes for remote or local programming.

Do not confuse the term *command* as used here with some device instructions, which can also be called commands. These device-specific instructions are actually data messages.

Talkers, Listeners, and Controllers

GPIB devices can be Talkers, Listeners, and/or Controllers. A digital voltmeter, for example, is a Talker and may be a Listener as well. A Talker sends data messages to one or more Listeners. The Controller manages the flow of information on the GPIB by sending commands to all devices.

The GPIB is like an ordinary computer bus, except that the computer has its circuit cards interconnected via a backplane bus, whereas the GPIB has stand-alone devices interconnected via a cable bus.

The role of the GPIB Controller is similar to the role of the CPU of a computer, but a better analogy is to the switching center of a city telephone system. The switching center (Controller) monitors the communications network (GPIB). When the center (Controller) notices that a party (device) wants to make a call (send a data message), it connects the caller (Talker) to the receiver (Listener).

The Controller addresses a Talker and a Listener before the Talker can send its message to the Listener. After the Talker transmits the message, the Controller may unaddress both devices.

Some bus configurations do not require a Controller. For example, one device may always be a Talker (called a Talk-only device) and there may be one or more Listen-only devices.

A Controller is necessary when you must change the active or addressed Talker or Listener. A computer usually handles the Controller function.

With the GPIB board and its software, your personal computer plays all three roles:

- Controller—to manage the GPIB
- Talker—to send data
- Listener—to receive data

The Controller-In-Charge and System Controller

Although there can be multiple Controllers on the GPIB, only one Controller at a time is active or Controller-In-Charge (CIC). You can pass active control from the current CIC to an idle Controller. Only one device on the bus–the System Controller–can make itself the CIC. The GPIB board is usually the System Controller.

GPIB Signals and Lines

The interface system consists of 16 signal lines and 8 ground-return or shield-drain lines. The 16 signal lines are divided into three groups:

- Eight data lines
- Three handshake lines
- Five interface management lines

Data Lines

The eight data lines, DIO1 through DIO8, carry both data and command messages. All commands and most data use the 7-bit ASCII or International Standards Organization (ISO) code set, in which case the eighth bit, DIO8, is unused or is used for parity.

Handshake Lines

Three lines asynchronously control the transfer of message bytes among devices. This process is called a three-wire interlocked handshake, and it guarantees that message bytes on the data lines are sent and received without transmission error.

NRFD (Not Ready for Data). NRFD indicates whether a device is ready to receive a message byte. All devices drive NRFD when they receive commands, and Listeners drive it when they receive data messages.

NDAC (Not Data Accepted). NDAC indicates whether a device has accepted a message byte. All devices drive NDAC when they receive commands, and Listeners drive it when they receive data messages.

DAV (Data Valid). DAV tells whether the signals on the data lines are stable (valid) and whether devices can accept them safely. The Controller drives DAV when sending commands, and the Talker drives it when sending data messages.

Interface Management Lines

Five lines manage the flow of information across the interface.

ATN (Attention). The Controller drives ATN true when it uses the data lines to send commands and drives ATN false when a Talker can send data messages.

IFC (Interface Clear). The System Controller drives the IFC line to initialize the bus and become CIC.

REN (Remote Enable). The System Controller drives the REN line, which places devices in remote or local program mode.

SRQ (Service Request). Any device can drive the SRQ line to asynchronously request service from the Controller.

EOI (End or Identify). The EOI line has two purposes. The Talker uses the EOI line to mark the end of a message string. The Controller uses the EOI line to tell devices to respond in a parallel poll.

Physical and Electrical Characteristics

You usually connect devices with a cable assembly consisting of a shielded 24-conductor cable that has both a plug and a receptacle connector at each end. With this design, you can link devices in either a linear or a star configuration, or a combination of the two. See the following figures.

The standard connector is the Amphenol or Cinch Series 57 *Microribbon* or *Amp Champ* type. You can use an adapter cable with a non-standard cable and/or connector for special interconnection applications.

The GPIB uses negative logic with standard transistor-transistor logic (TTL) level. When DAV is true, for example, it is a TTL low level (≤ 0.8 V), and when DAV is false, it is a TTL high level (≥ 2.0 V).

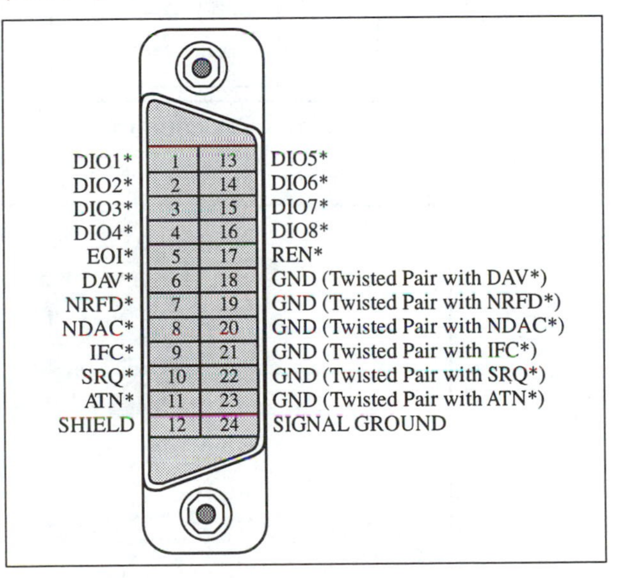

GPIB Connector Showing Signal Assignment

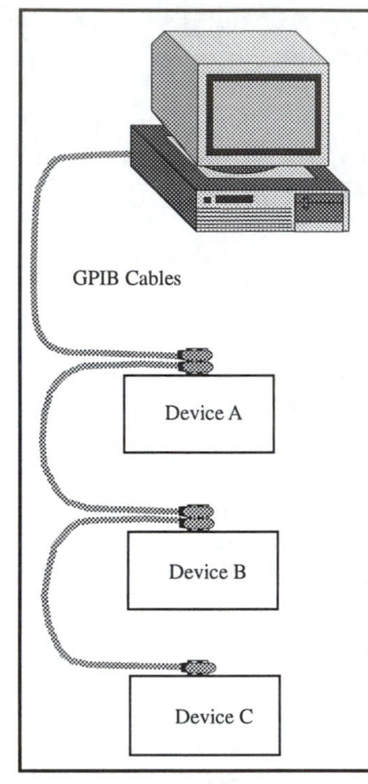

Linear Configuration

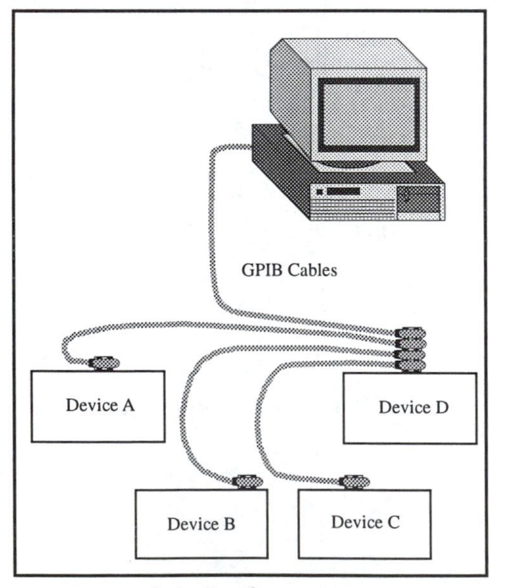

Star Configuration

Configuration Requirements

To achieve the high data transfer rate for which the GPIB was designed, the physical distance between devices and the number of devices on the bus must be limited. The following restrictions are typical.

- A maximum separation of 4 m between any two devices and an average separation of 2 m over the entire bus.
- A maximum total cable length of 20 m.
- No more than 15 devices connected to each bus, with at least two-thirds powered on.

Contact National Instruments for bus extenders if your requirements exceed these limits.

GPIB VIs

This section describes the GPIB VIs. To access these VIs, select **GPIB** from the **Functions** menu. You can observe and run an example GPIB program by opening **LabVIEW<->GPIB.vi**, located in EXAMPLES\GPIB\SMPLGPIB.LLB. The following figure shows the **GPIB** palette.

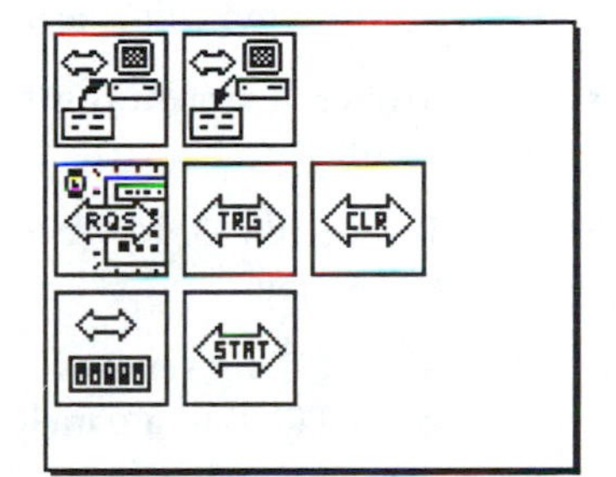

Most of the GPIB VIs use the following parameters

- **address string** contains the address of the GPIB device with which the VI communicates. You can input both the primary and secondary addresses in **address string** by using the form *primary+secondary.* Both *primary* and *secondary* are decimal values, so if *primary* is 2 and *secondary* is 3, **address string** is 2+3.

  If you do not specify an address, the VIs do not perform addressing before they attempt to write the string. The VIs assume that you have either sent these commands in another way or that another Controller is in charge and therefore responsible for the addressing. If the Controller is supposed to address the device but does not do so before the time limit expires, the VIs terminate with GPIB error 6 (timeout) and set bit 14 in **status**. If the GPIB is not the Controller-In-Charge, you must not specify an **address string**.

  When there are multiple GPIB Controllers that LabVIEW can use, a prefix to the **address string** in the form ID:address (or ID: if no address is necessary) determines the Controller that a specific VI uses. If a Controller ID is not present, the VIs assume Controller (or bus) 0.

- **status** is a 16-bit Boolean array in which each bit describes a state of the GPIB Controller. If an error occurs, bit 15 is set. **GPIB error** is valid only if bit 15 of **status** is set. Refer to the **GPIB Status** VI description for status bit error codes.

GPIB Initialization

Configures the GPIB device at **address string**.

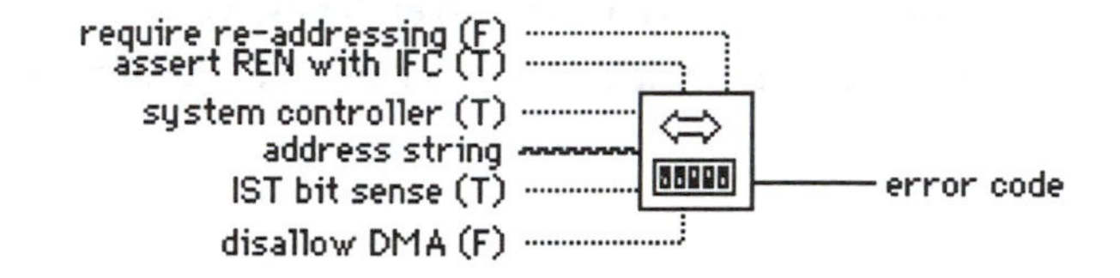

require re-addressing. If **require re-addressing** is TRUE, the VI addresses the device before every read or write. If FALSE, the device must be able to retain addressing from one read or write to the next.

assert REN with IFC. If **assert REN with IFC** is TRUE, and if this Controller (specified by the ID in **address string**) is the System Controller, the VI asserts the Remote Enable line.

TF

system controller. If **system controller** is TRUE, this Controller acts as the System Controller.

abc

address string. See the *GPIB VIs* section at the beginning of this chapter for more information.

IST bit sense. If **IST bit sense** is TRUE, the Individual Status bit of the device responds TRUE to a parallel poll; if **IST bit sense** is FALSE, the Individual Status bit of the device responds FALSE to a parallel poll.

disallow DMA. If **disallow DMA** is TRUE, this device uses programmed I/O for data transfers.

error code. Refer to the list of error codes in *Appendix C* or under the **GPIB Status** VI.

GPIB Write

Writes **data** to the GPIB device identified by **address string**.

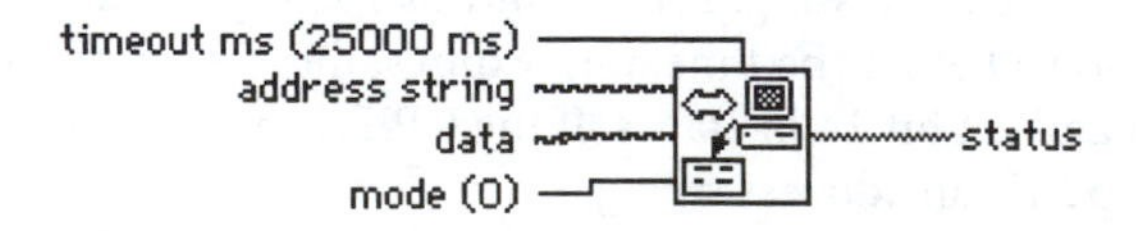

timeout ms. The operation aborts if it does not complete within **timeout ms.** If a timeout occurs, bit 14 of **status** is set. Timeout defaults to 25,000 ms. To disable timeouts, set **timeout ms** to 0.

address string. See the *GPIB VIs* section at the beginning of this chapter for more information.

data is the data the VI writes to the GPIB device.

mode indicates how to terminate the GPIB Write.

0: Send EOI with the last character of the string.

1: Append CR to the string and send EOI with CR.

2: Append LF to the string and send EOI with LF.

3: Append CR LF to the string and send EOI with LF.

4: Append CR to the string but do not send EOI.

5: Append LF to the string but do not send EOI.

6: Append CR LF to the string but do not send EOI.

7: Do not send EOI.

status. See the *GPIB VIs* section at the beginning of this chapter for more information.

GPIB Read

Reads **byte count** number of bytes from the GPIB device at **address string**.

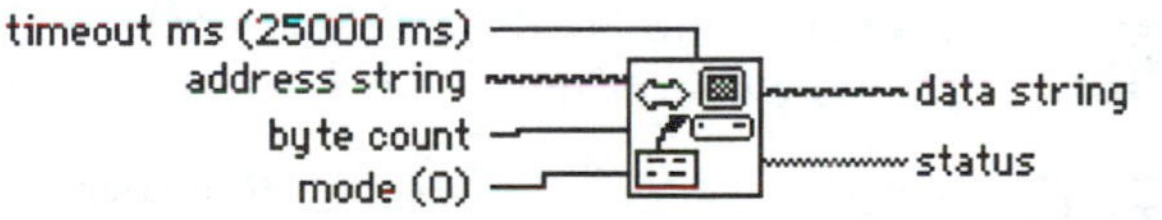

timeout ms. The operation aborts if it does not complete within **timeout ms.** If a timeout occurs, bit 14 of **status** is set. Timeout defaults to 25,000 ms. To disable timeouts, set **timeout ms** to 0.

address string. See the *GPIB VIs* section at the beginning of this chapter for more information.

byte count specifies the number of bytes the VI reads from the GPIB device.

mode specifies conditions other than reaching **byte count** for terminating the read.

0: No EOS character. The EOS termination mode is disabled.

1: EOS character is CR. Read terminated on EOI, **byte count**, or CR.

2: EOS character is LF. Read terminated on EOI, **byte count**, or LF.

x: Any other mode indicates the number (decimal) of the desired EOS character.

data string. The VI returns the data read in **data string**.

status. See the *GPIB VIs* section at the beginning of this chapter for more information.

The GPIB Read VI terminates when it:

- Reads the number of bytes requested.
- Detects an error.
- Exceeds the time limit.
- Detects the END message (EOI asserted).
- Detects the EOS character (if this option is enabled by the value supplied to **mode**).

Note
The VI compares only the lowest seven bits when it checks for the EOS character.

GPIB Trigger

Sends GET (Group Execute Trigger) to the device indicated by **address string**.

address string. See the *GPIB VIs* section at the beginning of this chapter for more information.

status. See the *GPIB VIs* section at the beginning of this chapter for more information.

GPIB Clear

Sends either SDC (Selected Device Clear) or DCL (Device Clear).

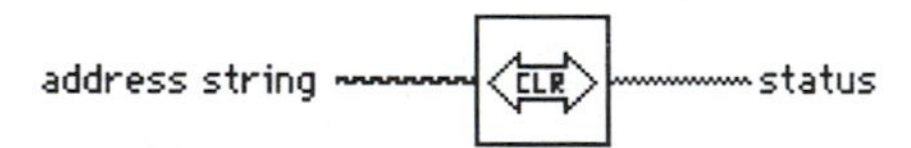

address string. If you specify an address in **address string**, the VI sends SDC to that address. If **address string** is unwired, the VI sends DCL. See the *GPIB VIs* section at the beginning of this chapter for more information.

status. See the *GPIB VIs* section at the beginning of this chapter for more information.

Wait for GPIB RQS

Waits for the device indicated by **address string** to assert SRQ.

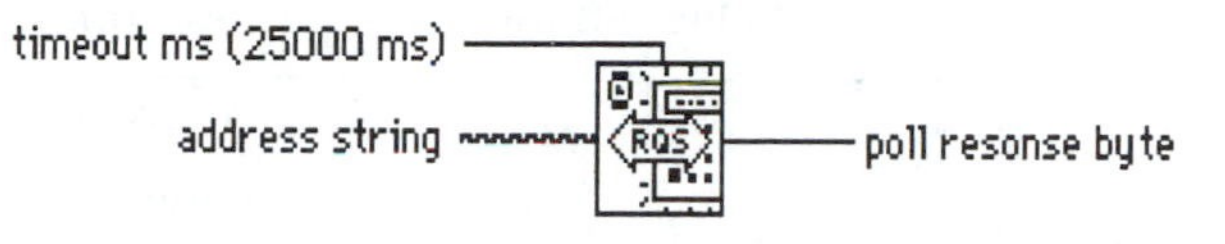

timeout ms. The operation aborts if it does not complete within **timeout ms**. If a timeout occurs, **poll response byte** is –1. Timeout defaults to 25,000 ms. To disable timeouts, set **timeout ms** to 0.

address string. See the *GPIB VIs* section at the beginning of this chapter for more information.

poll response byte. If SRQI is set, the VI polls the device at the specified address to see if it requested service. When the specified device requests service (bit 6 is set in **poll response byte**), the VI returns the serial poll response.

If the device indicated by **address string** does not respond within the timeout limit, **poll response byte** is –1, and the VI indicates GPIB error 6 by setting bit 14 of status, a front panel Boolean array indicator for this VI.

GPIB Status

Shows the status of the GPIB Controller indicated by **address string** after the previous GPIB operation.

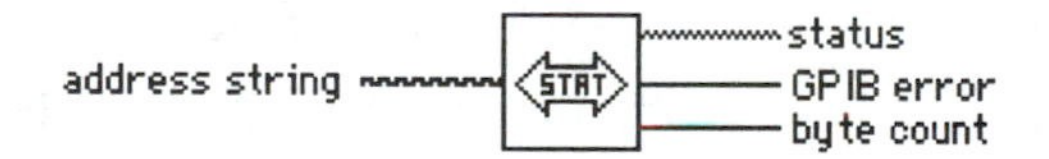

address string. See the *GPIB VIs* section at the beginning of this chapter for more information.

status. The following table shows the numeric value and symbolic status of each bit in **status**. This table also includes a description of each bit.

Status Bits

Status Bit	Numeric Value	Symbolic Status	Description
0	1	DCAS	Device Clear state
1	2	DTAS	Device Trigger State
2	4	LACS	Listener Active
3	8	TACS	Talker Active
4	16	ATN	Attention Asserted

Status Bits (Continued)

Status Bit	Numeric Value	Symbolic Status	Description
5	32	CIC	Controller-In-Charge
6	64	REM	Remote State
7	128	LOK	Lockout State
8	256	CMPL	Operation Completed
12	4096	SRQI	SRQ Detected while CIC
13	8192	END	EOI or EOS Detected
14	16384	TIMO	Timeout
15	-32768	ERR	Error Detected

NUM **GPIB error** contains the most recent error code reported by any of the GPIB VIs. The following table shows the possible values for **GPIB error** if bit 15 of **status** is set.

GPIB Error Bits

GPIB Error	Symbolic Status	Description
0	EDVR	Error Connecting to Driver
1	ECIC	Command Requires GPIB Controller to be CIC
2	ENOL	Write Detected No Listeners
3	EADR	GPIB Controller Not Addressed Correctly
4	EARG	Invalid Argument or Arguments
5	ESAC	Command Requires GPIB to be System Controller
6	EABO	I/O Operation Aborted
7	ENEB	Non-existent Board
8	EDMA	DMA Hardware Not Detected
9	EBTO	DMA Hardware uP Bus Timeout
11	ECAP	No Capability
12	EFSO	File System Operation Error
13	EOWN	Shareable Board Exclusively Owned
14	EBUS	GPIB Bus Error
15	ESTB	Serial Poll Byte Queue Overflow
16	ESRQ	SRQ Stuck On

GPIB Error Bits (Continued)

GPIB Error	Symbolic Status	Description
17	ECMD	Unrecognized Command
19	EBNP	Board Not Present
20	ETAB	Table Error
30	NADDR	No GPIB Address Input
31	NSTRG	No String Input (Write)
32	NCNT	No Count Input (Read)

NUM **Byte count** is the number of bytes the previous GPIB operation sent.

Serial Port VIs

This section describes the VIs for serial port operations, which you access by selecting **Serial** from the **Functions** menu. You can observe and run a serial example by opening **Serial Communication.vi**, located in EXAMPLES\SERIAL\SMPLSERL.LLB. The following figure shows the **Serial** palette.

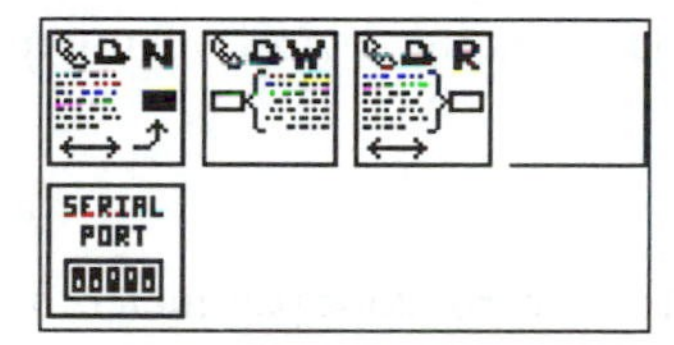

Common Serial Port VI Parameters

Port Number

When you use the serial port VIs under Windows, the **port number** parameter can have the following values

0: COM1 5: COM6 10: LPT1

1: COM2 6: COM7 11: LPT2

2: COM3 7: COM8 12: LPT3

3: COM4 8: COM9 13: LPT4

4: COM5

On the Macintosh, port 0 is the modem, using the drivers .ain and .aout. Port 1 is the printer, using the drivers.bin and .bout. To get more ports on a Macintosh, you must install other boards, with the accompanying drivers.

Handshaking Modes

A common problem in serial communications is ensuring that both sender and receiver keep up with data transmission. The serial port driver can buffer incoming/outgoing information, but that buffer is of a finite size. When it becomes full, the computer ignores new data until you have read enough data out of the buffer to make room for new information.

Handshaking helps prevent this buffer from overflowing. With handshaking, the sender and the receiver notify each other when their buffers fill up. The sender can then stop sending new information until the other end of the serial communication is ready for new data.

You can perform two kinds of handshaking in LabVIEW—software handshaking and hardware handshaking. You can turn both of these forms of handshaking on or off using the Serial Port Init VI. By default, the VIs do not use handshaking.

Software Handshaking—XON/XOFF

XON/XOFF is a software handshaking protocol you can use to avoid overflowing serial port buffers. When the receive buffer is nearly full, the receiver sends XOFF (<control-S> [decimal 19]) to tell the other device to stop sending data. When the receive buffer is sufficiently empty, the receiver sends XON (<control-Q> [decimal 17]) to indicate that transmission can begin again. When you enable XON/XOFF, the devices always interpret <control-Q> and <control-S> as XON and XOFF characters, never as data. When you disable XON/XOFF, you can send <control-Q> and <control-S> as data. Do not use XON/XOFF with binary data transfers because <control-Q> or <control-S> may be embedded in the data, and the devices will interpret them as XON and XOFF instead of data.

Error Codes

You can connect the **error code** parameter to one of the error handler VIs. These VIs can furnish you with a description of the error, and give you options on how to proceed when an error occurs. Error codes are listed under the **Serial Port Init** VI as well as in *Appendix C.*

Some error codes returned by the serial port VIs are platform specific. Please refer to your system documentation for a list of these error codes.

Note

Some peripheral devices require one or two characters to terminate strings of data sent to them, or they may return a timeout error or hang the machine. Common termination characters are a carriage return, a line feed, or a semicolon. Please consult the device manual to determine if you need a termination character.

Serial Port VI Descriptions

Serial Port Init

Initializes the selected serial port to the specified settings.

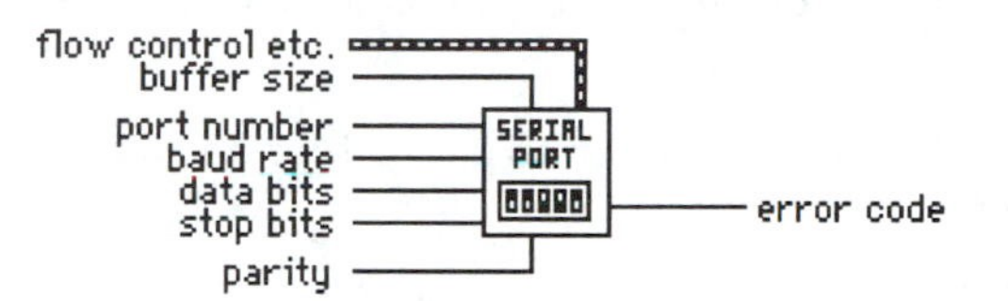

flow control etc. contains the following parameters.

Input XON/XOFF. See the *Common Serial Port VI Parameters* section of this chapter for more information.

Input HW Handshake. On computers running Windows, this parameter corresponds to Request To Send (RTS) handshaking.

Input alt HW Handshake. On computers running Windows, this parameter corresponds to Data Terminal Ready (DTR) handshaking.

Output XON/XOFF. See the *Common Serial Port VI Parameters* section of this chapter for more information.

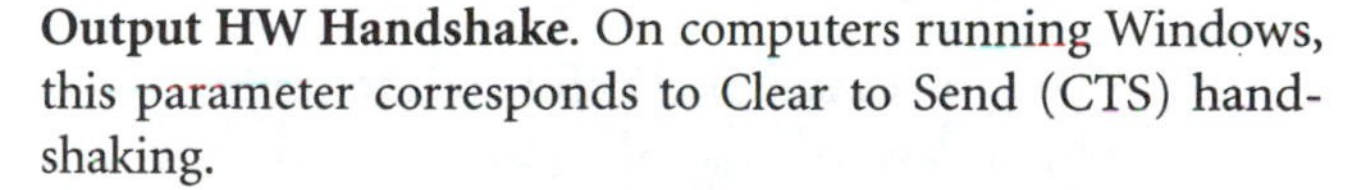

Output HW Handshake. On computers running Windows, this parameter corresponds to Clear to Send (CTS) handshaking.

Output alt HW Handshake. On computers running Windows, this parameter corresponds to Data Set Ready (DSR) handshaking.

XOFF byte is the byte used for XOFF (^S).

XON byte is the byte used for XON (^Q).

Parity Error Byte. If the high byte is non zero, the low byte is the character that is used to replace any parity errors found when **parity** is enabled.

buffer size indicates the size of the input and output buffers the VI allocates for communication through the specified port. If **buffer size** is less than or equal to 1 K, the VI uses 1 K as the **buffer size.** You may need to use larger buffers for large data transfers.

port number. See the *Common Serial Port VI Parameters* section of this chapter for more information.

baud rate is the rate of transmission.

data bits is the number of bits in the incoming data. The value of **data bits** is between five and eight.

stop bits is 0 for one stop bit, or 1 for two stop bits.

parity is 0 for no parity, 1 for odd parity, or 2 for even parity.

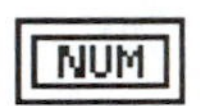

error code is –1 if **baud rate**, **data bits**, **stop bits**, **parity**, or **port number** is out of range, or if the serial port could not be initialized. Check the values of **baud rate**, **data bits**, **stop bits**, **parity**, and **port number**. If these values are valid, verify that the serial port has been initialized.

Serial Port Error Codes

Code	Name	Description
61	EPAR	Serial port parity error.
62	EORN	Serial port overrun error.
63	EOFL	Serial port receive buffer overflow.
64	EFRM	Serial port framing error.
65	SPTMO	Serial port timeout, bytes not received at serial port

You can connect **error code** to the **Simple Error Handler** VI, which can furnish you with a description of the error, and give you options on how to proceed when an error occurs.

Some error codes returned by the serial port VIs are platform-specific. Please refer to your system documentation for a list of these error codes.

Serial Port Write

Writes the data in **string to write** to the serial port indicated in **port number**.

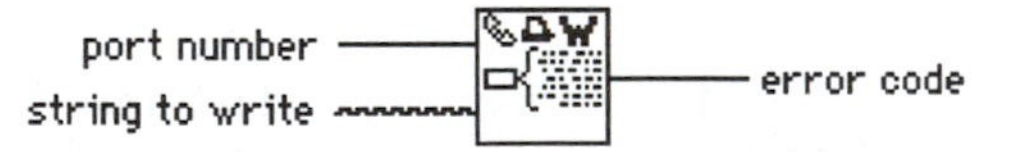

port number. See the *Common Serial Port VI Parameters* section of this chapter for a list of valid port numbers.

string to write.

error code. If **error code** is non-zero, an error occurred. Refer to the **Serial Port Init** section or *Appendix C* for a list of error codes.

Serial Port Read

Reads the number of characters specified by **requested byte count** from the serial port indicated in **port number**.

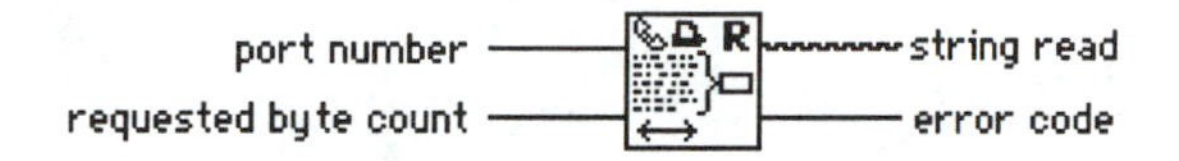

port number. See the *Common Serial Port VI Parameters* section of this chapter for a list of valid port numbers.

requested byte count specifies the number of characters to be read. If you want to read all of the characters currently at the serial port, first execute the **Bytes at Serial Port** VI to determine the exact number of bytes ready to be read. Then use the **byte count** output of that VI as the **requested byte count** input to the **Serial Port Read** VI.

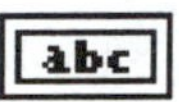

string read. The VI returns the bytes read in **string read.**

error code. If **error code** is non-zero, an error occurred. Refer to the **Serial Port Init** section or *Appendix C* for a list of error codes.

Bytes at Serial Port

Returns in **byte count** the number of bytes in the input buffer of the serial port indicated in **port number.**

port number. See the *Common Serial Port VI Parameters* section of this chapter for a list of valid port numbers.

byte count is the number of bytes currently queued up in the serial port buffer.

error code. If **error code** is non-zero, an error occurred. Refer to the **Serial Port Init** section or *Appendix C* for a list of error codes.

The Formula Node

This chapter describes how to use the Formula Node to execute mathematical formulas on the block diagram.

The Formula Node is available from the **Structs & Constants** palette of the **Functions** menu. It is a resizable box similar to the four structures, but instead of containing a subdiagram, the Formula Node contains one or more formula statements delimited by a semicolon, as in the following example.

```
y=3*x^2+x*log(x);
```

Formula statements use a syntax similar to most text-based programming languages for arithmetic expressions. You can add comments by enclosing them inside a slash-asterisk pair (/**comment**/).

The pop-up menu on the border of the Formula Node contains options to add input and output variables, as shown in the illustration below.

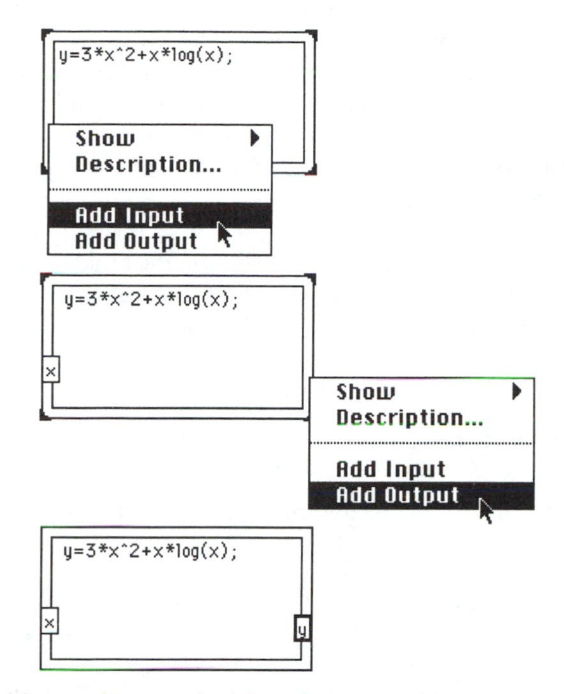

Output variables are distinguished from input variables by their thicker border.

There is no limit to the number of variables or formulas you can place in a Formula Node. No two inputs and no two outputs can have the same name. However, an output can have the same name as an input.

You can change an input to an output by selecting **Change to Output** from the pop-up menu, as shown in the following illustration.

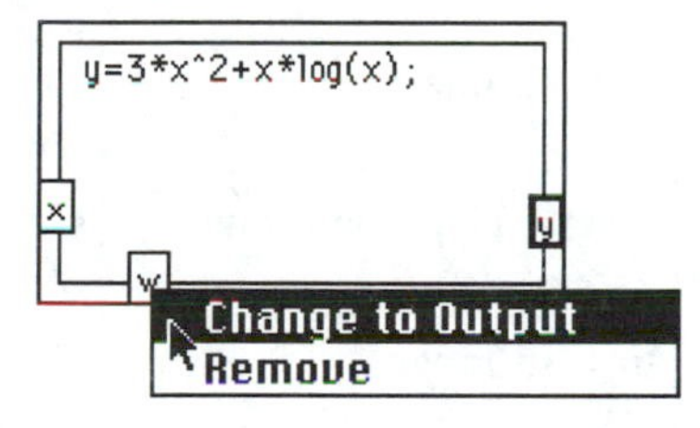

You can change an output to an input by selecting **Change to Input** from the pop-up menu, as shown below.

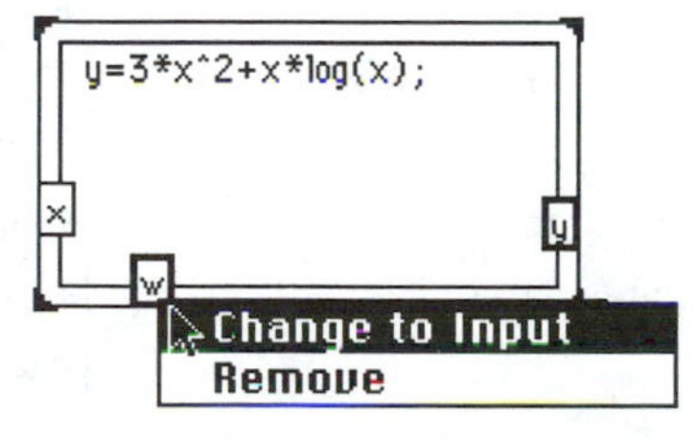

All variables are floating-point numeric scalars, whose precision depends on the configuration of your computer. All input variables that appear in the formulas must be wired. All output variables that are wired must be assigned in at least one statement; that is, they must be on the left side of an equal sign. An output variable may appear in an expression on the right side of an equal sign, but LabVIEW does not check to see whether it has been assigned in a previous statement. When an assignment occurs as a subexpression, the value of the subexpression is the value assigned; for example

```
x = sin(y = pi/3);
```

assigns pi/3 to y, and then assigns sin(pi/3) to x.

If a syntax error occurs you can click on the broken Run button to get the error listing. In the listing, the Formula Node displays a portion of the formula with a # symbol marking the point at which the error was detected.

Formula Node Functions

All function names must be lowercase. The following table shows the names of the Formula Node functions.

Formula Node Functions

Function	Corresponding LabVIEW Function Name	Description
abs(x)	Absolute Value	Returns the absolute value of x.
acos(x)	Inverse Cosine	Computes the inverse cosine of x in radians.
acosh(x)	Inverse Hyperbolic Cosine	Computes the inverse hyperbolic cosine of x in radians.
asin(x)	Inverse Sine	Computes the inverse sine of x in radians.
asinh(x)	Inverse Hyperbolic Sine	Computes the inverse hyperbolic sine of x in radians.
atan(x,y)	Inverse Tangent	Computes the inverse tangent of y/x in radians.
atanh(x)	Inverse Hyperbolic Tangent	Computes the inverse hyperbolic tangent of x in radians.
ceil(x)	Round to +Infinity	Rounds x to the next higher integer (smallest int ≥ x).
cos(x)	Cosine	Computes the cosine of x in radians.
cosh(x)	Hyperbolic Cosine	Computes the hyperbolic cosine of x in radians.
cot(x)	Cotangent	Computes the cotangent of x in radians [1/tan(x)].

Formula Node Functions (Continued)

Function	Corresponding LabVIEW Function Name	Description
csc(x)	Cosecant	Computes the cosecant of x in radians [1/sin(x)].
exp(x)	Exponential	Computes the value of e raised to the x power.
expm1(x)	Exponential (Arg) - 1	Computes the value of e raised to the x power minus one (e^x-1).
floor(x)	Round to -Infinity	Truncates x to the next lower integer (largest int ≤ x).
getexp(x)	mantissa and exponent	Returns the exponent.
getman(x)	mantissa and exponent	Returns the mantissa.
intrz(x)	round toward 0	Rounds x to the nearest integer between x and zero.
ln(x)	Natural Logarithm	Computes the natural logarithm of x (to the base e).
lnpl(x)	Natural Logarithm (Arg +1)	Computes the natural logarithm of (x + 1).
log(x)	Logarithm Base 10	Computes the logarithm of x (to the base of 10).
log2(x)	Logarithm Base 2	Computes the logarithm of x (to the base 2).
max(x,y)	maximum and minimum	Compares x and y and returns the larger value.
min(x,y)	maximum and minimum	Compares x and y and returns the smaller value
mod(x,y)	quotient and remainder	Computes the remainder of x/y, when the quotient is rounded toward -Infinity.
rand( )	Random Number (0- 1)	Produces a floating-point number between 0 and 1 exclusively.
rem(x,y)	remainder	Same as mod except quotient is rounded to the nearest integer.
sec(x)	Secant	Computes the secant of x radians [1/cos(x)].

Formula Node Functions (Continued)

Function	Corresponding LabVIEW Function Name	Description
sign(x)	Sign	Returns 1 if x is greater than 0, returns 0 if x is equal to 0, and returns –1 if x is less than 0.
sin(x)	Sine	Computes the sine of x radians.
sinc(x)	Sinc	Computes the sine of x divided by x radians [sin(x)/x].
sinh(x)	Hyperbolic Sine	Computes the hyperbolic sine of x in radians.
sqrt(x)	Square Root	Computes the square root of x.
tan(x)	Tangent	Computes the tangent of x in radians.
tanh(x)	Hyperbolic Tangent	Computes the hyperbolic tangent of x in radians.
x^y	x^y	Computes the value of x raised to the y power.

The Formula Node syntax is summarized below using Bakus-Naur Form (BNF) notation. Square brackets enclose optional items.

```
<assignlst>:=<outputvar> = <aexpr> ;
                              [ <assignlst> ]
<aexpr>:=  <expr> | <outputvar> = <aexpr>
<expr>:=   <expr> <binaryoperator> <expr>
        |  <unaryoperator> <expr>
        |  <expr> ? <expr> : <expr>
        |  ( <expr>)
        |  <inputvar>
        |  <outputvar>
        |  <const>
        |  <function> ( <arglist> )
<binaryoperator>:= + | - | * | / | ^ | != | ==
| > | < | >= | <= | && | ||
<unaryoperator>:=+ | - | !
<arglist>:= <aexpr>[ , <arglist> ]
<const>:=  pi | <number>
```

The precedence of operators is as follows, from lowest to highest.

=	assignment
? :	conditional
||	logical or
&&	logical and

!= == inequality, equality

< > <= >= other relational: less than, greater than, less than or equal, greater than or equal

+ - addition, subtraction

* / multiplication, division

+ - ! unary: positive, negative, logical not

^ exponentiation

Exponentiation and the assignment operator are right-associative (groups right to left). All other binary operators are left-associative. The numeric value of TRUE is 1 and FALSE is 0 (for output). The logical value of 0 is FALSE, and any nonzero number is TRUE. The logical value of the conditional expression

<lexpr> ? <texpr>: <fexpr>

is <texpr> if the logical value of <lexpr> is TRUE and <fexpr> otherwise.

The table below lists errors detected by the Formula Node.

Formula Node Errors

Error Message	Error Message Meaning
syntax error	Misused operator, and so on.
bad token	Unrecognized character.
output variable required	Cannot assign to an input variable.
missing output variable	Attempt to assign to a nonexistent output variable.
missing variable	References a nonexistent input or output variable.
too few arguments	Not enough arguments to a function.
too many arguments	Too many arguments to a function.
unterminated argument list	Formula ended before argument list close parenthesis seen.
missing left parenthesis	Function name not followed by argument list.
missing right parenthesis	Formula ended before all matching close parenthesis seen.
missing colon	Improper use of conditional ternary operator.
missing semicolon	Formula statement not terminated by a semicolon.
missing equal sign	Formula statement is not a proper assignment.

DAQ and GPIB Configuration

This appendix will help you become familiar with the DAQ and GPIB software architecture for your platform. You can find more information about DAQ and GPIB in the manuals that accompany the hardware.

DAQ and GPIB Drivers

The LabVIEW DAQ and GPIB drivers do not accompany the *LabVIEW Student Edition*, but they are available on a *LabVIEW Student Edition* Driver Setup disk provided in the instructor's package. While DAQ and GPIB drivers do come with your hardware, they may be incompatible; you should always use the drivers especially designated for the *LabVIEW Student Edition*. To install the drivers, simply insert the setup disk. Run SETUP.EXE on Windows machines; follow the instructions given on Macintoshes. The drivers will be installed in the proper locations.

Data Acquisition Driver Software Architecture

The *LabVIEW Student Edition* will support only the National Instruments Lab-NB, Lab-LC, NB-DIO-24, and all NB-MIO boards on the Macintosh. It supports the Lab-PC, Lab-PC+, PC-LPM-16, PC-DIO-24, DAQcard-700, DAQpad-1200, and all AT-MIO boards on ISA computers running Windows.

You must take several steps before you can use the LabVIEW DAQ VIs. First, you must install a National Instruments DAQ board and have access to a cable to connect your board to the signals you want to measure. Then you must install the correct driver software for the *LabVIEW Student Edition.* Then you must check the software configuration to be sure it is correct.

Watch Out!
You should not install the drivers that come with your DAQ board because they may not be compatible with your version of LabVIEW. You should install only the drivers especially designated for use with the LabVIEW Student Edition.

When you install the student edition drivers, you will overwrite any existing LabVIEW driver files in your WINDOWS and WINDOWS\SYSTEM directories, such as GPIB.DLL. This Setup disk installs the same DLLs as the LabVIEW Professional Version 3.1. However, if you have an older version of LabVIEW on your machine, you may have problems running it with the new DLLs installed by the student edition if you are doing data acquisition.

For example, LabVIEW 2.5.2 cannot use the ATWDAQ.DLL. that the student edition setup disk installs. If you want to switch between LabVIEW versions and do data acquisition, you will have to load the appropriate DLL each time you run a different version.

Windows Configuration

The *LabVIEW Student Edition* Driver Setup program copies the required files for LabVIEW DAQ onto your computer. While these files also come with your board, you should use the files that the Driver Setup disk installs to avoid compatibility problems. The installer loads the NIDAQ.DLL file, the high-level interface to your board, into the WINDOWS\SYSTEM directory. It also copies the WDAQCONF.EXE file, used for saving your configuration parameters, into the LabVIEW directory. The DAQ software architecture for computers running Windows is shown in the following illustration.

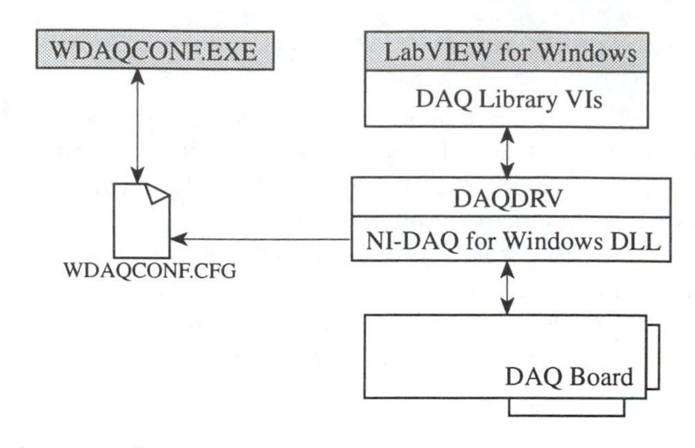

WDAQCONF

In order to acquire data in LabVIEW, you must have a configuration utility to establish parameters for your board. This utility, called WDAQCONF, saves the configuration parameters in a file called WDAQCONF.CFG. You start WDAQCONF by double-clicking on its icon in Windows. The following illustration shows the primary WDAQCONF window. It may appear slightly different on your machine, depending on the boards you have installed. WDAQCONF will let you test your board interactively and also let you configure an SCXI system. For more information on the DAQ software, please see the manual that comes with your board.

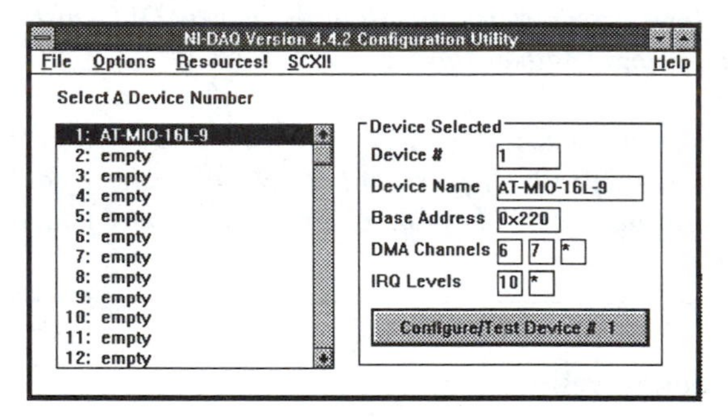

Note

The board's device number in WDAQCONF is the same device number that you will need to address the board in the LabVIEW VIs.

EXERCISE A.1
Exploring WDAQCONF (Windows Only)

You will open the WDAQCONF utility and study the current settings. You also will test the board interactively with the utility. If you don't have a board, obviously you can't test one, but you *can* learn about the configuration utility.

1. Start the WDAQCONF utility by double-clicking on the WDAQCONF icon in the LabVIEW group window.

Note

If the Windows Program Manager is not the active window, click on that window with the mouse or press <ctrl-esc> and select Program Manager from the Task List. If the LabVIEW group window is not active, double-click on the LabVIEW group window icon.

---

2. Examine the board parameters detailed in the WDAQCONF window, shown in the following illustration. You use this window to select board types and display settings for each board.

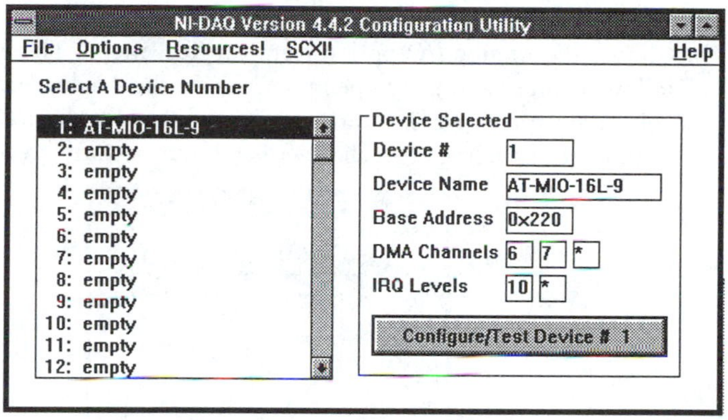

3. Open the Device Number 1 window, shown in the following illustration, by clicking on **Configure/Test Device# 1**. You use the **Devices**, **DMA**, and **IRQ** menu choices to select the device type, the DMA channels, and the interrupt level, respectively, that are jumper-set on your board. You can alter the base address (0x220 in this example) by toggling the switch settings with the mouse.

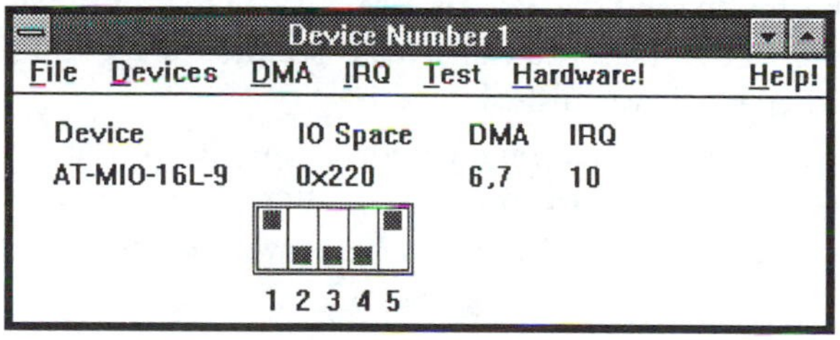

Note

You must configure WDAQCONF to use the same base address, DMA, and IRQ levels that you have selected using the jumpers on your board.

---

4. The **Hardware** menu, which is only visible if you've saved your configuration, opens a Hardware Configuration window, which you use to specify default settings such as polarity and mode on a per-device basis. Your settings must match the settings configured by the jumpers on the board if the board has jumpers for those settings.
5. Use the **Test** menu to verify the operation of your DAQ board. You should use the test routines to verify hardware operation before attempting to use the LabVIEW Data Acquisition VIs.
   a. Choose the **Configuration** option from the **Test** menu. Verify that the utility returns the message, "Device Configuration has been verified," which confirms the hardware setup. Click on **OK** in the dialog box to make it go away. If you receive any other message, make sure your base address and DMA and IRQ levels match those jumpered on your board. You may also have to change the jumpers and try other settings if you have a conflict. If you still can't verify device configuration, talk to your instructor.
   b. Choose the **Analog I/O** option from the **Test** menu. The window shown in the following illustration will open up. Click on the **Analog Input** button to acquire voltage from channel 0. (You must have channel 0 connected to a signal if you want to get a reading other than zero and thus perform a valid test.)

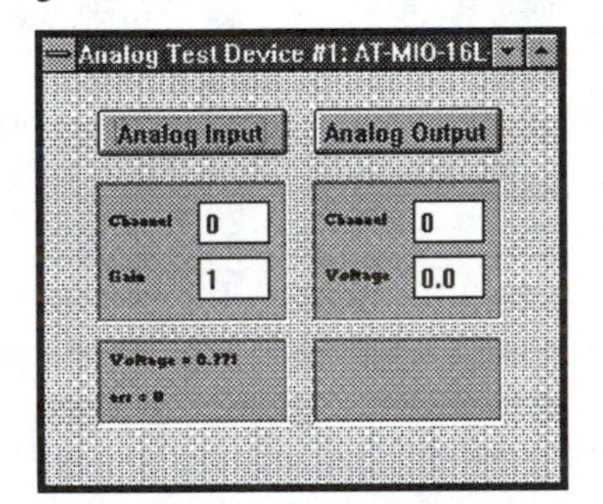

   c. Choose **Digital I/O** from the **Test** menu to display the Digital Test window. Toggle Lines 0 to 3 on Port 1 in the Digital Output section and press the **Digital Output** button to send the pattern. If you have LEDs or other devices hooked up to your digital channels, you can actually see the voltage you send out. Close the window when you have finished.
6. Use the **Help!** menu if you get lost in WDAQCONF. You can access all possible DAQ and LabVIEW DAQ-related error messages from the **Help!** menu.
7. Close the remaining WDAQCONF windows and return to LabVIEW.

Mac Configuration

The driver installation program copies the files NI-488 INIT (the GPIB driver) and NI-DAQ (the DAQ driver) into the Control Panels folder of your Mac. It also places NI-DMA/DSP, the DMA and DSP driver, into the Extensions folder in your System Folder. These files provide full support for the National Instruments NB and Lab Series boards. While the

Mac data acquisition software also comes with your board, you should use the files that LabVIEW installs to avoid compatibility problems.

NI-DAQ

You can use the Data Acquisition library once NI-DAQ and NI-DMA/DSP, the DAQ driver files, are correctly installed. The following illustration shows the DAQ software architecture for the Macintosh.

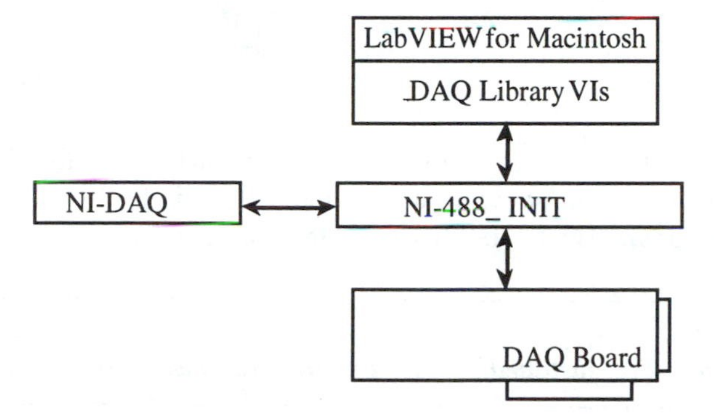

You use NI-DAQ, shown in the following illustration, to configure your hardware and view a list of boards installed in your Macintosh. Under the **Devices** screen, all National Instruments boards are listed in boldface; all others are dimmed. You can configure the hardware settings for each board using the **Device Configuration** option. You can set up SCXI hardware using the **SCXI Configuration** option. The **Errors** option contains a list of all DAQ errors. Unlike the Windows configuration utility, NI-DAQ does not allow you to test your board interactively. For more information on the DAQ software, please look at the manual that comes with your DAQ board.

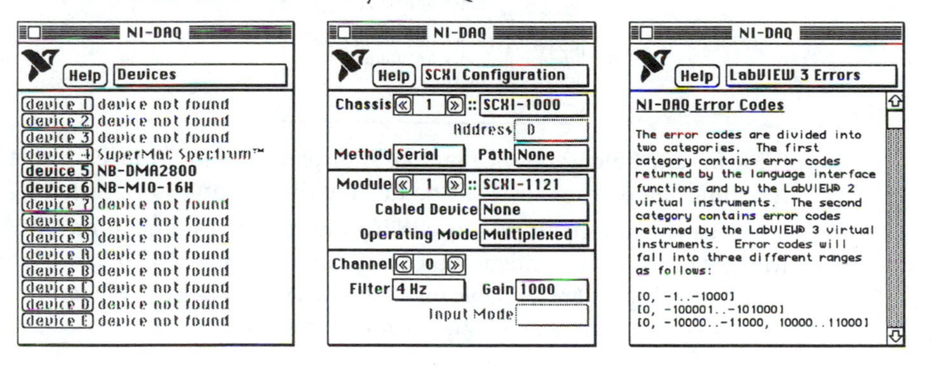

Note

The slot number occupied by your board, shown in the Devices screen of NI-DAQ, is the device number used by the LabVIEW DAQ VIs to address your board.

EXERCISE A.2
Exploring NI-DAQ (Mac Only)

In this exercise, you will open NI-DAQ and study the current settings. You will also view a list of possible DAQ error messages.

1. Start NI-DAQ by clicking on the icon shown below from the Control Panels (found in the menu).

2. Notice that the boards installed in your machine are listed in the NI-DAQ window shown in the following illustration. All National Instruments NB and Lab Series boards are listed in boldface; all others are dimmed.

Note
The boards installed in your computer may be different than those shown here.)

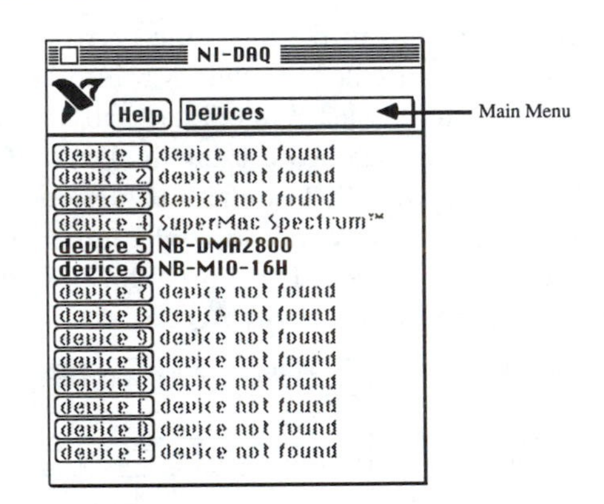

3. Click on any boldface device button to display the pin arrangement of the corresponding board. Click anywhere on the resulting window to close it.
4. Select **Device Configuration** by clicking on the NI-DAQ main menu. You use this window to select the configuration parameters for your DAQ board, as shown in the next figure. Inputs are available for changing Polarity/Range, Mode, and Number of Mux boards, as well as individual channel parameters.

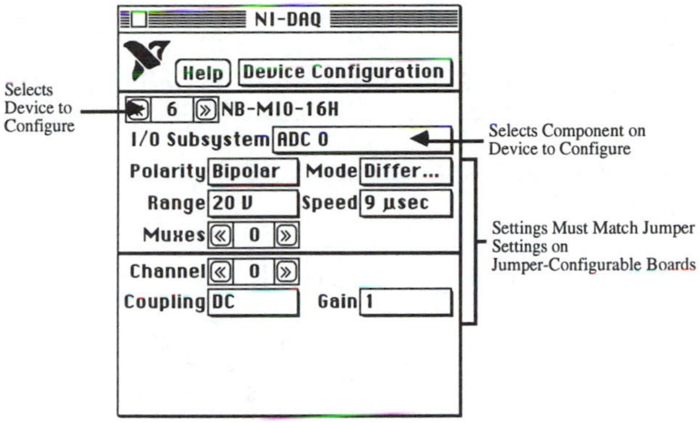

5. Select **LabVIEW 3 Errors** by clicking on the main menu. NI-DAQ displays a list of all possible DAQ error messages here, as shown in the following illustration.

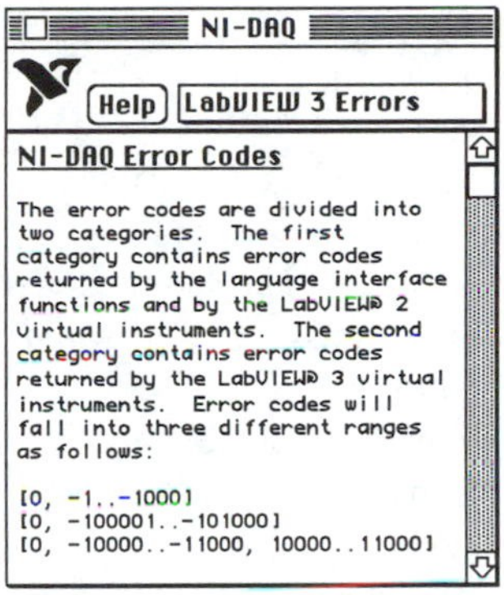

6. Close NI-DAQ.

GPIB Driver Software Architecture

Windows Configuration

LabVIEW GPIB VIs use the National Instruments standard NI-488.2 for Windows GPIB dynamic link library (DLL), via an intermediate driver, GPIBDRV. You must install the driver software yourself from the *LabVIEW Student Edition* Driver Setup disk. As a rule, you will want to use the driver on this disk to make sure you have a compatible version, *not* the driver that comes with your GPIB board.

You can use a utility program called WIBCONF.EXE to specify configuration parameters for devices on the GPIB. WIBCONF modifies the file GPIB.INI, which specifies the parameters for GPIB.DLL.

Note
Parameters specified in LabVIEW VIs take priority over those in WIBCONF.

The following illustration shows the GPIB software architecture on a computer running Windows.

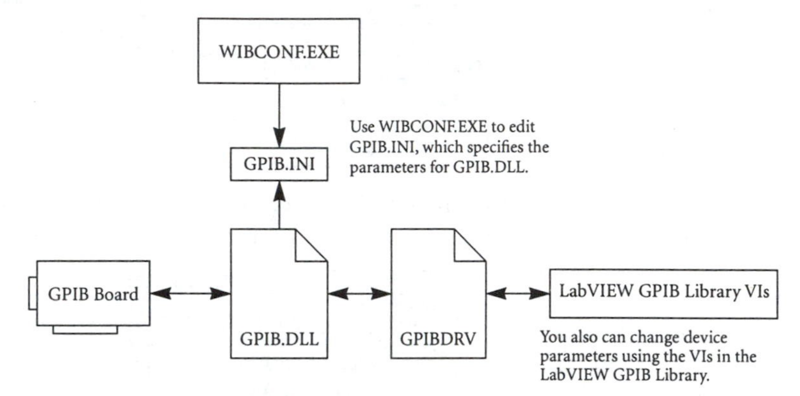

If you use LabVIEW VIs, you probably do not need to make any changes using WIBCONF, *as long as the base address and DMA and IRQ levels set by jumpers on the board match the settings in the software.*

You may want to open WIBCONF, the Windows GPIB configuration utility, so that you can examine the options available. When you are finished, close it and do not save any changes.

For more information on the GPIB driver files, please see the manual that comes with your board.

Macintosh Configuration

Before you can use any GPIB VI from the GPIB library, you must make sure the file NI-488 INIT resides in the Control Panels folder inside your System Folder (the setup disk should install it there for you). This file provides full support for the National Instruments GPIB interface boards. The following illustration shows the GPIB software architecture on the Macintosh.

Note
Parameters specified in LabVIEW VIs take priority over those in NI-488 INIT.

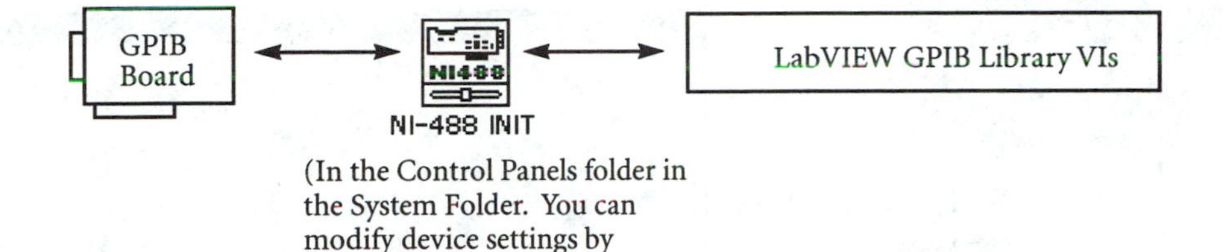

(In the Control Panels folder in the System Folder. You can modify device settings by double-clicking on the icon and using the GPIB Configuration)

You can use the NI-488 INIT GPIB configuration program, accessed by double-clicking on the NI-488 INIT icon, to specify configuration parameters for devices on the GPIB.

If you use LabVIEW VIs, you probably do not need to make any changes using NI-488 INIT. However, you may wish to open NI-488 INIT, shown in the following illustration so that you can examine the options available. When you are finished, close it and do not save any changes.

NI-488 INIT
National Instruments GPIB Configuration
Interface: System ...
4 5 6
Auto
NI-488 Bus#: 0
Bus/Device: gpib0
Primary: 0
Secondary: None
Timeout: 10 sec
System Controller
REN when SC
DMA
Write END on Last Byte
Write END on EOS
Read END on EOS
EOS Byte 0 8 Bit
Repeat Addressing
Unaddressing
NI488.2 V5.0
©1994 National Instruments

For more information on GPIB files, please see the manual that comes with your board.

Troubleshooting

Murphy's Law states that if anything can go wrong, it will. This premise applies to installations as much as anything else, if not more so. To help you around any possible troubles, we've included this list of common problems and questions, mostly related to installation free of charge. If you have other questions, please talk to your professor or lab instructor.

[Windows & Mac] I just received a new DAQ or GPIB board and a copy of LabVIEW. What do I have to install and in what order?

You need install the LabVIEW package as well as the LabVIEW Student Edition Driver Setup program. The *LabVIEW Student Edition* installation does not include the necessary files to control DAQ and GPIB boards with LabVIEW. While some driver files do come with your boards, they are very likely to be incompatible with LabVIEW, so you should only install drivers from the Student Edition Driver Setup disk, available from your instructor. Installation order doesn't matter at all.

[Windows & Mac] I have a previous version of LabVIEW and am starting the installation process. Should I do a full installation?

We suggest that you start from scratch with an empty directory. If you install the drivers from the Setup disk, the old data acquisition and GPIB drivers will automatically be overwritten during the installation process.

[Windows & Mac] The installation process aborts prematurely.

On Mac and Windows, virus protection software can prevent installation of the LabVIEW package; the error message you get depends on the virus protection software. We suggest that you first check the disks for viruses, and then turn off the virus-protection software while installing LabVIEW.

On both platforms, this symptom may indicate a problem with the disk. You can obtain a replacement disk by contacting Prentice Hall at (201) 592-3096.

[Windows] I am installing the LabVIEW for Windows software and get the error, "`Can't install daq.llb`" or some other file.

You may have a bad disk. Go to the DOS prompt and use the standard DOS command "expand" to try to extract the individual file. On the installation diskettes, all of the files have the last character in the names substituted with "_". For example, if the suspect file is `daq.llb` on disk 3, then put disk 3 in the drive and type "`expand b:\daq.ll_ daq.llb`". If the file cannot be read, then the disk is bad. You can obtain a replacement disk by contacting Prentice Hall at (201) 592-3096.

[Mac] While installing the LabVIEW for Macintosh software, I receive an error message stating that the disk is unreadable.

You may have a bad disk. We suggest aborting the installation procedure with <Command>-<.>. After you are out of the installer, insert the wayward disk and see if the Mac recognizes it. If so, then there is some other problem involving installation. If it does not recognize the disk, then the disk is bad. You can obtain a replacement disk by contacting Prentice Hall at (201) 592-3096.

[Mac] I'm having trouble with installation, but the disks all seem fine and I have turned off my virus protection software. Any other suggestions?

System INITs, extensions, or control panel documents may conflict with the installer. If you have problems during installation, first check to see that you have the necessary hard disk space available. A full installation takes about 10 MB of space. If you have plenty of space and installation still fails, restart the Mac while holding the shift key down to start up with extensions off, and then install LabVIEW. This procedure keeps the Mac from loading its `Control Panels` or `Extensions` folder utilities at start-up so they will not cause problems. Sometimes it may also help to rebuild your desktop before you install. As a last resort, run Norton Utilities on your hard drive.

If you are using MacOS 6.0.x, you will need 32-bit QuickDraw in the `System Folder` to both install and run LabVIEW (32-bit QuickDraw is built into System 7). It can be found on and installed from your System 6.0.x disks.

[Windows] I'm seeing inconsistencies or problems in the LabVIEW graphics or I'm seeing strange results when I print.

Make sure you have the Standard VGA video driver installed under Windows Setup. Other video drivers may not properly meet specifications that LabVIEW requires.

Error Codes

This appendix contains a table of the LabVIEW error codes.

Function Error Codes

Code	Description
0	No error.
1	Manager argument error.
2	Argument error.
3	Out of zone.
4	End of file.
5	File already open.
6	Generic file I/O error.
7	File not found.
8	File permission error.
9	Disk full.
10	Duplicate path.
11	Too many files open.
12	System feature not enabled.
13	Resource file not found.
14	Cannot add resource
15	Resource not found.

Function Error Codes (Continued)

Code	Description
16	Image not found.
17	Image memory error.
18	Pen does not exist.
19	Config type invalid.
20	Config token not found.
21	Config parse error.
22	Config memory error.
23	Bad external code format.
24	Bad external code offset.
25	External code not present.
26	Null window.
27	Destroy window error.
28	Null menu.
29	Print aborted.
30	Bad print record.
31	Print driver error.
32	Windows error during printing.
33	Memory error during printing.
34	Print dialog error
35	Generic print error.
36	Invalid device refnum.
37	Device not found.
38	Device parameter error.
39	Device unit error.
40	Cannot open device.
41	Device call aborted.
42	Generic error.
43	Canceled by user.
44	Object ID too low.
45	Object ID too high.
46	Object not in heap.
47	Unknown heap.
48	Unknown object (invalid DefProc).
49	Unknown object (DefProc not in table).
50	Message out of range.
51	Invalid (null) method.

Function Error Codes (Continued)

Code	Description
52	Unknown message.
53	Manager call not supported.
54	Bad address.
55	Connection in progress.
56	Connection timed out.
57	Connection is already in progress.
58	Network attribute not supported.
59	Network error.
60	Address in use.
61	System out of memory.
62	Connection aborted.
63	Connection refused.
64	Not connected.
65	Already connected.
66	Connection closed.
67	Initialization error (interapplication manager)
68	Bad occurrence.
69	Wait on unbound occurrence handler.
70	Occurrence queue overflow.
71	Datalog type conflict.
72	Unused.
73	Unrecognized type (interapplication manager).
74	Memory corrupt.
75	Failed to make temporary DLL.
76	Old CIN version.
77	Unknown error code

Analysis Error Codes

Code	Name	Description
0	NoErr	No error; the call was successful.
-20001	OutOfMemErr	There is not enough memory left to perform the specified routine.
-20002	EqSamplesErr	The input sequences must be the same size.
-20003	SamplesGTZeroErr	The number of samples must be greater than zero.

Analysis Error Codes (Continued)

Code	Name	Description
-20004	SamplesGEZeroErr	The number of samples must be greater than or equal to zero.
-20005	SamplesGEOneErr	The number of samples must be greater than or equal to one.
-20006	SamplesGETwoErr	The number of samples must be greater than or equal to two.
-20007	SamplesGEThreeErr	The number of samples must be greater than or equal to three.
-20008	ArraySizeErr	The input arrays do not contain the correct number of data values for this VI.
-20009	PowerOfTwoErr	The size of the input array must be a power of two: size = 2^m, $0<m<23$.
-20010	MaxXformSizeErr	The maximum transform size has been exceeded.
-20011	DutyCycleErr	The duty cycle must meet the condition: 0≤duty cycle≤100.
-20012	CyclesErr	The number of cycles must be greater than zero and less than or equal to the number of samples.
-20013	WidthLTSamplesErr	The width must meet the condition: 0 < width < samples.
-20014	DelayWidthErr	The following condition must be met: 0≤(delay+width) < samples.
-20015	DtGEZeroErr	dt must be greater than or equal to zero.
-20016	DtGTZeroErr	dt must be greater than zero.
-20017	IndexLTSamplesErr	The index must meet the condition: 0≤index < samples.
-20018	IndexLengthErr	The following condition must be met: 0≤(index+length) < samples.
-20019	UpperGELowerErr	The upper value must be greater than or equal to the lower value.
-20020	NyquistErr	The cutoff frequency, f_c, must meet the condition $0 \le f_c \le \frac{f_s}{2}$.
-20021	OrderGTZeroErr	The order must be greater than zero.
-20022	DecFactErr	The decimating factor must meet the condition: 0 < decimating≤samples.

Analysis Error Codes (Continued)

Code	Name	Description
-20023	BandSpecErr	The following condition must be met: $0 \le f_{flow} \le f_{high} \le \frac{f_s}{2}$
-20024	RippleGTZeroErr	The ripple amplitude must be greater than zero.
-20025	AttenGTZeroErr	The attenuation must be greater than zero.
-20026	WidthGTZeroErr	The width must be greater than zero.
-20027	FinalGTZeroErr	The final value must be greater than zero.
-20028	AttenGTRippleErr	The attenuation must be greater than the ripple amplitude.
-20029	StepSizeErr	The step-size, μ, must meet the condition: $0 \le \mu \le 0.1$.
-20030	LeakErr	The leakage coefficient must meet the condition: $0 \le leak \le \mu$.
-20031	EqRplDesignErr	The filter cannot be designed with the specified input values.
-20032	RankErr	The rank of the filter must meet the condition: $1 \le (2\ rank + 1) \le size$.
-20033	EvenSizeErr	The number of coefficients must be odd for this filter.
-20034	OddSizeErr	The number of coefficients must be even for this filter.
-20035	StdDevErr	The standard deviation must be greater than zero for normalization.
-20036	MixedSignErr	The elements of the Y Values array must be nonzero and either all positive or all negative.
-20037	SizeGTOrderErr	The number of data points in the Y Values array must be greater than two.
-20038	IntervalsErr	The number of intervals must be greater than zero.
-20039	MatrixMulErr	The number of columns in the first matrix is not equal to the number of rows in the second matrix or vector.
-20040	SquareMatrixErr	The input matrix must be a square matrix.
-20041	SingularMatrixErr	The system of equations cannot be solved because the input matrix is singular.
-20042	LevelsErr	The number of levels is out of range.

Analysis Error Codes (Continued)

Code	Name	Description
-20043	FactorErr	The level of factors is out of range for some data.
-20044	ObservationsErr	Zero observations were made at some level of a factor.
-20045	DataErr	The total number of data points must be equal to the product of the levels for each factor and the observations per cell.
-20046	OverflowErr	There is an overflow in the calculated F-value.
-20047	BalanceErr	The data is unbalanced. All cells must contain the same number of observations.
-20048	ModelErr	The Random Effect model was requested when the Fixed Effect model was required.
-20049	DistinctErr	The x values must be distinct.
-20050	PoleErr	The interpolating function has a pole at the requested value.
-20051	ColumnErr	All values in the first column in the X matrix must be one.
-20052	FreedomErr	The degrees of freedom must be one or more.
-20053	ProbabilityErr	The probability must be between zero and one.
-20054	InvProbErr	The probability must be greater than or equal to zero and less than one.
-20055	CategoryErr	The number of categories or samples must be greater than one.
-20056	TableErr	The contingency table must not contain a negative number.
-20061	InvSelectionErr	One of the input selections is invalid.

Data Acquistion Error Codes

This section lists the error codes returned by the LabVIEW data acquisition VIs and the configuration utility, including the error number, name, and description.

The following table lists the negative error codes returned by the data acquisition VIs. Each data acquisition VI returns an error code that indicates whether the VI executed successfully. When a VI returns a code that is a negative number, it means that the VI did not execute.

Note

All error codes and descriptions are also included in the configuration utility help panels.

Data Acquisition Error Codes

Code	Name	Description
-10001	syntaxErr	An error was detected in the input string; the arrangement or ordering of the characters in the string is not consistent with the expected ordering.
-10002	semanticsErr	An error was detected in the input string; the syntax of the string is correct, but certain values specified in the string are inconsistent with other values specified in the string.
-10003	invalidValueErr	The value of a numeric parameter is invalid.
-10004	valueConflictErr	The value of a numeric parameter is inconsistent with another parameter, and the combination is therefore invalid.
-10005	badDeviceErr	The device parameter is invalid.
-10006	badLineErr	The line parameter is invalid.
-10007	badChanErr	A channel is out of range for the board type or input configuration, the combination of channels is not allowed, or you must reverse the scan order so that channel 0 is last.
-10008	badGroupErr	The group parameter is invalid.
-10009	badCounterErr	The counter parameter is invalid.
-10010	badCountErr	The count parameter is too small or too large for the specified counter.
-10011	badIntervalErr	The interval parameter is too small or too large for the associated counter or I/O channel.
-10012	badRangeErr	The analog input or analog output voltage range is invalid for the specified channel.
-10013	badErrorCodeErr	The driver returned an unrecognized or unlisted error code.
-10014	groupTooLargeErr	The group size is too large for the board.
-10015	badTimeLimitErr	The time limit parameter is invalid.
-10016	badReadCountErr	The read count parameter is invalid.

Data Acquisition Error Codes (Continued)

Code	Name	Description
-10017	badReadModeErr	The read mode parameter is invalid.
-10018	badReadOffsetErr	The offset is unreachable.
-10019	badClkFrequencyErr	The frequency parameter is invalid.
-10020	badTimebaseErr	The timebase parameter is invalid.
-10021	badLimitsErr	The limits are beyond the range of the board.
-10022	badWriteCountErr	Your data array contains an incomplete update or you are trying to write past the end of the internal buffer or your output operation is continuous and the length of your array is not a multiple of one-half of the internal buffer size.
-10023	badWriteModeErr	The write mode is out of range or is disallowed.
-10024	badWriteOffsetErr	The write offset plus the write mark is greater than the internal buffer size or it must be set to 0.
-10025	limitsOutOfRangeErr	The voltage limits are out of range for this board in the current configuration. Alternate limits were selected.
-10026	badInputBufferSpecification	The input buffer specification is invalid. This error results if, for example, you try to configure a multiple-buffer acquisition for a board that does not support multiple-buffer acquisition.
-10027	badDAQEventErr	For DAQEvents 0 and 1, general value A must be greater than 0 and less than the internal buffer size. If DMA is used for DAQEvent 1, general value A must divide the internal buffer evenly. If the TIO-10 is used for DAQEvent 4, general value A must be 1 or 2.
-10028	badFilterCutoffErr	The cutoff frequency specified is not valid for this device.
-10080	badGainErr	The gain parameter is invalid.
-10081	badPretrigCountErr	The pretrigger sample count is invalid.
-10082	badPosttrigCountErr	The posttrigger sample count is invalid.
-10083	badTrigModeErr	The trigger mode is invalid.
-10084	badTrigCountErr	The trigger count is invalid.

Data Acquisition Error Codes (Continued)

Code	Name	Description
-10085	badTrigRangeErr	The trigger range or trigger hysteresis window is invalid.
-10086	badExtRefErr	The external reference value is invalid.
-10087	badTrigTypeErr	The trigger type parameter is invalid.
-10088	badTrigLevelErr	The trigger level parameter is invalid.
-10089	badTotalCountErr	The total count specified is inconsistent with the buffer configuration and pretrigger scan count or with the board type.
-10090	badRPGErr	The individual range, polarity, and gain settings are valid but the combination specified is not allowed for this board.
-10091	badIterationsErr	The analog output buffer iterations count is not allowed. It must be 0 (for indefinite iterations) or 1.
-10100	badPortWidthErr	The requested digital port width is not a multiple of the hardware port width.
-10240	noDriverErr	The driver interface could not locate or open the driver.
-10241	oldDriverErr	The driver is out of date.
-10242	functionNotFoundErr	The specified function is not located in the driver.
-10243	configFileErr	The driver could not locate or open the configuration file, or the format of the configuration file is not compatible with the currently installed driver.
-10244	deviceInitErr	The driver encountered a hardware-initialization error while attempting to configure the specified device.
-10245	osInitErr	The driver encountered an operating system error while attempting to perform an operation, or the operating system does not support an operation performed by the driver.
-10246	communicationsErr	The driver is unable to communicate with the specified external device.
-10247	cmosConfigErr	The CMOS configuration memory for the computer is empty or invalid, or the configuration specified does not agree with the current configuration of the computer.

Data Acquisition Error Codes (Continued)

Code	Name	Description
-10248	dupAddressErr	The base addresses for two or more devices are the same; consequently, the driver is unable to access the specified device.
-10249	intConfigErr	The interrupt configuration is incorrect given the capabilities of the computer or device.
-10250	dupIntErr	The interrupt levels for two or more devices are the same.
-10251	dmaConfigErr	The DMA configuration is incorrect given the capabilities of the computer/DMA controller or device.
-10252	dupDMAErr	The DMA channels for two or more devices are the same.
-10253	switchlessBoardErr	NI-DAQ was unable to find one or more switchless boards you have configured using `WDAQCONF`.
-10254	DAQCardConfigErr	Cannot configure the DAQCard because: 1. The correct version of card and socket services software is not installed. 2. The card in the PCMCIA socket is not a DAQCard. 3. The base address and/or interrupt level requested are not available according to the card and socket services resource manager. Try different settings or use AutoAssign in the configuration utility.
-10340	noConnectErr	No RTSI signal/line is connected, or the specified signal and the specified line are not connected.
-10341	badConnectErr	The RTSI signal/line cannot be connected as specified.
-10342	multConnectErr	The specified RTSI signal is already being driven by a RTSI line, or the specified RTSI line is already being driven by a RTSI signal.
-10343	SCXIConfigErr	The specified SCXI configuration parameters are invalid, or the function cannot be executed given the current SCXI configuration.
-10360	DSPInitErr	The DSP driver was unable to load the kernel for its operating system.

Data Acquisition Error Codes (Continued)

Code	Name	Description
-10370	badScanListError	The scan list is invalid. This error can result if, for example, you mix AMUX-64T channels and onboard channels, or if you scan multiplexed SCXI channels out of order.
-10400	userOwnedRsrcErr	The specified resource is owned by the user and cannot be accessed or modified by the driver.
-10401	unknownDeviceErr	The specified device is not a National Instruments product, or the driver does not support the device (for example, the driver was released before the device was supported).
-10402	deviceNotFoundErr	No device is located in the specified slot or at the specified address.
-10403	deviceSupportErr	The specified device does not support the requested action (the driver recognizes the device, but the action is inappropriate for the device).
-10404	noLineAvailErr	No line is available.
-10405	noChanAvailErr	No channel is available.
-10406	noGroupAvailErr	No group is available.
-10407	lineBusyErr	The specified line is in use.
-10408	chanBusyErr	The specified channel is in use.
-10409	groupBusyErr	The specified group is in use.
-10410	relatedLCGBusyErr	A related line, channel, or group is in use; if the driver configures the specified line, channel, or group, the configuration, data, or handshaking lines for the related line, channel, or group will be disturbed.
-10411	counterBusyErr	The specified counter is in use.
-10412	noGroupAssignErr	No group is assigned, or the specified line or channel cannot be assigned to a group.
-10413	groupAssignErr	A group is already assigned, or the specified line or channel is already assigned to a group.
-10414	reservedPinErr	Selected signal indicates a pin reserved by NI-DAQ. You cannot configure this pin yourself.

Data Acquisition Error Codes (Continued)

Code	Name	Description
-10440	sysOwnedRsrcErr	The specified resource is owned by the driver and cannot be accessed or modified by the user.
-10441	memConfigErr	No memory is configured to support the current data transfer mode, or the configured memory does not support the current data transfer mode. (If block transfers are in use, the memory must be capable of performing block transfers.)
-10442	memDisabledErr	The specified memory is disabled or is unavailable given the current addressing mode.
-10443	memAlignmentErr	The transfer buffer is not aligned properly for the current data transfer mode. For example, the memory buffer is at an odd address, is not aligned to a 32-bit boundary, is not aligned to a 512-bit boundary, and so on. Alternatively, the driver is unable to align the buffer because the buffer is too small.
-10444	memFullErr	No more system memory is available on the heap, or no more memory is available on the device.
-10445	memLockErr	The transfer buffer cannot be locked into physical memory.
-10446	memPageErr	The transfer buffer contains a page break; system resources may require reprogramming when the page break is encountered.
-10447	memPageLockErr	The operating environment is unable to grant a page lock.
-10448	stackMemErr	The driver is unable to continue parsing a string input due to stack limitations.
-10449	cacheMemErr	A cache-related error occurred, or caching is not supported in the current mode.
-10450	physicalMemErr	A hardware error occurred in physical memory, or no memory is located at the specified address.

Data Acquisition Error Codes (Continued)

Code	Name	Description
-10451	virtualMemErr	The driver is unable to make the transfer buffer contiguous in virtual memory and therefore cannot lock the buffer into physical memory; thus, you cannot use the buffer for DMA transfers.
-10452	noIntAvailErr	No interrupt level is available for use.
-10453	intInUseErr	The specified interrupt level is already in use by another device.
-10454	noDMACErr	No DMA controller is available in the system.
-10455	noDMAAvailErr	No DMA channel is available for use.
-10456	DMAInUseErr	The specified DMA channel is already in use by another device.
-10457	badDMAGroupErr	DMA cannot be configured for the specified group because it is too small, too large, or misaligned. Consult the user manual for the device in question to determine group ramifications with respect to DMA.
-10459	DLLInterfaceErr	The DLL could not be called due to an interface error.
-10460	interfaceInteractionErr	You have attempted to mix LabVIEW 2.2 VIs and LabVIEW 3.0 VIs. You may run an application consisting only of 2.2 VIs, then run the 2.2 Board Reset VI, before you can run any 3.0 VIs. You may run an application consisting of only 3.0 VIs, then run the 3.0 Device Reset VI, before you can run any 2.2 VIs.
-10560	invalidDSPhandleError	The DSP handle input to the VI is not a valid handle.
-10600	noSetupErr	No setup operation has been performed for the specified resources.
-10601	multSetupErr	The specified resources have already been configured by a setup operation.
-10602	noWriteErr	No output data has been written into the transfer buffer.
-10603	groupWriteErr	The output data associated with a group must be for a single channel or must be for consecutive channels.

Data Acquisition Error Codes (Continued)

Code	Name	Description
-10604	activeWriteErr	Once data generation has started, only the transfer buffers originally written to can be updated. If DMA is active and a single transfer buffer contains interleaved channel-data, new data must be provided for all output channels currently using the DMA channel.
-10605	endWriteErr	No data was written to the transfer buffer because the final data block has already been loaded.
-10606	notArmedErr	The specified resource is not armed.
-10607	armedErr	The specified resource is already armed.
-10608	noTransferInProgErr	No transfer is in progress for the specified resource.
-10609	transferInProgErr	A transfer is already in progress for the specified resource.
-10610	transferPauseErr	A single output channel in a group cannot be paused if the output data for the group is interleaved.
-10611	badDirOnSomeLinesErr	Some of the lines in the specified channel are not configured for the transfer direction specified. For a write transfer, some lines were configured for input. For a read transfer, some lines were configured for output.
-10612	badLineDirErr	The specified line does not support the specified transfer direction.
-10613	badChanDirErr	The specified channel does not support the specified transfer direction.
-10614	badGroupDirErr	The specified group does not support the specified transfer direction.
-10615	masterClkErr	The clock configuration for the clock master is invalid.
-10616	slaveClkErr	The clock configuration for the clock slave is invalid.
-10617	noClkSrcErr	No source signal has been assigned to the clock resource.
-10618	badClkSrcErr	The specified source signal cannot be assigned to the clock resource.
-10619	multClkSrcErr	A source signal has already been assigned to the clock resource.

Data Acquisition Error Codes (Continued)

Code	Name	Description
-10620	noTrigErr	No trigger signal has been assigned to the trigger resource.
-10621	badTrigErr	The specified trigger signal cannot be assigned to the trigger resource.
-10622	preTrigErr	The pretrigger mode is not supported or is not available in the current configuration, or no pretrigger source has been assigned.
-10623	postTrigErr	No posttrigger source has been assigned.
-10624	delayTrigErr	The delayed trigger mode is not supported or is not available in the current configuration, or no delay source has been assigned.
-10625	masterTrigErr	The trigger configuration for the trigger master is invalid.
-10626	slaveTrigErr	The trigger configuration for the trigger slave is invalid.
-10627	noTrigDrvErr	No signal has been assigned to the trigger resource.
-10628	multTrigDrvErr	A signal has already been assigned to the trigger resource.
-10629	invalidOpModeErr	The specified operating mode is invalid, or the resources have not been configured for the specified operating mode.
-10630	invalidReadErr	An attempt was made to read 0 bytes from the transfer buffer, or an attempt was made to read past the end of the transfer buffer.
-10631	noInfiniteModeErr	Continuous input or output transfers are not allowed in the current operating mode.
-10632	someInputsIgnoredErr	Certain inputs were ignored because they are not relevant in the current operating mode.
-10633	invalidRegenModeError	This board does not support the specified analog output regeneration mode.
-10680	badChanGainErr	All channels must have an identical setting for this board.
-10681	badChanRangeErr	All channels of this board must have the same range.
-10682	badChanPolarityErr	All channels of this board must have the same polarity.
-10683	badChanCouplingErr	All channels of this board must have the same coupling.

Data Acquisition Error Codes (Continued)

Code	Name	Description
-10684	badChanInputModeErr	All channels of this board must have the same input range.
-10685	clkExceedsBrdsMaxConvRate	The clock rate selected exceeds the recommended maximum rate for this board.
-10686	scanListInvalidErr	A configuration change has invalidated the scan list.
-10687	bufferInvalidErr	A configuration change has invalidated the allocated buffer.
-10688	noTrigEnabledErr	The total number of scans and pretrigger scans implies that a trigger start is intended, but no trigger is enabled.
-10689	digitalTrigBErr	Digital trigger B is illegal for the total scans and pretrigger scans specified.
-10690	digitalTrigAandBErr	This board does not allow digital triggers A and B to be enabled at the same time.
-10691	extConvRestrictionErr	This board does not allow an external sample clock with an external scan clock, start trigger, or stop trigger.
-10692	chanClockDisabledErr	Cannot start the acquisition because the channel clock is disabled.
-10693	extScanClockError	Cannot use an external scan clock when performing a single scan of a single channel.
-10694	unsafeSamplingFreqError	The sampling frequency exceeds the safe maximum rate for the ADC, gains, and filters you are using.
-10695	DMANotAllowedErr	You must use interrupts. DMA is not allowed.
-10696	multiRateModeErr	Multirate scanning can not be used with AMUX-64, SCXI, or pretriggered acquisitions.
-10697	rateNotSupportedErr	NI-DAQ was unable to convert your timebase/interval pair to match the actual hardware capabilities of the specified board.
-10698	timebaseConflictErr	You cannot use this combination of scan and sample clock timebases for the specified board.
-10699	polarityConflictErr	You cannot use this combination of scan and sample clock source polarities for this operation, for the specified board.

Data Acquisition Error Codes (Continued)

Code	Name	Description
-10700	signalConflictErr	You cannot use this combination of scan and convert clock signal sources for this operation, for the specified board.
-10740	SCXITrackHoldErr	A signal has already been assigned to the SCXI track-and-hold trigger line, or a control call was inappropriate because the specified module is not configured for one-channel operation.
-10780	sc2040InputModeErr	When you have an SC2040 attached to your device, all analog input channels must be configured for differential input mode.
-10800	timeOutErr	The operation could not complete within the time limit.
-10801	calibrationErr	An error occurred during the calibration process.
-10802	dataNotAvailErr	The requested amount of data has not yet been acquired, or the acquisition has completed and no more data is available to read.
-10803	transferStoppedErr	The transfer has been stopped to prevent regeneration of output data.
-10804	earlyStopErr	The transfer stopped prior to reaching the end of the transfer buffer.
-10805	overRunErr	The clock source for the input transfer is faster than the maximum input-clock rate; the integrity of the data has been compromised. Alternatively, the clock source for the output transfer is faster than the maximum output-clock rate; a data point was generated more than once since the update occurred before new data was available.
-10806	noTrigFoundErr	No trigger value was found in the input transfer buffer.
-10807	earlyTrigErr	The trigger occurred before sufficient pretrigger data was acquired.
-10808	LPTCommunicationErr	An error occurred in the parallel port communication with the SCXI-1200.
-10809	gateSignalError	Attempted to start a pulse width measurement with the pulse in the active state.

Data Acquisition Error Codes (Continued)

Code	Name	Description
-10840	softwareErr	The contents or the location of the driver file was changed between accesses to the driver.
-10841	firmwareErr	The firmware does not support the specified operation, or the firmware operation could not complete due to a data-integrity problem.
-10842	hardwareErr	The hardware is not responding to the specified operation, or the response from the hardware is not consistent with the functionality of the hardware.
-10843	underFlowErr	The update rate exceeds your system's capacity to supply data to the output channel.
-10844	underWriteErr	At the time of the update for the device-resident memory, insufficient data was present in the output transfer buffer to complete the update.
-10845	overFlowErr	At the time of the update clock for the input channel, the device-resident memory was unable to accept additional data—one or more data points may have been lost.
-10846	overWriteErr	New data was written into the input transfer buffer before the old data was retrieved.
-10847	dmaChainingErr	New buffer information was not available at the time of the DMA chaining interrupt; DMA transfers will terminate at the end of the currently active transfer buffer.
-10848	noDMACountAvailErr	The driver could not obtain a valid reading from the transfer-count register in the DMA controller.
-10849	openFileError	Unable to open a file.
-10850	closeFileError	Unable to close a file.
-10851	fileSeekError	Unable to seek within a file.
-10852	readFileError	Unable to read from a file.

Data Acquisition Error Codes (Continued)

Code	Name	Description
-10853	writeFileError	Unable to write to a file.
-10854	miscFileError	An error occurred accessing a file.
-10880	updateRateChangeError	A change to the update rate is not possible at this time because: 1. When waveform generation is in progress, you cannot change the interval timebase. 2. When you make several changes in a row, you must give each change enough time to take effect before requesting further changes.
-10920	gpctrDataLossError	One or more data points may have been lost during buffered GPCTR operations due to the speed limitations of your system.

Configuration Utility Error Codes

Error Code	Error Name	Description
-60	notOurBrdErr	The board in the specified slot is not an MC Series, AT Series, EISA Series, or Lab-PC board.
-61	badBrdNumErr	The board parameter is out of range.
-62	badGainErr	The gain parameter is out of range.
-63	badChanErr	The channel parameter is out of range.
-64	noSupportErr	Function cannot be executed by the specified board.
-65	badPortErr	The port parameter is out of range or the port is busy.
-66	badOutPortErr	The specified port has not been configured as an output port.
-67	noLatchModeErr	The specified port has not been configured for handshaking.
-69	badInputValErr	One or more input parameters are out of range.
-70	timeOutErr	A/D conversion did not complete or timeout period has expired.
-71	outOfRangeErr	Scaled input value is out of range.
-72	daqInProgErr	Data acquisition is in progress; therefore, call was not executed.

Configuration Utility Error Codes (Continued)

Error Code	Error Name	Description
-75	overFlowErr	A/D FIFO memory has overflowed as a result of a DAQ or SCAN operation.
-76	overRunErr	Minimum sample interval has been exceeded as a result of a DAQ or SCAN operation.
-81	portAssignToGrp	The specified port is currently assigned to a group and can be accessed only through digital group calls until unassigned.
-197	incompatibleVISRDErr	Incorrect version of NIVISRD.386 is installed.

GPIB Error Codes

Code	Name	Description
0	EDVR	Error connecting to driver.
1	ECIC	Command requires GPIB Controller to be CIC.
2	ENOL	Write detected no Listeners.
3	EADR	GPIB Controller not addressed correctly.
4	EARG	Invalid argument or arguments.
5	ESAC	Command requires GPIB Controller to be SC.
6	EABO	I/O operation aborted.
7	ENEB	Nonexistent board.
8	EDMA	DMA hardware error detected.
9	EBTO	DMA hardware µP bus timeout.
11	ECAP	No capability.
12	EFSO	File system operation error.
13	EOWN	Shareable board exclusively owned.
14	EBUS	GPIB bus error.
15	ESTB	Serial poll byte queue overflow.
16	ESRQ	SRQ stuck on.
17	ECMD	Unrecognized command.
19	EBNP	Board not present.
20	ETAB	Table error.
30	NADDR	No GPIB address input.
31	NSTRG	No string input (write).
32	NCNT	No count input (read).

Serial Port Error Codes

Code	Name	Description
61	EPAR	Serial port parity error.
62	EORN	Serial port overrun error.
63	EOFL	Serial port receive buffer overflow.
64	EFRM	Serial port framing error.
65	SPTMO	Serial port timeout, bytes not received at serial port.

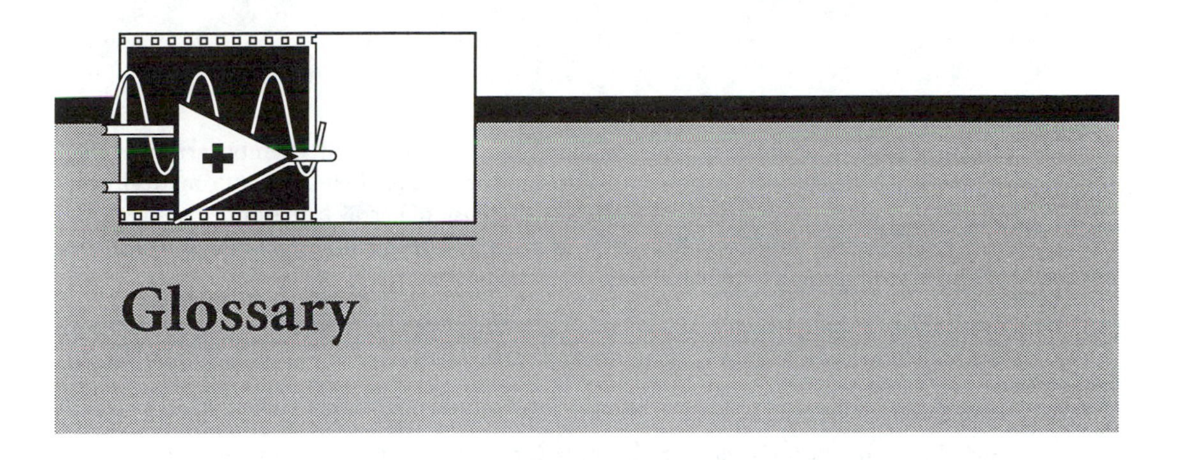

Glossary

Prefix	Meaning	Value
μ-	micro-	10^{-6}
m-	milli-	10^{-3}
k-	kilo-	10^{3}

∞ Infinity.

π Pi.

A Amperes.

active window Window that is currently set to accept user input, usually the frontmost window. The title bar of an active window is highlighted. You make a window active by clicking on it, or by selecting it from the **Windows** menu.

ANSI American National Standards Institute.

array Ordered, indexed set of data elements of the same type.

array shell Front panel object that houses an array. It consists of an index display, a data object window, and an optional label. It can accept various data types.

artificial data dependency Condition in a dataflow programming language in which the arrival of data rather than its value triggers execution of a node.

ASCII American Standard Code for Information Interchange.

asynchronous execution Mode in which multiple processes share processor time. For example, one process executes while others wait for interrupts during device I/O or while waiting for a clock tick.

ATE Automatic test equipment.

auto-indexing Capability of loop structures to disassemble and assemble arrays at their borders. As an array enters a loop with auto-indexing enabled, the loop automatically disassembles it with scalars extracted from one-dimensional arrays, one-dimensional arrays extracted from two-dimensional arrays, and so on. Loops assemble data into arrays as they exit the loop according to the reverse of the same procedure.

autoscaling Ability of scales to adjust to the range of plotted values. On graph scales, this feature determines maximum and minimum scale values, as well.

autosizing Automatic resizing of labels to accommodate text that you enter.

block diagram Pictorial description or representation of a program or algorithm. In LabVIEW, the block diagram, which consists of executable icons called nodes and wires that carry data between the nodes, is the source code for the VI. The block diagram resides in the Diagram window of the VI.

Boolean controls and indicators Front panel objects used to manipulate and display or input and output Boolean (TRUE or FALSE) data. Several styles are available, such as switches, buttons, and LEDs.

breakpoint Mode that halts execution when a subVI is called. You set a breakpoint by clicking on the Breakpoint button in the execution palette.

broken VI VI that cannot be compiled or run; signified by a broken arrow in the Run button.

Bundle Function that creates clusters from various types of elements.

c Speed of light.

case One subdiagram of a Case structure.

Case structure Conditional branching control structure, which executes one and only one of its subdiagrams based on its input. It is the combination of the IF, THEN, ELSE, and CASE statements in control flow languages.

chart *See* waveform chart.

clone To duplicate graphics by selecting and dragging while pressing a control key.

cloning To make a copy of a control or some other LabVIEW object by clicking the mouse button while pressing the <option> key on the Mac or the <Ctrl> key under Windows and dragging the copy to its new location.

cluster A set of ordered, unindexed data elements of any data type including numeric, Boolean, string, array, or cluster. The elements must be all controls or all indicators.

cluster shell Front panel object that contains the elements of a cluster.

Coloring tool Tool you use to color objects and backgrounds.

compile Process that converts high-level code to machine-executable code. LabVIEW automatically compiles VIs before they run for the first time after creation or alteration.

conditional terminal The terminal of a While Loop that contains a Boolean value that determines whether the VI performs another iteration.

connector Part of the VI or function node that contains its input and output terminals, through which data passes to and from the node.

connector pane Region in the upper-right corner of a Panel window that displays the VI terminal pattern. It underlies the Icon pane.

constant *See* universal and user-defined constants.

continuous run Execution mode in which a VI is run repeatedly until the operator stops it. You enable it by clicking on the Continuous Run button.

control Front panel object for entering data to a VI interactively or to a subVI programmatically.

control flow Programming system in which the sequential order of instructions determines execution order. Most conventional text-based programming languages, such as C, Pascal, and BASIC, are control-flow languages.

Controls menu Menu of controls and indicators.

conversion Changing the type of a data element.

count terminal The terminal of a For Loop whose value determines the number of times the For Loop executes its subdiagram.

CPU Central processing unit.

current VI VI whose Panel window, Diagram window, or icon editor window is the active window.

DAQ, data acquisition Process of acquiring data, typically from A/D or digital input plug-in boards.

data dependency Condition in a dataflow programming language in which a node cannot execute until it receives data from another node. *See also* artificial data dependency.

dataflow Programming system consisting of executable nodes in which nodes execute only when they have received all required input data and produce output automatically when they have executed. LabVIEW is a dataflow system.

Description box Online documentation for a LabVIEW object.

destination terminal *See* sink terminal.

Diagram window VI window that contains the block diagram code.

dialog box An interactive screen with prompts in which you specify additional information needed to complete a command.

dimension Size and structure attribute of an array.

drag To drag the mouse cursor on the screen to select, move, copy, or delete objects.

e Electronic charge.

edit mode The mode in which you create or edit a VI.

empty array Array that has zero elements, but has a defined data type. For example, an array that has a numeric control in its data display window but has no defined values for any element is an empty numeric array.

EOF End-of-File. Character offset of the end of file relative to the beginning of the file (that is, the EOF is the size of the file).

execution highlighting Feature that animates VI execution to illustrate.the data flow in the VI.

FFT Fast Fourier transform.

file refnum An identifier that LabVIEW associates with a file when you open it. You use the file refnum to specify that you want a function or VI to perform an operation on the open file.

For Loop Iterative loop structure that executes its subdiagram a set number of times. Equivalent to conventional code: For i=0 to n-1, do ....

Formula Node Node that executes formulas that you enter as text. Especially useful for lengthy formulas that would be cumbersome to build in block diagram form.

frame Subdiagram of a Sequence structure.

free label Label on the front panel or block diagram that does not belong to any other object.

front panel The interactive user interface of a VI. Modeled from the front panel of physical instruments, it is composed of switches, slides, meters, graphs, charts, gauges, LEDs, and other controls and indicators.

function Built-in execution element, comparable to an operator, function, or statement in a conventional language.

G The LabVIEW graphical programming language.

g Gram.

global variable Nonreentrant subVI with local memory that uses an uninitialized shift register to store data from one execution to the next. The memory of copies of these subVIs is shared and thus can be used to pass global data between them.

GPIB General Purpose Interface Bus is the common name for the communications interface system defined in ANSI/IEEE Standard 488.1-1987 and ANSI/IEEE Standard 488.2-1987. Hewlett-Packard, the inventor of the bus, calls it the HP-IB.

graph control Front panel object that displays data in a Cartesian plane.

handle Pointer to a pointer to a block of memory; handles reference arrays and strings. An array of strings is a handle to a block of memory containing handles to strings.

Help window Special window that displays the names and locations of the terminals for a function or subVI, the description of controls and indicators, the values of universal constants, and descriptions and data types of control attributes.

hex Hexadecimal. A base-16 number system.

hierarchical menu Menu that contains submenus or palettes.

hierarchy The ability in LabVIEW to have subVIs within subVIs within VIs, each subVI being its own stand-alone unit.

housing Nonmoving part of front panel controls and indicators that contains sliders and scales.

Hz Hertz. Cycles per second.

icon Graphical representation of a node on a block diagram.

icon editor Interface similar to that of a paint program for creating VI icons.

icon pane Region in the upper-right corner of the Panel and Diagram windows that displays the VI icon.

IEEE Institute for Electrical and Electronic Engineers

IEEE 488 The IEEE standard governing GPIB transfers.

indicator Front panel object that displays output.

Inf Digital display value for a floating-point representation of infinity.

inplace execution Ability of a function or VI to reuse memory instead of allocating more.

instrument driver VI that controls a programmable instrument.

I/O Input/Output. The transfer of data to or from a computer system involving communications channels, operator input devices, and/or data acquisition and control interfaces.

iteration terminal The terminal of a For Loop or While Loop that contains the current number of completed iterations.

J Joule. Absolute unit of work or energy equal to 10^7 ergs.

label Text object used to name or describe other objects or regions on the front panel or block diagram.

Labeling tool Tool used to create labels and enter text into text windows.

LabVIEW Laboratory Virtual Instrument Engineering Workbench.

latch action Mechanical action setting of a Boolean control in which the Boolean returns to its default setting after you have clicked on it and a value has been read.

LED Light-emitting diode.

legend Object owned by a chart or graph that display the names and plot styles of plots on that chart or graph.

MB Megabytes of memory.

marquee A moving, dashed border that surrounds selected objects.

matrix Two-dimensional array.

mechanical action Setting on a Boolean control that determines how it behaves when you click on it. You have the choice of latch action or switch action.

menu bar Horizontal bar that contains names of main menus.

modular programming Programming that uses interchangeable computer routines.

NaN Digital display value for a floating-point representation of *not a number*, typically the result of an undefined operation, such as log(-1).

nodes Execution elements of a block diagram consisting of functions, structures, and subVIs.

nondisplayable characters ASCII characters that cannot be displayed, such as newline, tab, and so on.

not-a-path A predefined value for the path control that means the path is invalid.

not-a-refnum A predefined value that means the refnum is invalid.

numeric controls and indicators Front panel objects used to manipulate and display or input and output numeric data.

object Generic term for any item on the front panel or block diagram, including controls, nodes, wires, and imported pictures. Described in your tutorial manual.

Operating tool Tool used to enter data into controls as well as operate them. Resembles a pointing finger.

palette Menu that displays a palette of pictures that represent possible options.

Panel window VI window that contains the front panel, the execution palette and the icon/connector pane.

path controls and indicators Front panel objects used to manipulate and display or input and output file path data.

platform Computer and operating system.

plot A graphical representation of an array of data shown either on a graph or a chart.

polymorphism Ability of a node to automatically adjust to data of different representation, type, or structure.

pop up To call up a special menu by clicking (usually on an object) with the right mouse button under Windows and while holding down the command key on Macintosh.

pop-up menus Menus accessed by popping up, usually on an object. Menu options pertain to that object specifically.

Positioning tool Tool used to move and resize objects.

probe Debugging feature for checking intermediate values in a VI.

programmatic printing Automatic printing of a VI front panel after execution.

pseudocode Simplified language-independent representation of programming code.

pull-down menus Menus accessed from a menu bar. Pull-down menu options are usually general in nature.

reentrant execution Mode in which calls to multiple instances of a subVI can execute in parallel with distinct and separate data storage.

refnum A file refnum is an identifier of open files that can be referenced by other VIs.

resizing handles Angled handles on the corner of objects that indicate resizing points.

ring control Special numeric control that associates numbers, starting at 0 and increasing sequentially, with a series of text labels or graphics.

run mode The mode in which you execute a VI.

scalar Number capable of being represented by a point on a scale. A single value as opposed to an array. Scalar Booleans, strings, and clusters are explicitly singular instances of their respective data types.

scale Part of mechanical-action, chart, and graph controls and indicators that contains a series of marks or points at known intervals to denote units of measure.

scope mode Mode of a waveform chart numeric indicator modeled on the operation of an oscilloscope.

selector terminal The input terminal of a Case structure that determines which case will execute. The selector terminal can be either Boolean or numeric, depending on the data type of what you wire to it.

sequence local Terminal that passes data between the frames of a Sequence structure.

Sequence structure Program control structure that executes its subdiagrams in numeric order. Commonly used to force nodes that are not data dependent to execute in a desired order.

serial port A port on the back of the computer that you can use to transfer data one bit at a time.

shift register Optional mechanism in loop structures used to pass the value of a variable from one iteration of a loop to a subsequent iteration.

single-step mode Feature that controls VI execution to illustrate the dataflow in the VI and slow-down execution for debugging purposes. In single-step mode, the VI executes one step at a time, each time the user pushes the single step button.

sink terminal Terminal that absorbs data. Also called a destination terminal.

slider Movable part of slide controls and indicators.

stub VI A nonfunctional prototype of a subVI. It has inputs and outputs, but is incomplete. It is used during early planning stages of a VI design as a place holder for future VI development.

source terminal Terminal that emits data.

spreadsheet file A text file formatted so that you can open it with a spreadsheet program. Usually columns are separated by tabs and rows are separated by carriage returns.

string controls and indicators Front panel objects used to manipulate and display or input and output text.

strip mode Mode of a waveform chart numeric plotting indicator modeled after a paper strip chart recorder, which scrolls as it plots data.

structure Program control element, such as a Sequence, Case, For Loop, or While Loop.

subdiagram Block diagram within the border of a structure.

subVI VI used in the block diagram of another VI; comparable to a subroutine.

SubVI Node Setup Option that allows you to configure behavioral characteristics of a subVI. Modifications in subVI node setup affect only that particular subVI call.

sweep mode Mode of a waveform chart that is similar to scope mode; except a line sweeps across the display to separate old data from new data.

switch action Mechanical action setting of a Boolean control in which the control changes value each time you click on it.

terminal Object or region on a node through which data passes.

text ring *See* ring control.

tool Special LabVIEW cursor you can use to perform specific operations.

top-level VI VI at the top of the VI hierarchy. This term distinguishes the VI from its subVIs.

tunnel Data entry or exit terminal on a structure.

Unbundle Function that disassembles clusters into their component parts.

universal constant Uneditable block diagram object that emits a particular ASCII character or standard numeric constant, for example, pi.

user-defined constant Block diagram object that emits a value you set.

UUT Unit under test.

V Volts.

VI *See* virtual instrument.

VI library Special file that contains a collection of related VIs for a specific use. VI libraries, recognizable by their .LLB extensions, are special LabVIEW structures that can only be accessed from within LabVIEW.

VI Setup Option that allows you to configure behavioral characteristics of a VI when it is used as a subVI. Modifications in VI Setup affect all calls to that VI.

virtual instrument LabVIEW program, so called because it models the appearance and function of a physical instrument.

waveform chart A numeric plotting indicator that updates interactively. A waveform chart has three update modes; *see* scope mode, strip mode, and sweep mode.

waveform graph A numeric plotting indicator that plots entire arrays of data in which points are evenly distributed.

While Loop Post-iterative test loop structure that repeats a section of code until a condition is met. Comparable to a Do loop or a Repeat-Until loop in conventional programming languages.

wire Data path between nodes.

Wiring tool Tool used to define data paths between source and sink terminals.

XY graph A numeric plotting indicator that plots multivalued functions such as circular shapes or waveforms with varying timebases.

Index

B

D

E

F

G

H

I

J

K

L

M

N

R

S

T

U

V

W

X

Y

Z